OXFORD

Endorsed by edexcel

GCSE Maths

Higher

Marguerite Appleton
Head of Mathematics, Queen Mary's High School, Walsall

Derek Huby
Mathematics Consultant, West Sussex

Jayne Kranat
Mathematics Consultant & Senior Examiner

OXFORD
UNIVERSITY PRESS

OXFORD
UNIVERSITY PRESS

Great Clarendon Street, Oxford OX2 6DP

Oxford University Press is a department of the University of Oxford.
It furthers the University's objective of excellence in research, scholarship,
and education by publishing worldwide in

Oxford New York

Auckland Cape Town Dar es Salaam Hong Kong Karachi
Kuala Lumpur Madrid Melbourne Mexico City Nairobi
New Delhi Shanghai Taipei Toronto

With offices in

Argentina Austria Brazil Chile Czech Republic France Greece
Guatemala Hungary Italy Japan South Korea Poland Portugal
Singapore Switzerland Thailand Turkey Ukraine Vietnam

Oxford is a registered trade mark of Oxford University Press
in the UK and in certain other countries

British Library Cataloguing in Publication Data

Data available

ISBN 0 19 915089 3

ISBN 978 019 915089 2

10 9 8 7 6 5 4

Typeset by MCS Publishing Services Ltd, Salisbury, Wiltshire

Printed and bound by Rotolito Lombarda

Acknowledgements

The publisher would like to thank Edexcel for their kind permission to reproduce
past exam questions. Edexcel Ltd accepts no responsibility whatsoever for the
accuracy or method of working in the answers given.

This high quality material is endorsed by Edexcel and has been through a rigorous
quality assurance programme to ensure that it is a suitable companion to the
specification for both learners and teachers.
This does not mean that its contents will be used verbatim when setting
examinations nor is it to be read as being the official specification – a copy of
which is available at www.edexcel.org.uk

The Publisher would also like to thank Peter Hind for his authoritative guidance in
preparing this book.

The image on the cover is reproduced courtesy of ImageDJ.

The Publisher would like to thank the following for permission to
reproduce photographs:

p11 Oxford University Press; **p17** Paul Doyle/Alamy; **p81** Science Photo Library;
p86 Brendan Regan/Corbis; **p127** TopFoto UK Ltd; **p135** Oxford University Press;
p149 Hemera/Oxford University Press; **p150** Photodisc/Oxford University Press; **p155**
Oxford University Press; **p176** NASA/Oxford University Press; **p182** Oxford University
Press; **p195** Oxford University Press; **p197** The Photolibrary Wales/Alamy; **p199**
Araldo de Luca/Corbis; **p243** John La Gette/Alamy; **p285** India Images/Dinodia
Images/Alamy; **p287** Stockfolio/Alamy; **p304** World Pictures Ltd/Alamy **p308** Oxford
University Press; **p315** Andrew Syred/Science Photo Library; **p316l** Bob
Croxford/Atmosphere Picture Library/Alamy; **p316r** Ordnance Survey.

Figurative artwork is by Dylan Gibson.
Technical artwork is by MCS Publishing Services Ltd.

About this book

This book has been specifically written to help you get your best possible grade in your Edexcel GCSE Mathematics examinations. It is designed for students who have achieved level 6 or a strong level 5 at Key Stage 3 and are looking to progress to a grade C or B at GCSE.

The authors are experienced teachers and examiners who have an excellent understanding of the Edexcel two-tier specification and so are well qualified to help you successfully meet your objectives.

The book is made up of units that are based on the Edexcel specification, and provide coverage of the National Curriculum strands at Key Stage 4.

The units are:

Each unit contains double page spreads for each lesson. These are shown on the full contents list.

Problem solving is integrated throughout the material as suggested in the National Curriculum.

How to use this book

This book is made up of units of work that are colour-coded:
Algebra (green), Data (blue), Number (orange) and Shape, space and measures (pink).

Each unit starts with an overview of the content, so that you know exactly what you are expected to learn.

This unit will show you how to

- Use place value to round and order whole numbers and decimals
- Multiply and divide any number by powers of 10
- Add and subtract with positive and negative numbers
- Multiply and divide with negative numbers
- Understand the concepts of prime numbers, factors and prime factor decomposition

The first page of a unit also provides prior knowledge questions to help you revise before you start – then you can apply your knowledge later in the unit:

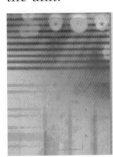

Before you start ...

You should be able to answer these questions.

1 Write in words the value of the digit 4 in each of these numbers.

a 4506   b 23 409

c 200.45   d 13.054

Review

Key stage 3

Inside each unit, the content develops in double page spreads that all follow the same structure.

Each spread starts with a list of the learning outcomes and a summary of the keywords:

N1.1 Place value and ordering numbers

This spread will show you how to:

- Use place value to round and order whole numbers and decimals
- Multiply and divide any number by powers of 10

Keywords
Ascending
Digit
Order

Key points are highlighted in the text so you can see the facts you need to learn:

- Area of trapezium $= \frac{1}{2} \times (a + b) \times h$

Examples showing the key skills and techniques you need to develop are shown in boxes. Also, margin notes show tips and reminders you may find useful:

Find the prime factor decomposition of 990.

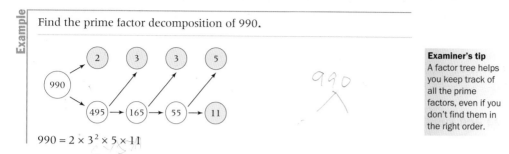

$990 = 2 \times 3^2 \times 5 \times 11$

Examiner's tip
A factor tree helps you keep track of all the prime factors, even if you don't find them in the right order.

Each exercise is carefully graded, set at three levels of difficulty:
- The first few questions are mainly repetitive to give you confidence, and simplify the content of the spread
- The questions in the middle of the exercise consolidate the topic, focusing on the main techniques of the spread
- Later questions extend the content of the spread – some of these questions may be problem-solving in nature, and may involve different approaches.

At the end of the unit is an exam review page so that you can revise the learning of the unit before moving on. The key Edexcel objectives are identified:

N1

Exam review

Key objectives
- Understand highest common factor, least common multiple, prime number and prime number decomposition
- Multiply and divide by a negative number

1 **a** Using the information that $7.4 \times 3.5 = 25.9$, write down the value of:

   i 7.4×-3.5

   ii -7.4×3.5

   iii -7.4×-3.5 (3)

Summary questions, including past exam questions, are provided to help you check your understanding of the key concepts covered and your ability to apply the key techniques.

An Edexcel formula page is provided near the end of the book so you can see what information you will be given in the exam.

You will find the answers to all exercises at the back of the book so that you can check your own progress and identify any areas that need work.

Contents

N1

This unit will show you how to

- Use place value to round and order whole numbers and decimals
- Multiply and divide any number by powers of 10
- Add and subtract with positive and negative numbers
- Multiply and divide with negative numbers
- Understand the concepts of prime numbers, factors and prime factor decomposition
- Find the highest common factor and lowest common multiple of two numbers

Before you start ...

You should be able to answer these questions.

Review

1 Write in words the value of the digit 4 in each of these numbers.

    a 4506   **b** 23 409

    c 200.45   **d** 13.054

Key stage 3

2 **a** Sketch a number line showing values from −5 to +5 and mark these numbers on it.

    i −3   **ii** +4   **iii** −2.5   **iv** 0

  b Write this set of directed numbers in ascending order.

    +5, −2.4, −3, +6, 0, +1.5, −1.8

Key stage 3

3 **a** Write all of the factors of these numbers.

    i 6   **ii** 12   **iii** 28   **iv** 36

  b Write all of the prime numbers between 1 and 50.

Key stage 3

This spread will show you how to:

- Use place value to round and order whole numbers and decimals
- Multiply and divide any number by powers of 10

Keywords

Ascending
Descending
Digit
Integer
Order
Round

- You can **round** large numbers to the nearest hundred, thousand or any other power of ten and round decimal numbers to any number of decimal places.
  - Identify the final **digit** required
  - Round it up if the following digit is a 5 or more
  - Write the rounded number, including any zeros needed to make the place value correct.

Example

Round 72 456 to the nearest   **a** 10   **b** 100   **c** 1000.

a 72 456 = 72 460 to the nearest 10
b 72 456 = 72 500 to the nearest 100
c 72 456 = 72 000 to the nearest 1000

Example

Round 6.0374 to the nearest   **a** tenth   **b** hundredth   **c** thousandth.

a 6.0374 = 6.0 to the nearest tenth
b 6.0374 = 6.04 to the nearest hundredth
c 6.0374 = 6.037 to the nearest thousandth

To **order** decimals, look at the tenths digit first, then the hundredths digit, then the thousandths and so on.

Example

Write these numbers in **ascending** order.

0.3   0.275   0.28   0.3005   0.269 997

0.275, 0.28 and 0.269 997 are smaller than 0.3 and 0.3005.
0.269 997 is smallest.
0.275 is smaller than 0.28.
0.3 is smaller than 0.3005.
In ascending order, the numbers are 0.269 997, 0.275, 0.28. 0.3, 0.3005.

Do not be misled by the number of digits. 0.28 is equal to 0.280, and is larger than 0.275.

- Multiplying a number by 10 moves the digits one place to the left. Multiplying by 100 moves the digits two places to the left.
- Dividing a number by 10 moves the digits one place to the right. Dividing by 100 moves the digits two places to the right.

Example

Work out   **a** 3.72 ÷ 100   **b** 0.0349 × 10 000   **c** 17.3 ÷ 1000

a 3.72 ÷ 100 = 0.0372   Move the digits 2 places right.

b 0.0349 × 10 000 = 349   Move the digits 4 places left.

c 17.3 ÷ 1000 = 0.0173   Move the digits 3 places right.

1 Write these sets of numbers in ascending order.

   a 0.3, 3.1, 1.3, 2, 1, 0.1   **b** 607, 77.2, 27.6, 7.06, 6.07

2 Write these sets of numbers in descending order.

   a 6008, 682.8, 862.6, 6000.8, 8000.6   **b** 47.9, 94.7, 49.7, 79.4, 74.9, 97.4

3 Multiply these numbers by 10.

   a 16.7   **b** 24.8   **c** 0.716   **d** 1.095   **e** 243   **f** 281.3

4 Divide these numbers by 10.

   a 214   **b** 67.3   **c** 4106   **d** 200.7   **e** 6.025   **f** 86

5 Decide which number in each pair is bigger.
   Explain your answers.

   a 4.52 and 4.05   **b** 5.5 and 5.05   **c** 16.8 and 16.75   **d** 16.8 and 16.15

6 Write these sets of numbers in ascending order.

   a 7.83, 7.3, 7.8, 7.08, 7.03, 7.38   **b** 4.2, 8.24, 8.4, 4.18, 2.18, 2.4

7 Write these sets of numbers in descending order.

   a 16.7, 18.16, 16.18, 17.16, 18.7, 17.6   **b** 1.06, 13.145, 1.1, 2.38, 13.2, 2.5

8 Round these numbers to the nearest   **i** 10   **ii** 100.

   a 3048   **b** 1763   **c** 294   **d** 51   **e** 43   **f** 743

9 Round these numbers to the nearest 1000.

   a 2964   **b** 1453   **c** 17   **d** 24 598   **e** 16 344   **f** 167 733

10 Round these numbers to   **i** 1 decimal place   **ii** 2 decimal places.

   a 39.114   **b** 7.068   **c** 5.915   **d** 512.715

   e 4.259   **f** 12.007   **g** 0.833   **h** 26.8813

11 Round these numbers to the nearest
   i tenth   **ii** hundredth   **iii** thousandth.

   a 0.07   **b** 15.9184   **c** 127.9984

   d 887.172   **e** 55.144 55   **f** 0.007 49

12 Calculate.

   a 13.06×100 **b** $208.5 \div 100$ **c** 1.085×1000

   d $2487 \div 1000$ **e** $0.008 \div 10$ **f** $0.006\ 19 \times 1000$

   g $45.13 \div 1000$ **h** $0.000\ 045 \times 100$

Adding and subtracting negative numbers

This spread will show you how to:
- Add and subtract with positive and negative numbers

Keywords
Directed number
Negative
Number line
Positive

A number with a plus or minus sign is a **directed number**.
You can extend the basic rules of addition and subtraction to
include negative numbers.

- Adding a **positive** number counts as addition. Move right along the **number line**.
- Subtracting a positive number counts as subtraction. Move left along the number line.
- Adding a **negative** number counts as subtraction. Move left along the number line.
- Subtracting a negative number counts as addition. Move right along the number line.

For subtracting a
negative number,
think of reducing
an overdraft, or
taking ice cubes
out of a cold drink.

Example

Work out   **a** −5 + −6   **b** +4 − −2   **c** −7 − +2   **d** −5 + +8

a Start at −5 on the number line and
move 6 places to the left.
The answer is −11.

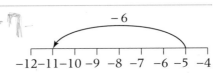

b Start at +4 on the number line and move 2
places to the right. The answer is +6.

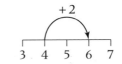

c Start at −7 on the number line and
move 2 places to the left. The answer is −9.

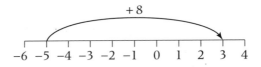

d Start at −5 on the number line
and move 8 places to the right.
The answer is +3.

Example

Ben writes: −5 + −2 = +7
Is he correct?

> Two minuses make a plus.
> I've got −5 and −2, so the
> answer must be positive.

Examiner's tip
Avoid simple rules
like 'two minuses
make a plus',
which can be
misleading. Use
the number line.

No.
Using the number line, you start at −5 and move
2 places to the left.
The correct answer is −5 + −2 = −7.

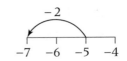

1 Calculate.

   a $+7 - +9$ **b** $+5 - +6$ **c** $+8 - +10$

   d $-7 + +5$ **e** $-11 + +6$ **f** $-7 + +2$

2 Calculate.

   a $-7 + +8$ **b** $-9 + +12$ **c** $-6 + +10$

   d $+3 - -6$ **e** $+5 - -7$ **f** $+2 - -3$

3 Calculate.

   a $-9 - -4$ **b** $-8 - -6$ **c** $-5 - -1$

   d $-6 - -8$ **e** $-5 - -9$ **f** $-3 - -10$

4 Copy and complete these calculations by replacing the boxes with the correct number or sign.

   a $-3 + \square = -5$ **b** $+7 \,\square\, -5 = +2$

   c $\square 8 + \square 5 = +3$ **d** $\square 2 + \square 11 = -13$

5 Calculate.

   a $+8 - -14$ **b** $-1 + -11$ **c** $-9 - -7$

   d $+3 + -17$ **e** $+8 - -4$ **f** $+13 + -1$

   g $+48 - +29$ **h** $-19 + +4$ **i** $+34 + -23$

   j $-104 + +43$ **k** $+208 - -136$ **l** $+347 + -298$

6 Calculate.

   a $-4.5 + -6.3$ **b** $+2.8 - -3.5$ **c** $+5.6 - -7.9$

   d $-9.4 + +8.7$ **e** $-26.5 + -11.7$ **f** $+45.9 - -66.8$

7 Find the balance in these bank accounts after the transactions shown.

   a Opening balance £133.45. Deposits of £45.55 and £63.99, followed by withdrawals of £17.50 and £220.

   b Opening balance is −£459.77. Deposit of £650, followed by a withdrawal of £17.85.

> A negative number represents an overdraft.

8 Find the final temperatures in these science experiments.

   a Starting temperature 55 °C. It goes up 32°, then down 100°.

   b Starting temperature −15 °C. It goes down 28°, increases by 75°, and then goes down 17°.

   c Starting temperature −22 °C. It goes up 12°, then down 2°, then increases by 53°.

This spread will show you how to:

- Multiply and divide with negative numbers

Keywords

Negative
Positive

For multiplication and division, these simple rules tell you the sign of the answer when negative numbers are multiplied or divided.

- positive number × positive number = positive number
- positive number × negative number = negative number
- negative number × negative number = positive number

The same rules apply to division.

Example

Calculate **a** $+4 \times +3$ **b** $-5 \times +4$ **c** $+7 \times -2$
  d -6×-2 **e** $-5 \times +7$

a $+4 \times +3 = +12$ **b** $-5 \times +4 = -20$ **c** $+7 \times -2 = -14$
d $-6 \times -2 = +12$ **e** $-5 \times +7 = -35$

Example

Calculate **a** $-12 \div +2$ **b** $+50 \div +2$ **c** $+24 \div -8$
  d $-18 \div -3$ **e** $-48 \div +4$

a $-12 \div +2 = -6$ **b** $+50 \div +2 = +25$ **c** $+24 \div -8 = -3$
d $-18 \div -3 = +6$ **e** $-48 \div +4 = -12$

These examples use both rules.

Example

Calculate **a** $+4 - +3 \times -2$ **b** $\dfrac{-2 \times +3}{+2 + -4}$

  c $+5 - -2 \times +3$ **d** $2(3 \times -4) \div 4(-5 \times 2)$

a $(+4) - (+3) \times (-2)$

$= (+4) - (-6)$

$= +10$

b $\dfrac{-2 \times +3}{+2 + -4}$

$= \dfrac{-6}{-2}$

$= +3$

c $(+5) - (-2) \times (+3)$

$= (+5) - (-6)$

$= +11$

d $2(3 \times -4) \div 4(-5 \times 2)$

$= 2(-12) \div 4(-10)$

$= \dfrac{-24}{-40}$

$= \dfrac{3}{5}$ Cancel by -8

Carry out multiplication and division before addition and subtraction.

When multiplying or dividing negative numbers, the combination of the signs gives the sign of the answer.

B
O
D
M
A
S

1 Calculate.

    a +5 × −3   **b** +2 × −9   **c** +7 × −3

    d −8 × +7   **e** −4 × +9   **f** −6 × +2

2 Calculate.

    a −4 × −4   **b** −2 × −8   **c** −3 × −5

    d −6 × −7   **e** −7 × −8   **f** −9 × −9

3 Calculate.

    a +5 × −5   **b** +4 × −8   **c** −8 × +9   **d** −4 × +5

    e −3 × −10   **f** −7 × −7   **g** +8 × +2   **h** +5 × −4

    i −2 × +9   **j** −13 × −2   **k** −7 × +6   **l** +12 × −4

4 Calculate.

    a −36 ÷ +12   **b** −16 ÷ +4   **c** +28 ÷ −4   **d** +18 ÷ −9

    e −38 ÷ −2   **f** −80 ÷ −16

5 Calculate.

    a −18 ÷ +9   **b** −20 ÷ +4   **c** −30 ÷ −6   **d** −12 ÷ −3

    e −66 ÷ +3   **f** +47 ÷ −47   **g** −80 ÷ −2   **h** +24 ÷ +6

    i −45 ÷ −9   **j** −51 ÷ +3   **k** +57 ÷ −19   **l** −81 ÷ −3

6 Copy and complete these calculations, replacing the boxes with the correct positive or negative number or sign.

    a −7 × □8 = −56   **b** +48 ÷ □ = −8

    c □ ÷ +45 = +1   **d** +108 ÷ □ = −9

7 Multiply these numbers by 10.

    a +45   **b** −15   **c** +6.3   **d** −2.5   **e** −0.073   **f** +0.0092

8 Multiply these numbers by −10.

    a +4.9   **b** −6.3   **c** −0.377   **d** +61.97   **e** −14.09   **f** −0.009

9 Divide these numbers by +10.

    a −360   **b** +1   **c** −9.8   **d** −0.087   **e** +0.073   **f** −0.0006

10 Divide these numbers by −10.

    a +550   **b** −4.8   **c** −52.66   **d** +1560   **e** −0.082   **f** +5.0005

11 Calculate.

    a +18 ÷ +100   **b** +9 × −3   **c** −14 ÷ +2   **d** −3.8 × +100

This spread will show you how to:

● Use the concepts of factors, prime numbers and prime factor decomposition

Keywords
Factor
Prime
Prime factor
     decomposition
Product

You can write a number as a **product** of **factors** in different ways.

● A **prime** number is a number with only two factors – itself and 1.
$17 = 17 \times 1$

● Any number greater than 1 can be written as a product of its prime factors. This is called the **prime factor decomposition** of the number.
$56 = 2 \times 2 \times 2 \times 7 = 2^3 \times 7$

● There is only one prime factor decomposition for any number.

The factor tree method is a good way to find factors systematically.

Example

Express 84 and 112 as a product of their prime factors.

Start by dividing by the smallest factor.

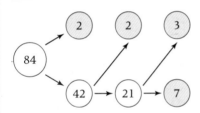

$84 = 2 \times 2 \times 3 \times 7 = 2^2 \times 3 \times 7$

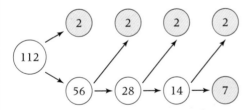

$112 = 2 \times 2 \times 2 \times 2 \times 7 = 2^4 \times 7$

Example

Find the prime factor decomposition of 990.

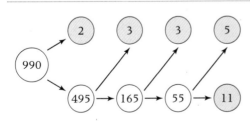

$990 = 2 \times 3^2 \times 5 \times 11$

Examiner's tip
A factor tree helps you keep track of all the prime factors, even if you don't find them in the right order.

1 Copy and complete these calculations to show the different ways that 24 can be written as a product of its factors.

   a $24 = \boxed{12} \times 2$ **b** $24 = 3 \times \boxed{8}$ **c** $24 = 2 \times 3 \times \boxed{3}$ **d** $24 = 4 \times \boxed{6}$

2 Each of these numbers has just two prime factors, which are not repeated. Write each number as the product of its prime factors.

   a 77   **b** 51   **c** 65   **d** 91   **e** 119   **f** 221

3 Copy and complete the factor tree to find the prime factor decomposition of 18.

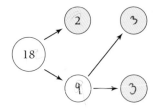

4 For each number, find its prime factors and write it as the product of powers of its prime factors.

   a 36   **b** 120   **c** 34   **d** 25

   e 48   **f** 90   **g** 27   **h** 60

5 Write the prime factor decomposition for each of these numbers.

   a 1052   **b** 2560   **c** 630   **d** 825

   e 715   **f** 1001   **g** 219   **h** 289

   i 2840   **j** 2695   **k** 1729   **l** 3366

   m 9724   **n** 11 830   **o** 2852   **p** 10 179

6 A cuboid is made from 210 small cubes.
   The prime factor decomposition of 210 is $2 \times 3 \times 5 \times 7$.
   One way of combining the prime factors is $(2 \times 3) \times 5 \times 7 = 6 \times 5 \times 7$.
   The dimensions of the cuboid could be $6 \times 5 \times 7$.

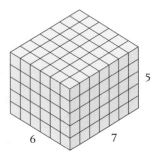

   a Find all the ways in which the prime factors of 210 can be combined to make 3 factors.

   b Using your answer to part **a**, list the dimensions of all the cuboids that could be made with 210 cubes.

This spread will show you how to:

- Find the highest common factor and lowest common multiple of two numbers

Keywords

HCF (highest common factor)
LCM (least common multiple)
Multiple
Prime
Venn diagram

- The **highest common factor (HCF)** of two numbers is the largest number that is a factor of both of them.

The HCF of 12 and 18 is 6.

- The **least common multiple (LCM)** of two numbers is the smallest number that is a **multiple** of both of them.

The LCM of 12 and 18 is 36.

To find the HCF of two numbers:

- list all the factors of each number
- identify the largest factor that is in both lists.

For example:

Factors of 18 = {1, 2, 3, 6, 9, 18}
Factors of 24 = {1, 2, 3, 4, 6, 8, 12, 24}
HCF of 18 and 24 = 6

To find the LCM of two numbers:

- list the first few multiples of each number
- identify the smallest multiple that is in both lists.

For example:

Multiples of 18 = 18, 36, 54, 72, 90, ...
Multiples of 24 = 24, 48, 72, 96, ...
LCM of 18 and 24 = 72

You can find the HCF and LCM by writing the **prime** factor decomposition for each number in a **Venn diagram**.

Note that there are other methods for finding LCM and HCF.

Example

Find the HCF and LCM of 36 and 28.

Find the prime factor decomposition of each number: $36 = 2^2 \times 3^2$ and $28 = 2^2 \times 7$.

Write these prime factors in a Venn Diagram.

The common factors are in the middle (the 'intersection').

The HCF of 36 and 28 is the product of the numbers in the intersection: $2 \times 2 = 4$.

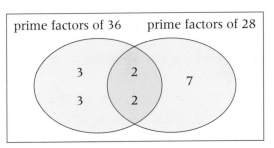

The LCM is the product of all of the numbers in the diagram:
$2^2 \times 3^2 \times 7 = 4 \times 9 \times 7 = 252$

1 The diagram shows how a student listed all the factors of 36. The lines show how the pairs of factors combine to make 36.

Factors of 36 = 1, 2, 3, 4, 6, 9, 12, 18, 36

a Copy the diagram, and explain why the factor 6 has not been joined to another factor.

b List the factors of 48 in the same way. Draw lines to show the factor pairs.

c The highest common factor (HCF) of 36 and 48 is the largest number that is in both lists. Write down the HCF of 36 and 48.

2 Using the method from question **1**, find the HCF of these pairs of numbers.

a 7 and 8   b 4 and 5   c 6 and 9

d 14 and 32   e 8 and 24   f 50 and 70

3 The diagram below shows how a student found the least common multiple (LCM) of 8 and 6.

multiples of 8 = 8, 16, 24, 32, 40, 48 . . .
multiples of 6 = 6, 12, 18, 24, 30, 36 . . .

a List the multiples of 12 and 9 in the same way.

b The least common multiple (LCM) of 12 and 9 is the smallest number that is in both lists.
Write the LCM of 12 and 9.

4 Using the method from question **3**, find the LCM of these pairs of numbers:

a 4 and 5   b 12 and 18   c 5 and 30

d 12 and 30   e 14 and 35   f 8 and 20

5 Use the Venn diagram method to find the HCF of 24 and 80.

6 Use the Venn diagram method to work out the HCF and LCM of these pairs of numbers.

a 25 and 120   b 16 and 108   c 60 and 144

d 42 and 56   e 15 and 35   f 20 and 110

DID YOU KNOW?
In 1880, John Venn, an English logician, introduced the logic diagrams that now bear his name.

Exam review

Key objectives

- Understand highest common factor, least common multiple, prime number and prime number decomposition
- Multiply and divide by a negative number

1 **a** Using the information that $7.4 \times 3.5 = 25.9$, write down the value of

    i 7.4×-3.5

    ii -7.4×3.5

    iii -7.4×-3.5 (3)

  b Use a mental method to work out

    i 5.32×-10

    ii $-53.2 \div 100$

    iii $-5320 \div -1000$ (3)

2 **a** Express the following numbers as products of their prime factors.

    i 60

    ii 96   (4)

  b Find the HCF of 60 and 96.   (1)

  c Work out the LCM of 60 and 96.   (2)

(Edexcel Ltd., 2003)

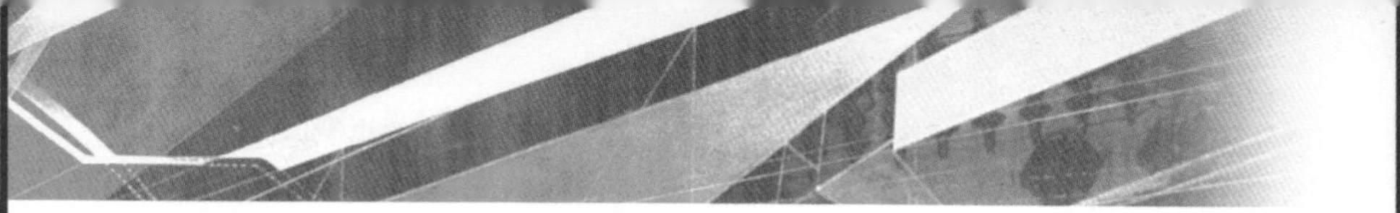

This unit will show you how to

- Calculate the perimeter and area of shapes made from rectangles and triangles
- Calculate the area of parallelograms and trapeziums
- Use the correct vocabulary to describe the parts of a circle
- Calculate the area and circumference of circles
- Calculate the length of an arc and the area of a semicircle
- Calculate the surface area of simple 3-D shapes

Before you start ...

You should be able to answer these questions.

Review

1 These shapes are drawn on a centimetre square grid.
Use shapes **a–c** to answer questions **i–iii**.

Key stage 3

a  **b**

c

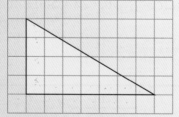

  i   Write the length of each side in cm.
      Hence find the perimeter in cm.
  ii   Find the perimeter in mm.
  iii   Calculate the area of the shape in cm².

Key stage 3

2 Work out the surface area of this cuboid.

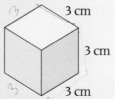

3 cm
3 cm
3 cm

This spread will show you how to:
- Calculate the perimeter and area of shapes made from rectangles and triangles

Keywords
Area
Perimeter
Rectangle
Triangle

The **area** of a shape is the amount of space it covers.

- You can use formulae to find the areas of **rectangles** and **triangles**.

Rectangle

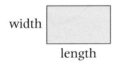

Area = length × width

The area of the triangle is half the rectangle.

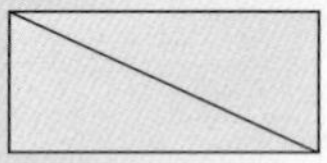

Triangle

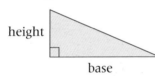

Area = $\frac{1}{2}$ × base × height

The height of a triangle is always at right angles to the base.

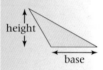

Example

Find the area of each shape.

a 9 cm, 2 cm

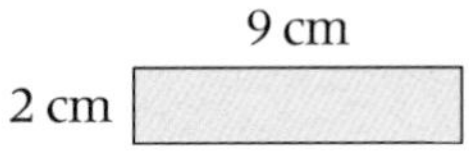

b 50 mm, 22 mm

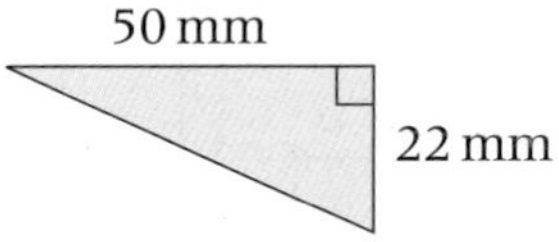

a Area = length × width
= 9 × 2 = 18 cm^2

b Area = $\frac{1}{2}$ base × height
= $\frac{1}{2}$ × 50 × 22 = 550 mm^2

Remember to write the units.
The ^{2} shows the measurement is an area.

- You can split compound shapes into rectangles and triangles.

Example

Find **a** the perimeter
b the area of this shape.

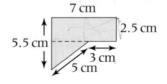

7 cm, 2.5 cm, 5.5 cm, 3 cm, 5 cm

Perimeter is the distance around a shape.

---

a Perimeter = 5.5 + 7 + 2.5 + 3 + 5
= 23 cm

b Area = area of rectangle + area of triangle
= (7 × 2.5) + ($\frac{1}{2}$ × 4 × 3)
= 17.5 + 6
= 23.5 cm^2

Triangle:
height = 7 − 3 = 4 cm
base = 5.5 − 2.5 = 3 cm

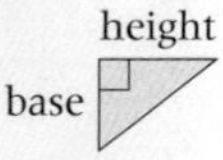

1 Calculate the areas of these rectangles.

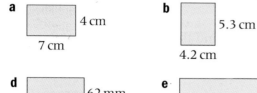

a 4 cm / 7 cm

b 5.3 cm / 4.2 cm

c 8.7 cm / 3 cm

d 62 mm / 120 mm

e 49 mm / 210 mm

2 Find the areas of these triangles.

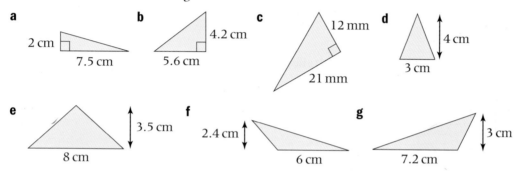

a 2 cm / 7.5 cm

b 4.2 cm / 5.6 cm

c 12 mm / 21 mm

d 4 cm / 3 cm

e 3.5 cm / 8 cm

f 2.4 cm / 6 cm

g 3 cm / 7.2 cm

3 Split these shapes into rectangles and triangles to work out
i the perimeter   **ii** the area.

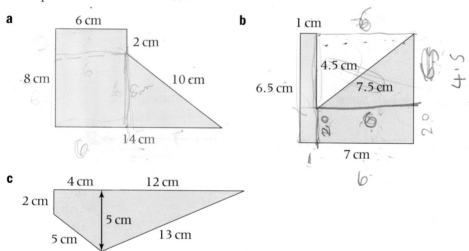

a 6 cm / 2 cm / 8 cm / 10 cm / 14 cm

b 1 cm / 4.5 cm / 6.5 cm / 7.5 cm / 7 cm

c 4 cm / 12 cm / 2 cm / 5 cm / 5 cm / 13 cm

4 Pete is making a mobile out of shapes like this:

He cuts the shape out of a piece of card that is
30 cm × 20 cm.

What is the area of the card left over?

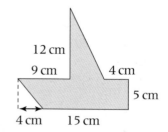

12 cm / 9 cm / 4 cm / 5 cm / 4 cm / 15 cm

15

Area of a parallelogram and a trapezium

This spread will show you how to:

● Calculate the area of parallelograms and trapeziums

Keywords

Area
Parallelogram
Trapezium

You can cut a triangle from one side and stick it on the other.

The height is at right angles to the base.

You can use the formula for the **area** of a rectangle to find the formula for the area of a **parallelogram**.

This parallelogram can be made into a rectangle.

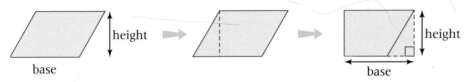

● **Area of parallelogram = base × height.**

Two congruent **trapeziums** make a parallelogram.

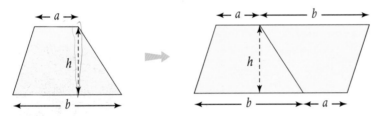

Area of the parallelogram = base × height

$$= (a + b) \times h$$

● **Area of trapezium** $= \frac{1}{2} \times (a + b) \times h$

The area of a trapezium is half the sum of the parallel sides times the distance between them.

Example

Find the areas of these shapes.

a

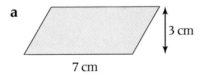

3 cm

7 cm

b

4 cm

3 cm

6 cm

a Area of parallelogram

= base × height

= 7 × 3

= 21 cm^2

b Area of trapezium

$= \frac{1}{2} \times (a + b) \times h$

$= \frac{1}{2} \times (4 + 6) \times 3$

= 5 × 3

= 15 cm^2

1 Find the area of each parallelogram.

a
2.5 cm
6 cm

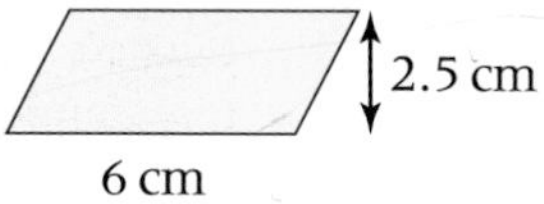

b
6.2 cm
5.4 cm

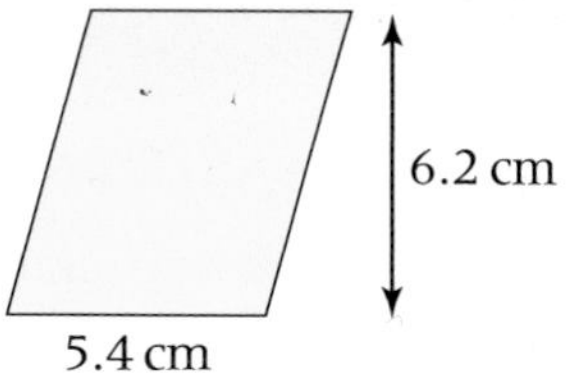

c
3.8 cm
12 cm

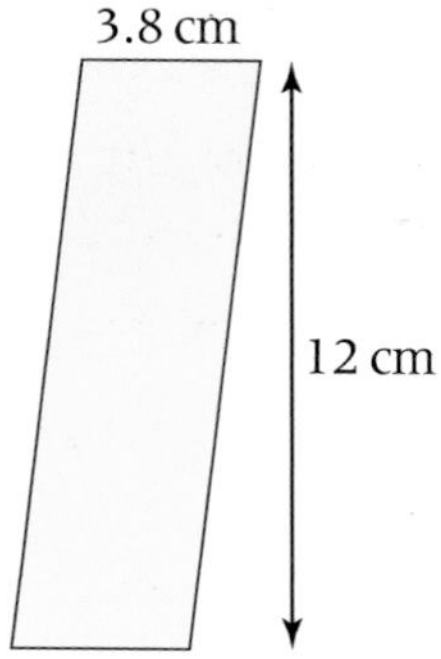

d
4.2 cm
7.5 cm

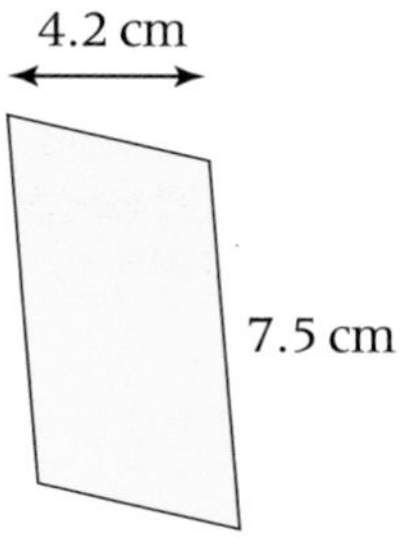

e
2.9 cm
4.6 cm

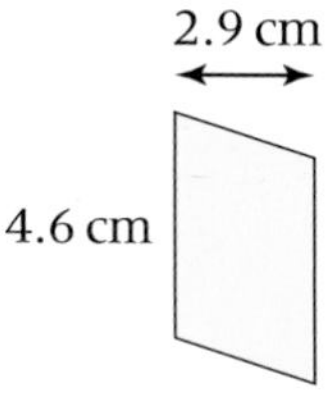

2 Find the area of each trapezium.

a
2 cm
4 cm
4 cm

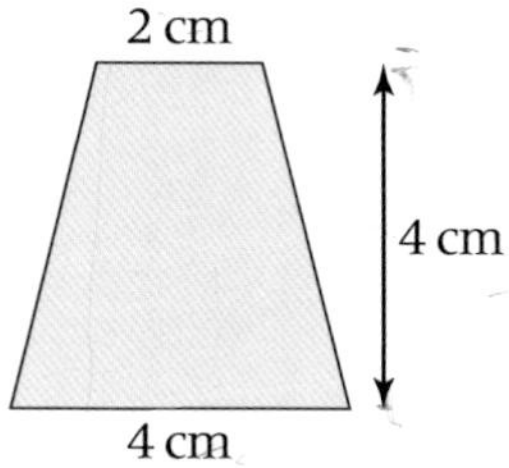

b
3 cm
4 cm
7 cm

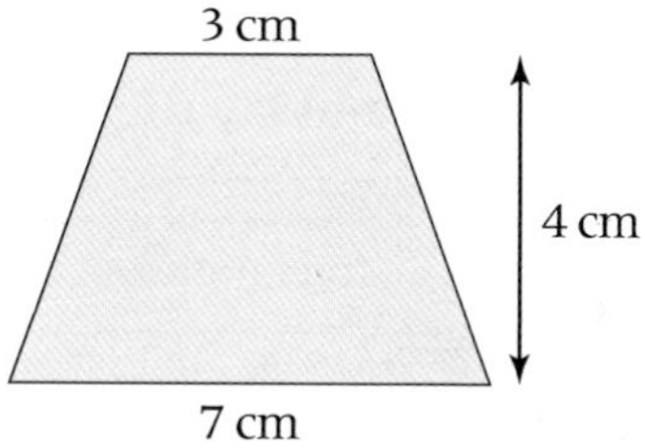

c
5 cm
5 cm
3.2 cm

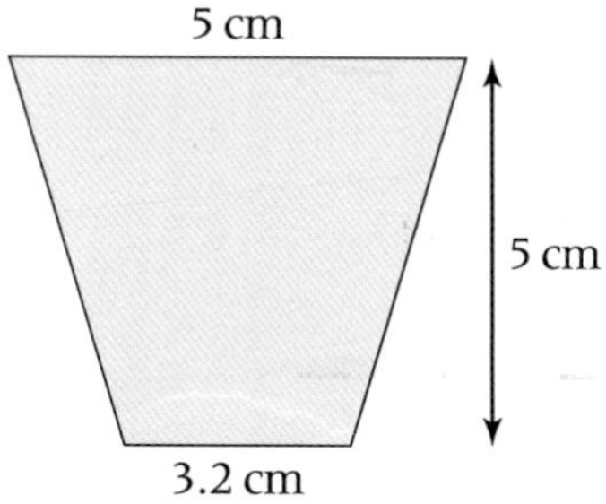

d 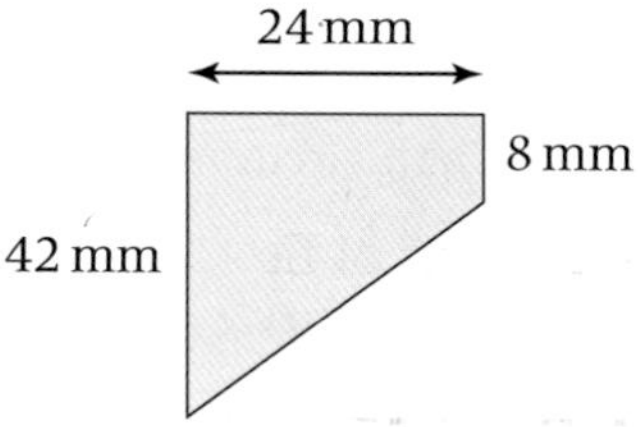
24 mm
8 mm
42 mm

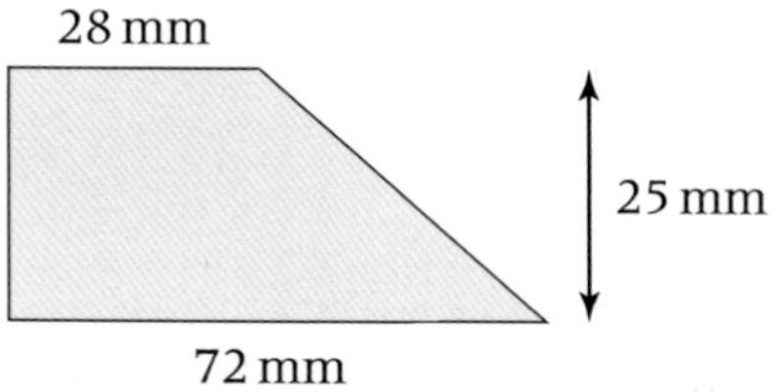

DID YOU KNOW?

The word 'trapezium' originates from the shape made by the ropes and bar of an old-fashioned flying trapeze.

e
28 mm
25 mm
72 mm

3 Caroline has drawn a sandcastle.
What is the area of her castle and flag?
Start by dividing the shape into parts.

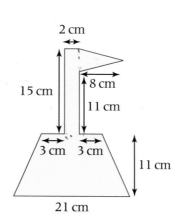

2 cm
15 cm
8 cm
11 cm
3 cm   3 cm
11 cm
21 cm

Area and circumference of a circle

This spread will show you how to:

Keywords
Area
Circle
Circumference
Diameter
Radius
Proportion

- Use the correct vocabulary to describe the parts of a circle
- Calculate the area and circumference of circles

In a **circle**:

- The **diameter**, d, is the distance across the circle through the centre.
- The **radius**, r, is the distance from the centre to the edge.
- The **circumference**, C, is the perimeter – the distance around the edge.

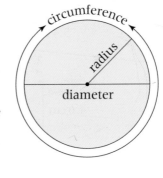

$d = 2 \times r$

The circumference of a circle is in **proportion** to its diameter: $C \approx 3d$

The actual proportion is a decimal. You use a symbol, π (pi).

π is about 3.14

- $C = \pi \times d$ or $C = 2 \times \pi \times r$

You can cut a circle into lots of sectors and lay them out side by side, to make a rectangle.

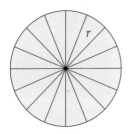
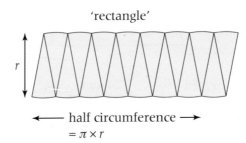

'rectangle'

← half circumference →
$= \pi \times r$

Area of 'rectangle' = length × width = $(\pi \times r) \times r = \pi \times r^2$

- Area of a circle = $\pi \times r^2$ or $A = \pi \times r^2$

Find   **i** the circumference   **ii** the area of each circle.

a

$r = 5\,\text{cm}$

b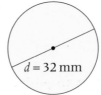

$d = 32\,\text{mm}$

Use the π key on your calculator.

a i $C = 2 \times \pi \times r = 2 \times \pi \times 5$
   $= 31.415\ldots$
   $= 31.4\,\text{cm}$ (to 3 sf)
ii $A = \pi \times r^2 = \pi \times 5^2$
   $= 78.539\ldots$
   $= 78.5\,\text{cm}^2$ (to 3 sf)

b i $C = \pi \times d = \pi \times 32$
   $= 100.530\ldots$
   $= 101\,\text{mm}$ (to 3 sf)
ii $A = \pi \times r^2 = \pi \times 16^2$
   $= 804.247\ldots$
   $= 804\,\text{mm}^2$ (to 3 sf)

Round your answers to a sensible degree of accuracy.
3 sf is usually good practice.

1 Find the circumferences of these circles.

a 4 cm

b 38 mm

c 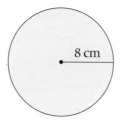 8 cm

d 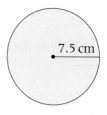 7.5 cm

e 24 mm

f  42 mm

g 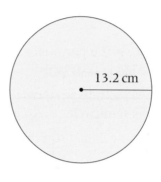 13.2 cm

Give your answers to 3 sf. Significant figures (sf) are covered on page 38.

2 Find the circumferences of these circles.

   a radius = 12 mm   **b** radius = 23 cm   **c** diameter = 105 mm

   d diameter = 1.2 cm   **e** radius = 3.6 cm   **f** diameter = 125 cm

3 Find the areas of the circles in question **1**.

4 Find the areas of the circles in question **2**.

5 A circular pond has radius 3 m.
Work out the circumference of the pond.

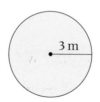 3 m

6 A round hole has circumference of 44 cm.
Work out the radius of the hole, to 1 decimal place.

7 Shamin is cutting out circles for an art project.
She has squares of card that are 4.2 cm wide.

   a What is the area of the biggest circle she can cut out?

   b What area of card is left?

4.2 cm

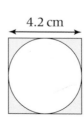

Area and perimeter of a semicircle

This spread will show you how to:

● Calculate the length of an arc and the area of a semicircle

Keywords
Area
Circumference
Diameter
Semicircle

A **semicircle** is half a circle.

🔹 **Area of a semicircle $= \frac{1}{2} \times$ area of whole circle**

Example

Calculate the area of this semicircle.

← 6 cm →

Area $= \frac{1}{2} \times (\pi \times r^2)$
$= \frac{1}{2} \times (\pi \times 6^2)$
$= \frac{1}{2} \times 113.097 \ldots$
$= 56.548 \ldots \text{ cm}^2$
$= 56.5 \text{ cm}^2$ (1 dp)

● **Perimeter of a semicircle $= \frac{1}{2} \times$ circumference of whole circle + diameter**

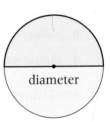

diameter

Example

Calculate the perimeter of this semicircle.

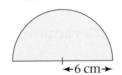

← 6 cm →

Circumference of whole circle $= 2 \times \pi \times r$
$= 37.699 \ldots \text{ cm}$
Perimeter of semicircle $= \frac{1}{2} \times$ circumference of whole circle + diameter
$= (\frac{1}{2} \times 37.699\ldots) + (2 \times 6)$
$= 18.849 \ldots + 12$
$= 30.8 \text{ cm}$ (to 1 dp)

Example

A door is shaped in an arch, with a semicircle on top.

Calculate the perimeter of the door, giving your answer to 1 decimal place.

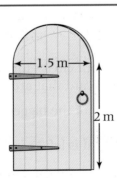

← 1.5 m →

2 m

Perimeter of arch
$= 2 \text{ m} + 1.5 \text{ m} + 2 \text{ m} + \frac{1}{2} \times$ circumference of circle
Circumference $= 2 \times \pi \times r = 1.5 \times \pi = 4.712\ldots$
Perimeter of arch $= 2 + 1.5 + 2 + \frac{1}{2} \times 4.712$
$= 7.9 \text{ m}$ (to 1 dp)

1 Find the area of each semicircle.

Give your answers to 1 dp.

a

5 cm

b
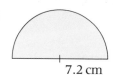
7.2 cm

c

24 mm

d

← 18 cm →

e
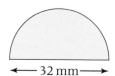
← 32 mm →

f

15.4 cm

g
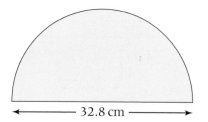
← 32.8 cm →

h
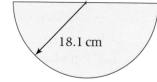
18.1 cm

2 Find the area of each semicircle.

   a radius = 12 cm   **b** radius = 2.3 cm   **c** diameter = 12.9 m

   d diameter = 22.3 cm   **e** radius = 9.5 mm   **f** diameter = 3.39 cm

3 Work out the perimeter of each semicircle in question **1**.

4 Work out the perimeter of each semicircle in question **2**.

5 A flowerbed in the park is semicircular.
It has a radius of 2 m.

   a Work out the area of the flowerbed.

Percy the park keeper wants to plant
flowers that each need an area of 0.3 m².

   b How many of these flowers can Percy plant in
the flowerbed? What space does he have left?

6 Viaduct arches have straight sides 50 m high.
The arch at the top is a semicircle with diameter 8 m.

A spider crawls from ground level on one side,
around the arch, and back down the other side.
Work out how far it crawls.

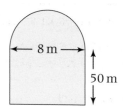
← 8 m →
50 m

7 A bathroom window is a semicircle with an internal
diameter of 80 cm.
Work out the area of the glass in the window.

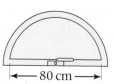
← 80 cm →

Surface area of 3-D shapes

This spread will show you how to:

- Find the surface area of simple 3-D shapes

Keywords
Cuboid
Cylinder
Prism
Surface area

- **Surface area** is the total area of all the faces of a 3-D shape.

To find the surface area, first imagine the net of the shape.

The faces of this **cuboid** are in pairs.

Front/back:	Area = 6 × 4 = 24
Side:	Area = 4 × 1.5 = 6
Top/bottom:	Area = 6 × 1.5 = 9

Surface area $= 2 \times (24 + 6 + 9)$
$= 2 \times 39$
$= 78$ cm^2

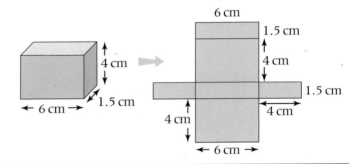

Example

Find the surface area of this triangular **prism**.

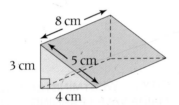

Triangle:	Area $= \frac{1}{2} \times 4 \times 3 = 6$
Triangle:	Area $= \frac{1}{2} \times 4 \times 3 = 6$
Side:	Area $= 3 \times 8 = 24$
Bottom:	Area $= 4 \times 8 = 32$
Sloping side:	Area $= 5 \times 8 = 40$

The two end faces of a prism are identical.

Surface area
$= 6 + 6 + 24 + 32 + 40 = 108$ cm^2

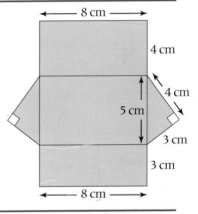

Example

Calculate the surface area of this **cylinder**.

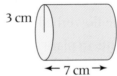

The curved surface of a cylinder is a rectangle.
Length of rectangle = circumference of circle.
Width of rectangle = height of cylinder.

Area of top circle
$\pi \times 3^2 = 28.27$
Area of bottom circle
$\pi \times 3^2 = 28.27$
Area of curved surface
$2 \times \pi \times 3 \times 7 = 131.95$

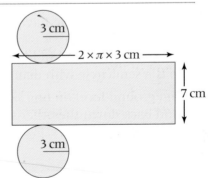

Surface area of cylinder

$= 28.27 + 28.27 + 131.95$
$= 188.5$ cm^2 (to 1 dp)

1 Work out the surface areas of these cuboids.

Give your answers to 1 dp.

a
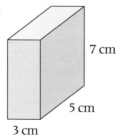
7 cm
5 cm
3 cm

b
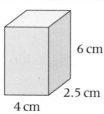
6 cm
2.5 cm
4 cm

c
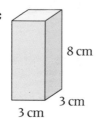
8 cm
3 cm
3 cm

d

7.2 cm
2 cm
2 cm

e
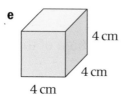
4 cm
4 cm
4 cm

f
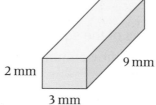
9 mm
2 mm
3 mm

2 Work out the surface areas of these cylinders.

a 2 cm

6 cm

b 5 cm
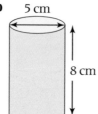
8 cm

c 4 cm

4 cm

d 3.2 cm

5 cm

The top measurements give the diameters.

3 Work out the surface areas of these prisms.

a
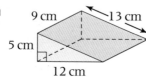
9 cm
13 cm
5 cm
12 cm

b
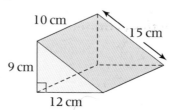
10 cm
15 cm
9 cm
12 cm

4 A scout troop is making a tent out of canvas and a groundsheet out of PVC.

a What area of canvas do they need for the cover (including front and back flaps)?

b What area of PVC do they need for the groundsheet?

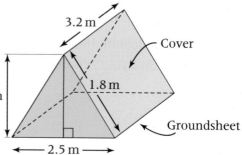
3.2 m
Cover
1.8 m
1.3 m
Groundsheet
2.5 m

Key objectives
- Calculate perimeters and areas of shapes made from triangles and rectangles
- Find circumferences of circles and areas enclosed by circles
- Solve problems involving surface areas and volumes of prisms and cylinders

1 The area of the square is 16 times the area of the triangle.

Work out the perimeter of the square. (5)

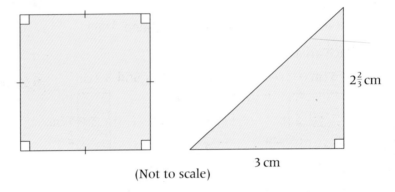

(Not to scale)

$2\frac{2}{3}$ cm

3 cm

2

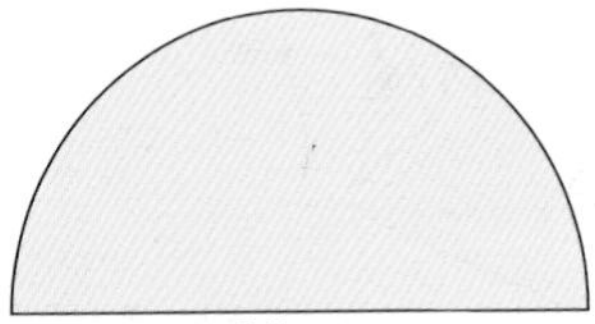

(Not to scale)

15 cm

The diagram shows a semicircle.

The diameter of the semicircle is 15cm.

Calculate the area of the semicircle.

Give your answer correct to 3 significant figures. (3)

> Significant figures are covered in N2.1 page 38.

(Edexcel Ltd., 2003)

This unit will show you how to

- Use the rules of algebra to write and manipulate algebraic expressions
- Simplify algebraic expressions by collecting like terms
- Use index notation and simple laws of indices
- Multiply a single term over a bracket
- Expand double brackets
- Factorise into single and double brackets

Before you start ...

You should be able to answer these questions.

Review

1 Work out these multiplications and divisions mentally.

    a 15×3 **b** 4×13

    c $(-2) \times 13$ **d** 14×14

    e $56 \div 8$ **f** $91 \div 7$

    g $150 \div (-3)$ **h** $1200 \div 40$

Key stage 3

2 Explain why the answer to each of these questions is 15.

    a $9 + 3 \times 2$ **b** $24 - 9 \div 3$

    c $(3 + 2) \times 3$ **d** $3^3 - 4 \times 3$

Key stage 3

3 Find all the factors of these values.

    a 24   **b** 36

    c 100   **d** 121

Key stage 3

4 Find the highest common factor of these number pairs.

    a 6 and 9   **b** 8 and 12

    c 20 and 30   **d** 12 and 18

    e 24 and 52   **f** 50 and 75

    g 99 and 132   **h** 7 and 14

Key stage 3

Keywords

Factorise
Index/indices
Like terms
Product
Simplify

This spread will show you how to:

- Use the rules of algebra to write and manipulate algebraic expressions
- Simplify algebraic expressions by collecting like terms
- Use index notation and simple laws of indices

There are rules for writing expressions in algebra

- Do not include the multiplication sign $\quad 3 \times p \rightarrow 3p$
- Write divisions as fractions $\quad 3 \div p \rightarrow \frac{3}{p}$
- Write numbers first in products $\quad p \times 3 \rightarrow 3p$
- Write letters in products in alphabetical order $\quad 4 \times q \times r \times p \rightarrow 4pqr$

- To **simplify** expressions involving addition or subtraction, you collect **like terms** together.

Like terms: $\qquad$ Unlike terms:
$3z, 9z, -4z \qquad 5p, 2p^2, 8q$

- To simplify expressions involving multiplication or division, you multiply the numbers then the letters.

You may need to use **indices**.

Example

a Write these using the rules of algebra.
  i $5 \times q \times 3 \times p$
  ii $y \times y \times y \times y \times y$
b Evaluate $3x + 2$ when $x = -4$

a i $5 \times q \times 3 \times p = 15pq$
  ii $y \times y \times y \times y \times y = y^5$
b $3x + 2 = 3 \times (-4) + 2 = -12 + 2 = -10$

Numbers first, letters in alphabetical order

$3x = 3 \times x$

Example

Simplify these algebraic expressions.

a $3p + 9q - 2p + 7q$ $\qquad$ b $5q - 7 + q^2$
c $7ab + 3ba$ $\qquad$ d $3t \times 4t \times 2t \times 2s$

e $\dfrac{15ab}{5b}$

a $3p + 9q - 2p + 7q = 3p - 2p + 9q + 7q$
$\qquad\qquad\qquad\qquad = p + 16q$
b $5q - 7 + q^2$ cannot be simplified as there are no like terms
c $7ab + 3ba = 7ab + 3ab = 10ab$
d $3t \times 4t \times 2t \times 2s = 48st^3$
e $\dfrac{15ab}{5b} = \dfrac{{}^3\cancel{15}a\cancel{b}}{\cancel{5}\cancel{b}} = 3a$

'1p' is simply 'p'

'q' and 'q^2' are not like terms

Cancelling – divide top and bottom by 5 and by b.

1 Write these expressions using the rules of algebra.

   a $5 \times w$ **b** $6 \div k$ **c** $y \times y$ **d** $ab6$ **e** $k \times k \times 8 \times k$

2 Evaluate these expressions, given that $x = 6$.

   a $3x + 2$ **b** $10 - x$ **c** x^2 **d** $\dfrac{10x - 16}{2}$ **e** $3x^2$

3 Simplify these expressions by collecting like terms.

   a $3a + 4b + 8a + 2b$ **b** $3t + 9 - t + 17$

   c $3x - 4y - 2x - 8y$ **d** $9p + p^2 + 5p$

   e $10xy + 10yx$ **f** $6ab + 2ba - ba$

4 Three students tried to simplify $3m + 5$. Which of them did it correctly?

Sara	Paul	Abdul
$3m + 5 = 8m$	$3m + 5 = 15m$	$3m + 5 = 3m + 5$

5 Simplify these expressions.

   a $4m \times 7n$ **b** $6m \times 2m$ **c** $\dfrac{20p}{2}$ **d** $\dfrac{14a}{7a}$

   e $2a \times 3b \times 4c$ **f** $k \times 2k \times 3k$ **g** $\dfrac{20ab}{5a}$ **h** $\dfrac{45c^2}{5c}$

6 Simplify the expressions in the grid and find the 'odd one out' for each row.

$3p + 2q + p + 5q$	$6p + 3q - 2p + 4q$	$5p - 3q - p + 5q$
$2m \times 3n$	$2 \times n \times m \times 5$	$6mn$
$\dfrac{24cd}{12c}$	$\dfrac{2d^2}{d^2}$	$\dfrac{2d^2}{d}$
$2n - 8$	$3m + 2n - m - 2m$	$3n - 2 - 6 - n$

7 **a** Write a simplified expression for the

     i perimeter   **ii** area

    of this rectangle

   b What are the measurements of a rectangle with perimeter $6x + 4y$ and area $6xy$?

8 Copy this grid, replacing each expression with its simplified form (where possible).

$3a + 7b - 5a + 2b$	$3a \times 4a$	$\dfrac{20b}{5}$
$\dfrac{16ab^2}{8b}$	$2p + 7p^2 + 5p^3 + 8p$	$11abc + 2cab$
$5m - 4$	$3m \times 4m \times 5m$	$\dfrac{4a}{2a^3}$

Expanding single brackets

This spread will show you how to:

- Multiply a single term over a bracket

Keywords

Bracket
Expand

A **bracket** in algebra means 'all multiplied by'.

$3(x + 5)$ means 'I think of a number, add 5 then **multiply it all** by 3'

To write an expression without brackets, multiply all the terms inside the bracket by the term outside.
This is called **expanding** the bracket.

$3(x + 5)$ expanded is $3x + 15$

For negative terms, use the rules for multiplying by negative numbers:

- negative term × positive term → negative term

$$-3(x + 5) = -3 \times x + -3 \times 5$$
$$= -3x - 15$$

- negative term × negative term → positive term

$$-5(y - 8) = -5 \times y + -5 \times -8$$
$$= -5y + 40$$

Example

Expand each of these.

a $5(m + 9)$ **b** $y(y - 7)$
c $3p(2p + 7 - q)$ **d** $-4m(m - 2)$

a $5(m + 9) = 5m + 45$
b $y(y - 7) = y^2 - 7y$
c $3p(2p + 7 - q) = 6p^2 + 21p - 3pq$
d $-4m(m - 2) = -4m^2 + 8m$

$y \times y$ is y^2

Be careful with negatives.

Example

Expand and simplify $3(t - 2) + 5(2 + t)$.

$$3(t - 2) + 5(2 + t) = 3t - 6 + 10 + 5t$$
$$= 3t + 5t - 6 + 10$$
$$= 8t + 4$$

Collect like terms.

Exam question

A cuboid with a square base of length x has height 1 cm more than the length. Its volume is 230 cm^3.
Show that $x^3 + x^2 = 230$

As the base is square, the length and width must both be x cm.
You are told that the height is 1 cm more than the length,
so height is $(x + 1)$ cm.
Volume of cuboid = length × width × height
$$230 = (x + 1) \times x \times x$$
$$230 = x^2(x + 1)$$
$$230 = x^3 + x^2 \text{ (as requested)}$$

(Edexcel Ltd., 2003)

Sketch a diagram to help:

Exercise A1.2

1 Expand these brackets.

a $4(n+5)$ **b** $6(b-7)$ **c** $a(a+3)$

d $a(b-c)$ **e** $4(2x+3y-4z)$ **f** $2h(h+9)$

2 Expand these brackets.

a $-3(k+9)$ **b** $-2(h-5)$ **c** $-(w-4)$

d $-(t-p)$ **e** $-k(k+7)$ **f** $-9(2m-k+4)$

g $-(x^2-x-8)$ **h** $-2(x^2+3)$ **i** $-3(1-x)$

> Be careful with negatives.

3 Expand and simplify these expressions.

a $3(c+2)+7(c+8)$ **b** $4(2x+8)+5(3x+7)$

c $x(x+8)+x(x+2)$ **d** $5t(3t+6)+2t(t+1)$

e $3(x-7)+4(x-6)$ **f** $5(2-x)+7(x-3)$

g $4(m-6)-2(m+1)$ **h** $3(g-3)-7(2g-6)$

i $2(p+5)-(p-4)$ **j** $(q-4)-(3-q)$

4 Expand and simplify $2x(x+7)+x(9-x)-3x(2x-7)$.

5 a Using brackets, write a formula for the area of this rectangle.

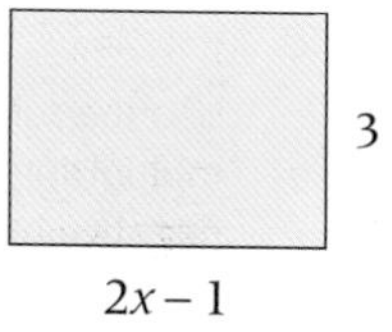

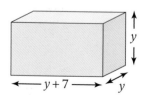

3

$2x-1$

b Expand the brackets.

c The area of the rectangle is 15 cm².
Show that $6x-18=0$.

6 An expression expands to give $24x+16$.

a What could the expression have been if it involved one pair of brackets?

b What could the expression have been if it involved two single brackets in succession?

7 Write an expression involving brackets for the volume of this cuboid.

y

$y+7$ y

Expand the brackets and simplify your expression.

Handwritten working (right margin):

$5(m+9)$
$= 5m+45$.

$y(y-7)$
y^2-7y

$3p(2p+7-q)$
$6p^2+21-3pq$.

$-4m(m-2)$
$-4m^2 + -8$

$3(t-2)+5(2+t)$
$3t-6 + 10+5t$
$3t+5t-6+10$
$8t-4$

Expanding double brackets

This spread will show you how to:

● Expand double brackets

You can **expand** a double bracket in algebra by multiplying pairs of terms.

Each term in the first bracket multiplies each term in the second bracket:

$$(p+7)\ (p+3) \longrightarrow p^2 + 3p + 7p + 21 \longrightarrow p^2 + 10p + 21$$

$p^2 + 10p + 21$ is the product of $(p+7)$ and $(p+3)$.

F... Firsts
O ... **O**uters
I ... **I**nners
L ... **L**asts

Example

Expand and simplify.

$(x+3)(x+4)$

$$(x+3)(x+4) = x^2 + 4x + 3x + 12$$
$$= x^2 + 7x + 12$$

F: $x \times x = x^2$
O: $x \times 4 = 4x$
I: $3 \times x = 3x$
L: $3 \times 4 = 12$

Example

Expand and simplify.

a $(x+3)(2x-2)$ **b** $(3x+2)^2$

a $(x+3)(2x-2) = 2x^2 - 2x + 6x - 6$
$$= 2x^2 + 4x - 6$$
b $(3x+2)^2 = (3x+2)(3x+2)$
$$= 9x^2 + 6x + 6x + 4$$
$$= 9x^2 + 12x + 4$$

Use the rules for multiplying negative terms:
O: $x \times -2 = -2x$
L: $3 \times -2 = -6$

Example

A rectangle has length $x+5$ and width $x-2$.
Show that $A = x^2 + 3x - 10$, where A is the area.

Sketch a diagram:

$x+5$

$x-2$

Area of rectangle = length × width
$$A = (x+5)(x-2)$$
$$= x^2 - 2x + 5x - 10$$
$$A = x^2 + 3x - 10$$

1 Expand and simplify these expressions involving double brackets.

    a $(x + 2)(x + 3)$ **b** $(p + 5)(p + 6)$

    c $(w + 1)(w + 4)$ **d** $(c + 5)^2$

    e $(x + 4)(x - 2)$ **f** $(y - 2)(y + 7)$

    g $(t + 6)(t - 2)$ **h** $(x - 2)(x - 5)$

    i $(y - 4)(y - 10)$ **j** $(w - 1)(w - 2)$

    k $(p - 5)^2$ **l** $(q - 12)^2$

2 Expand and simplify.

    a $(2x + 1)(3x + 7)$ **b** $(5p + 2)(2p + 3)$

    c $(3y + 4)(2y + 1)$ **d** $(2y + 6)^2$

    e $(5t - 4)(2t + 4)$ **f** $(5w - 1)(3w + 9)$

    g $(2x + 2y)(3x - 3y)$ **h** $(3m - 4)^2$

    i $(2p + 5q)(3p - 8q)$ **j** $(2m - 3n)^2$

3 Write an expression for the areas of this rectangle and square.

    a 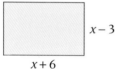 $x - 3$ **b**  $2m - 3$

        $x + 6$

4 **a** Expand $(a + b)^2$.

    b Hence, or otherwise, calculate $1.32^2 + 2 \times 1.32 \times 2.68 + 2.68^2$

    c Write another calculation that you could work out using this expansion.

5 **a** The diagram shows a rectangle with an area of 75 cm^2.
       Show that $6x^2 + 7x - 78 = 0$.

 $2x + 3$

       $3x - 1$

    b This triangle also has an area of 75 cm^2. Show that
       $15x^2 = 14x + 158$.

    $3x - 4$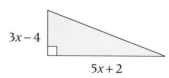

         $5x + 2$

This spread will show you how to:

- Factorise into single and double brackets

Keywords

Factorise
Highest common
factor (HCF)

The reverse of expanding a set of brackets is called **factorising**.
To factorise an expression, you put brackets in.

Expand

$12(x+2)$ $12x + 24$

Factorise

You divide the terms by their **highest common factor (HCF).**

HCF of $12x$ and 24 is 12

$12x + 24$ $12(x+2)$

Check your answer
by expanding

Example

Factorise fully.

a $6x + 9$ **b** $12pq - 4pw$ **c** $5x + 10x^2 - 25xy$

a The HCF of $6x$ and 9 is 3.

$6x + 9 = 3(2x + 3)$

b $12pq - 4pw$
Deal with numbers first, then letters.
HCF of 12 and 4 is 4.
HCF of pq and pw is p.
HCF of $12pq$ and $4pw$ is $4p$.
So, $12pq - 4pw = 4p(3q - w)$

c The HCF of $5x$, $10x^2$ and $25xy$ is $5x$.

$5x + 10x^2 - 25xy = 5x(1 + 2x - 5y)$

$5x \div 5x$ is 1

Example

a Factorise $y^2 - 3y$ **b** Factorise $(p + q)^2 - 2(p + q)$

a $y^2 - 3y = y(y - 3)$ **b** $(p + q)^2 - 2(p + q)$
$= (p + q)((p + q) - 2)$
$= (p + q)(p + q - 2)$

Each part in **b** has
$(p + q)$ in common

1 Factorise each of these fully by removing common factors.

 a $2x + 4$
  b $3y - 6$
  c $12p + 36q$

 d $25w - 5$
  e $6xy + xw$
  f $ab - 2bc$

 g $pqr + qrt - qsw$
  h $5xy - x$
  i $2xy + 6x$

 j $4ab - 6a^2$
  k $25p^2 - 10p$
  l $7x + 14xy$

 m $2ac + 4a^2 - 8a$
  n $15mn - 5m + 10m^3$
  o $6p^4 - 12p$

2 All three students have completed their factorisations incorrectly. Explain what they have done wrong.

  Clare   Ben   Vicky

$5x + 10xy$
$= 5x(0 + 2y)$

$6pq + 3p$
$= 3(2pq+1)$

$21p + 14pq$
$= 7p(14 + 7q)$

3 Factorise these expressions.

 a $10(x + y) + 13(x + y)$
  b $(a - b)^2 + 5(a - b)$

 c $6(q + r) - (q + r)^3$
  d $(pt - w) + 6(pt - w)$

4 Factorise fully.

 a $ax + bx + ay + by$
  b $cd + bd + cm + bm$

 c $a^2 + ab + 2a + 2b$
  d $cd + ce - me - md$

> First look at what each pair of terms has in common.

5 Write a factorised expression for

 a The perimeter of this rectangle

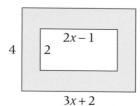

 b The perimeter of a square with sides $5b + 10$

6 Use factorisation to help you to evaluate these, without a calculator.

 a $2 \times 1.86 + 2 \times 1.14$
  b $3 \times 5.87 - 3 \times 0.37$

 c $5.86^2 + 5.86 \times 4.14$
  d $3.32 \times 6.68 + 3.32^2$

7 Show that the shaded area of this rectangle is $2(4x + 5)$.

This spread will show you how to:
- Factorise into single and double brackets

Keywords

Factor
Factorise
Quadratic

When you expand double brackets, you often get a **quadratic** expression with three terms. Therefore you can factorise a quadratic expression into double brackets.

A quadratic expression contains a squared term, such as x^2.

$$(x+2)(x+3) = x^2 + 3x + 2x + 6$$
$$= x^2 + 5x + 6$$
$$2 + 3 = 5 \qquad 2 \times 3 = 6$$

- The two numbers in the brackets **multiply** to give the number at the end and **add** to give the number of *x*s.

You can use this pattern to help you factorise a quadratic expression. You can check your answer by expanding the brackets.

Example

a Factorise $x^2 + 8x + 15$.
b Factorise $x^2 - 7x - 18$.

a Look for two numbers that multiply to give +15 and add to give +8. These are +3 and +5.
$$x^2 + 8x + 15 = (x+3)(x+5)$$
b Look for two numbers that multiply to give −18 and add to give −7.

Consider the factor pairs of 15
1 and 15
3 and 5

It can help to write out all the factor pairs.
Consider the factor pairs of −18:
−1 and 18
−18 and 1
−3 and 6
−6 and 3
−2 and 9
−9 and 2

The two numbers are −9 and +2.
$$x^2 - 7x - 18 = (x-9)(x+2)$$

You can check by expanding.

Example

Factorise $y^2 - 3y - 28$.

$$y^2 - 3y - 28 = (y-7)(y+4)$$

Two numbers that multiply to give −28 and add to give −3.
$-7 \times 4 = -28$
$-7 + 4 = -3$

1 Factorise each of these into double brackets.

a $x^2 + 6x + 8$ **b** $x^2 + 10x + 21$ **c** $x^2 + 11x + 28$

d $x^2 + 11x + 24$ **e** $x^2 - 8x + 12$ **f** $x^2 - 9x + 18$

g $x^2 - 13x + 36$ **h** $x^2 + x - 12$ **i** $x^2 - 2x - 35$

j $x^2 + 6x - 27$ **k** $x^2 - 14x + 32$ **l** $x^2 + 18x - 40$

2 a Write an expression for the perimeter of this triangle.

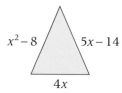

$x^2 - 8$ $5x - 14$

$4x$

b Factorise your expression.

3 Factorise each of these expressions.

a $x^2 + 6x - 72$ **b** $x^2 - 10x - 24$ **c** $x^2 - 20x + 75$

d $x^2 + 12x - 64$ **e** $x^2 - 64$ **f** $x^2 - 29x + 100$

4 Decide if each of these are single or double bracket factorisations. Factorise each fully.

a $x^2 + 21x + 38$ **b** $5x^2 + 5x + xy$ **c** $x^2 + 22x + 121$

d $x^2 + 7x - 18$ **e** $33 + p^2 + 14p$ **f** $2x^2 + 3xy$

5 Given that the area of this rectangle is 12 cm^2, show that $(x + 1)(x + 8) = 0$.

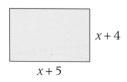

$x + 4$

$x + 5$

6 Use common factors and double brackets to factorise these expressions fully.

a $2x^2 + 22x + 56$

b $x^3 - 5x^2 - 24x$

c $x^3 - 16x$

Example:
$x^3 + 5x^2 + 6x$
$= x(x^2 + 5x + 6)$
$= x(x + 2)(x + 3)$

7 Factorise $2.3^2 + 2 \times 2.3 \times 1.7 + 1.7^2$ and use this to show that the calculation results in 16.

Exam review

Key objectives

- Expand the product of two linear expressions
- Manipulate algebraic expressions by collecting like terms, multiplying a single term over a bracket, taking out common factors and factorising quadratic expressions

1 a Factorise completely $\quad 4x + 2x^2$ (2)

b Expand $\quad (x + 5)(x - 2)$ (2)

c Simplify $\quad 3ab^2 \times 4b^3a$ (2)

2 The width of a rectangle is x centimetres.

The length of the rectangle is $(x + 4)$ centimetres.

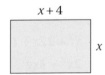

$x + 4$

x

a Find an expression, in terms of x, for the perimeter of the rectangle.

Give your expression in its simplest form. (2)

The perimeter of the rectangle is 54 centimetres.

b Work out the length of the rectangle. (3)

(Edexcel Ltd., 2005)

This unit will show you how to

- Round numbers to any number of significant figures
- Round numbers to a given power of 10 or number of decimal places
- Understand where to place the decimal point in calculations
- Use mental and written methods to calculate with decimal numbers
- Estimate answers to calculations involving decimals

Before you start ...

You should be able to answer these questions.

Review

1 Write notes to show how you could carry out these additions and subtractions mentally.

    a 23 + 97   **b** 234 − 96

    c 46 + 44   **d** 973 − 708

Key stage 3

2 Show how you could use a column method to work these out.

    a 3775 + 663   **b** 2886 − 909

    c 3775 − 2918   **d** 5549 + 8675

Key stage 3

3 Write notes to show how you could carry out these multiplications and divisions mentally.

    a 21 × 7   **b** 101 × 15

    c 8320 × 8   **d** 7350 × 7

Key stage 3

4 Show how you can calculate these multiplications and divisions using a written method.

    a 325 × 36   **b** 4356 × 18

    c 3485 ÷ 17   **d** 889 × 76

Key stage 3

This spread will show you how to:
- Round numbers to a given power of ten or number of decimal places
- Round numbers to any number of significant figures

Keywords
Decimal
Power
Round
Significant

Numbers are **rounded** when it is not appropriate to give an answer that is too precise.

- Numbers can be rounded:

    to **decimal places**   4.16 = 4.2 to 1 dp, and 5.663 = 5.66 to 2 dp

    to the nearest unit, 10, 100, 1000   32 559 = 33 000 to the nearest thousand

Always check the digit after the one you're rounding to: if it is a 5 or more, round up your final digit.

- When rounding to **significant figures**, count from the first non-zero digit

    to 2 sf: 712.4 = 710 and 0.00405 = 0.0041.

    to 3 sf: 6.339 = 6.34 and 0.000 000 338 754 = 0.000 000 339.

dp and sf are abbreviations for 'decimal places' and 'significant figures'.

Example

a Round these numbers to 2 dp.
  i 34.567   **ii** 3.887 126   **iii** 215.587 54

b Round 323 754.885 to the nearest:
  i unit   **ii** 10   **iii** 100   **iv** 1000   **v** 10 000

Unit, 10, 100, ... are **powers** of ten.

c Round these numbers to 2 sf.
  i 39.54   **ii** 217   **iii** 0.000 455   **iv** 12 019   **v** 25.505

a i 34.57   **ii** 3.89   **iii** 215.59
b i 323 755   **ii** 323 750   **iii** 323 800   **iv** 324 000   **v** 320 000
c i 40   **ii** 220   **iii** 0.000 46   **iv** 12 000   **v** 26

You should always round the original value.

Example

Round 3.447 to   **a** 2 dp   **b** 1 dp.

a 3.447 = 3.45 to 2 dp   **b** 3.447 = 3.4 to 1 dp

Do not round the 2 dp to get the 1 dp answer: use the original value 3.447.

Example

Find approximate answers to:

a 12.3 − 8.9   **b** 76.5 + 184.2   **c** 20 − 14.53

a $12.3 - 8.9 \approx 12 - 9 = 3$
b $76.5 + 184.2 \approx 80 + 200 = 280$
c $20 - 14.53 \approx 20 - 15 = 5$

Rounding to 1 sf is a useful way of finding a quick approximate answer to a calculation.

1 Round these numbers to the nearest 10.

   a 28   **b** 32   **c** 50   **d** 209   **e** 776   **f** 23 775

2 Round these decimal numbers to the nearest whole number.

   a 5.8   **b** 4.4   **c** 21.67   **d** 39.175   **e** 18.405   **f** 453.66

3 Round these numbers to the nearest 100.

   a 205   **b** 173   **c** 52   **d** 734   **e** 1389   **f** 134 545

4 Round these numbers to the nearest 1000.

   a 2239   **b** 12 563   **c** 7500   **d** 11 452   **e** 78 466   **f** 155 669

5 Round these numbers to one decimal place (nearest tenth).

   a 0.31   **b** 0.73   **c** 0.25   **d** 0.205   **e** 4.55   **f** 105.449

6 Round these whole numbers to two decimal places (nearest hundredth).

   a 0.317   **b** 0.455   **c** 15.304   **d** 104.675   **e** 16.445   **f** 0.0036

7 Round these whole numbers to two significant figures.

   a 483   **b** 1206   **c** 488   **d** 13 562   **e** 533   **f** 14 511

8 Round these numbers to two significant figures.

   a 0.355   **b** 0.421   **c** 0.0566   **d** 0.004 673   **e** 1.357   **f** 0.000 004 152

9 Round these numbers to one significant figure.

   a 157   **b** 2488   **c** 4.66   **d** 13.77   **e** 0.000 453   **f** 121 450

10 Use a calculator to work these out.
    Write your answers correct to two significant figures.

   a $8 \div 13$ **b** $4 \div 7$ **c** $5 \div 9$ **d** 24×16 **e** 7.8×71 **f** 2093×3493

11 By rounding all of the numbers to one significant figure,
    write a calculation that you could carry out mentally to
    estimate the answers to these calculations.

   a $355 \div 21$ **b** 39×43 **c** $1053 \div 92$ **d** $4385 + 11\ 655$
   e $108 + (2360 \div 52)$

12 Use mental calculations to work out the value of each
    of the estimates that you wrote for question **11**.

13 Use a calculator to work out an exact answer for each of the
    calculations from question **11**.
    For each one, write a sentence to say how well the calculator result
    agrees with the estimated answer that you wrote in question **12**.

This spread will show you how to:

- Understand where to place the decimal point in calculations
- Use mental methods to calculate with decimal numbers

Keywords

Compensation
method
Estimate

You already know several ways of adding and subtracting whole numbers, such as 36 + 59 or 326 − 138, in your head.
You can use the same techniques with decimal numbers.

- Start by working out an **estimate** in your head so that you can check that your final answer is reasonable.

  $15.1 − 3.8 ≈ 15 − 4$
  $= 11$

Then calculate the exact answer using the **compensation method**.

  15 − 4 = 11
  Add 0.1 (for 15.1) = 11.1
  Add 0.2 (for 3.8) = 11.3
  Answer is 11.3

Or treat the numbers as whole numbers, and adjust the place value afterwards. For 15.1 − 3.8, you could work out 151 − 38 = 113, and adjust this to 11.3.

  151 − 38 = 113
  so 15.1 − 3.8 = 11.3

Example

Work these out in your head.

a 4.7 + 9.8   **b** 16.9 − 7.3   **c** 108.4 − 37.1   **d** 6.07 + 5.3   **e** 13.9 − 6.75

a Initial estimate: 5 + 10 = 15.
One method is: 4.7 + 10 = 14.7, and then 14.7 − 0.2 = 14.5.
4.7 + 9.8 = 14.5

b Initial estimate: 17 − 7 = 10.
One method is: 169 − 73 = 169 − 69 − 4 = 96.
16.9 − 7.3 = 9.6

c Initial estimate: 100 − 40 = 60.
One method is to adjust the initial approximation:
60 + 8.4 + 2.9 = 71.3.
108.4 − 37.1 = 71.3

d Initial estimate: 6 + 5 = 11.
Now add the figures after the decimal point (0.07 and 0.3).
6.07 + 5.3 = 11.37

e Initial estimate: 14 − 7 = 7.
The difference between 6.75 and 14 is 7.25, so the difference between
6.75 and 13.9 is 0.1 less.
13.9 − 6.75 = 7.15

There is more than one way to solve each of these problems.

1 Write the answers to these additions.

   a 0.1 + 0.7   **b** 0.2 + 0.3   **c** 0.3 + 0.1

   d 0.7 + 0.4   **e** 0.3 + 0.1   **f** 0.5 + 0.5

2 Use your answers to question **1** to write the answers
  to these additions.

   a 5.1 + 0.7   **b** 6.2 + 0.3   **c** 7.3 + 0.1

   d 3.7 + 0.4   **e** 11.3 + 0.1   **f** 9.5 + 0.5

3 Work out these in your head and write the answers.

   a 4.7 + 5.3   **b** 4.7 + 5.4   **c** 3.6 + 6.7

   d 6.8 + 4.3   **e** 7.5 + 8.9   **f** 2.7 + 4.8

4 Work out these in your head and write the answers.

   a 3.55 + 4.22   **b** 2.13 + 3.12   **c** 3.18 + 0.42

   d 3.72 + 0.18   **e** 1.42 + 0.71   **f** 8.39 + 4.65

5 Work out these in your head and write the answers.

   a 3.35 + 0.8   **b** 0.15 + 6.7   **c** 0.7 + 3.88

   d 6.92 + 3.5   **e** 5.34 + 6.8   **f** 1.44 + 0.6

6 Write the answers to these subtractions.

   a 1 − 0.3   **b** 1 − 0.4   **c** 1 − 0.7

   d 4 − 0.6   **e** 11 − 0.8   **f** 15 − 0.9

7 Use your answers to question **6** to write the answers
  to these subtractions.

   a 1.1 − 0.3   **b** 1.1 − 0.4   **c** 1.5 − 0.7

   d 4.7 − 0.6   **e** 11.5 − 0.8   **f** 15.3 − 0.9

8 Work out these in your head and write the answers.

   a 2.16 − 1.42   **b** 1.51 − 0.46   **c** 6.39 − 4.88

   d 15.46 − 8.32   **e** 5.17 − 4.09   **f** 4.29 − 3.65

9 Write the answers to these subtractions.

   a 2 − 0.38   **b** 4 − 0.49   **c** 1 − 0.55

   d 5 − 0.63   **e** 15 − 0.48   **f** 12 − 0.34

10 Use your answers to question **9** to write the answers
   to these subtractions.

   a 2.1 − 0.38   **b** 4.1 − 0.49   **c** 1.2 − 0.55

   d 5.3 − 0.63   **e** 15.5 − 0.48   **f** 12.7 − 0.34

Written methods for adding and subtracting decimals

This spread will show you how to:

- Understand where to place the decimal point in calculations
- Use mental and written methods to calculate with decimal numbers

Keywords
Column
Decomposition
Digit

When adding or subtracting decimal numbers with more than one decimal place, use a written method to ensure accuracy.

Example

Calculate these using a written method.

a 102.773 + 28.47 **b** 26.44 − 1.105

a Initial estimate: 100 + 30 = 130

$$\begin{array}{r} 102.773 \\ +\ \ 28.470 \\ \hline 131.243 \\ \hline \end{array}$$

b Initial estimate: 30 − 1 = 29

$$\begin{array}{r} 26.4\,{}^{3}\not{4}\,{}^{10}0 \\ -\ \ 1.105 \\ \hline 25.335 \\ \hline \end{array}$$

You should still use an estimate to help check your answer.

Make sure that you line up the numbers correctly; put the decimal points above each other.

When the numbers in the calculation have different numbers of decimal digits, it can be useful to add extra zeros at the end of the number with fewer digits (shown above in red).

The 'carrying' and **decomposition** of numbers works in the same way as with whole numbers.

- **When you use a written method for adding or subtracting decimals, you should estimate first.**

Example

Here are some incorrect attempts at adding and subtracting with decimals. Try to decide what went wrong in each case.

a 10.03 − 2.55 **b** 38.53 + 2.474 **c** 100.773 − 28.782

a Initial estimate: 10 − 3 = 7

$$\begin{array}{r} 10.{}^{9}\not{0}{}^{1}3 \\ -\ 2.55 \\ \hline 8.48 \\ \hline \end{array} ✗$$

The top number has not been split up (decomposed) properly. Note that this calculation is actually quite easy to do mentally.

The correct answer is 7.48.

b Initial estimate: 39 + 2 = 41

$$\begin{array}{r} 38.53 \\ 2.474 \\ \hline 6.327 \\ \hline 1\ 1\ \ \ \end{array} ✗$$

In this example the decimal points, and therefore all of the columns, were not properly aligned.

The correct answer is 41.004.

c Initial estimate: 100 − 30 = 70

$$\begin{array}{r} 100.773 \\ -28.782 \\ \hline 128.011 \\ \hline \end{array} ✗$$

Here the smallest digit in each column has been subtracted from the bigger one. This seems silly, but it's easy to do when in a hurry!

The correct answer is 71.991

1 Calculate these using a mental method.

  a 4.3 + 8.1   **b** 6.4 + 5.6   **c** 9.2 + 3.9

  d 12.7 + 9.8   **e** 14.3 + 8.8   **f** 16.2 + 9.9

2 Work out each of the calculations from question **1**, using a standard written method.
  If you do not get the same answers by both methods, check to find your mistake.

3 Some of the calculations below are easy to do mentally.
  Others are best performed using a written method.
  Find the answer to each, showing your method clearly.

  a 5.9 + 7.1   **b** 0.673 + 1.198   **c** 94.834 + 106.487

  d 16.7 + 28.3   **e** 36.87 + 2.1   **f** 17.71 + 3.98

4 Work out these calculations using a standard written method.
  Remember to write down an estimate first.

  a 31.45 + 108.88   **b** 182.7 + 59.6   **c** 81.927 + 16.88

  d 104.7 + 98.89   **e** 57.784 + 103.218   **f** 61.386 + 40.614

5 Use a calculator to check your answers to question **4**.

6 Work out these using a mental method.

  a 8.4 − 6.2   **b** 9.7 − 0.6   **c** 17.9 − 2.9

  d 7.2 − 2.3   **e** 15.3 − 6.9   **f** 17.8 − 14.9

7 Work out the calculations from question **6**, using a standard written method.
  Check that you get the same answers by both methods. If not, find and correct your mistake.

8 Work out these calculations using an appropriate method (written or mental). Show your method clearly.

  a 5.8 − 3.2   **b** 16.73 − 8.87   **c** 9.6 − 3.7

  d 109.54 − 17   **e** 2.37 − 1.4   **f** 26.25 − 1.98

9 Work out these calculations using a standard written method.
  Remember to write an estimate before you do the calculation.

  a 21.864 − 7.968   **b** 104.87 − 85.42   **c** 417.48 − 57.69

  d 24.503 − 16.82   **e** 19.21 − 18.884   **f** 102.01 − 90.59

10 Use a calculator to check your answers to question **9**.

11 Work out these calculations using an appropriate method, and show your working clearly.

  a 8.6 − 4.5   **b** 26.4 + 13.8   **c** 18 − 6.712

  d 15.808 − 9.84   **e** 4.008 − 3.116   **f** 4.109 − 3.64

This spread will show you how to:

- Multiply and divide decimal numbers
- Understand where to place the decimal point in calculations

Keywords

Estimate
Place value

You already know some methods for mental multiplication and division of whole numbers. For example,

$7 \times 14 = 7 \times (10 + 4) = 7 \times 10 + 7 \times 4 = 70 + 28 = 98.$

You can extend these methods to work with decimal numbers.

- **One basic number fact can be extended to many different decimal calculations.**
  For example, $4 \times 3 = 12$ leads to $4 \times 0.3 = 1.2$, $0.4 \times 3 = 1.2$, $0.3 \times 0.4 = 0.12$, ...

Start with a basic calculation using whole numbers, and then to adjust the **place value**.

Example

Calculate mentally.   **a** 7×19 **b** 0.7×19 **c** 0.007×190

a $7 \times 20 = 7 \times 2 \times 10 = 14 \times 10 = 140$. So, $7 \times 19 = 140 - 7 = 133$.

b The original 7 (in 7×9) has changed to 0.7, so $0.7 \times 19 = 13.3$.
- A rough estimate is another way to get the correct answer.
  $0.7 \times 19 \approx 1 \times 20$. The answer must be 13.3, since the alternatives (1.33 or 133) are much further from 20.

c Compare this to part **a**. 0.007, not 7, so answer is 1000 times smaller. 190, not 19, so answer is 10 times bigger. So, overall, answer is 100 times smaller. The answer is 1.33.
- An estimate confirms this:
  $0.007 \times 190 \approx 0.007 \times 200 = 0.007 \times 100 \times 2 = 0.7 \times 2 = 1.4$

Often you know what the digits in the answers are, but you need to decide on the place value. A rough estimate is usually good enough to decide.

Example

Given that $238 \times 17 = 4046$, find the value of
a $2.38 \times 17\,000$ **b** $404.6 \div 170$

a $2.38 \times 17\,000 \approx 2 \times 20\,000$, so the answer must be 40 460.

b The original calculation gives $4046 \div 17 = 238$.
$404.6 \div 170 \approx 400 \div 200 = 2$. The answer must be 2.38.

Example

Davinder buys 15 bottles of cola for a party. Each bottle costs 79p.
Work out mentally the total cost.

Estimate: $15 \times 80 = 10 \times 80 + 5 \times 80$
$= 800 + 400 = 1200$

$15 \times 79 = 15 \times 80 - 15 \times 1$
$= 1200 - 15 = 1185$

The answer is £11.85.

1   Write the answer to the calculation 4×12.
    Use this answer to work out
    a 0.4×12 **b** 4×1.2 **c** 0.4×1.2 **d** 40×1.2 **e** 400×0.12

2   Use a mental method to work out 7×13. Then use your answer to work out
    a 0.7×13 **b** 7×1.3 **c** 0.7×1.3 **d** 70×1.3 **e** 700×0.13

3   Write the answers to
    a 14×3 **b** $64 \div 4$ **c** $35 \div 5$ **d** $208 \div 2$ **e** $48 \div 24$ **f** 4×7

4   Use your answers from question **3** to work out
    a 1.4×3 **b** $6.4 \div 4$ **c** $3.5 \div 5$ **d** $20.8 \div 2$ **e** $4.8 \div 2.4$ **f** 0.4×0.7

5   Any multiplication (such as $2 \times 3 = 6$) is part of a larger family of
    related facts: the equivalent multiplication $3 \times 2 = 6$, and the
    divisions $6 \div 2 = 3$ and $6 \div 3 = 2$. Write three other related facts for
    each of these multiplications.
    a $7 \times 9 = 63$ **b** $6 \times 8 = 48$ **c** $13 \times 7 = 91$ **d** $15 \times 18 = 270$
    e $5 \times 3.5 = 17.5$ **f** $2.4 \times 3.9 = 9.36$

6   You can use your knowledge of place value to extend families of
    multiplication facts further. For example, starting with $8 \times 7 = 56$,
    you can write $7 \times 80 = 560$, $56 \div 7 = 8$, $5.6 \div 7 = 0.8$, $5.6 \div 8 = 0.7$,
    $5.6 \div 80 = 0.07$, and so on.
    Use these ideas to write some number facts related to these
    calculations, and check your answers with a calculator.
    a $4 \times 5 = 20$ **b** $7 \times 9 = 63$ **c** $32 \div 8 = 4$ **d** $24 \div 8 = 3$ **e** $81 \div 9 = 9$

7   Given that $5 \times 9 = 45$, calculate
    a 0.5×9 **b** 50×90 **c** 0.5×90 **d** 0.5×0.9
    e $45 \div 9$ **f** $450 \div 5$ **g** $450 \div 0.9$

8   Given that $38 \times 91 = 3458$, calculate
    a 3.8×91 **b** 38×9.1 **c** $3458 \div 91$ **d** 0.38×910
    e $345.8 \div 38$ **f** $0.3458 \div 38$

9   Given that $2.91 \times 350 = 1018.5$, write the value of
    a 29.1×350 **b** 29.1×3.5 **c** 0.0291×3.5 **d** $101.85 \div 291$
    e $10.185 \div 2.91$ **f** $1.0185 \div 350$

10 Given that $4.7 \times 6.3 = 29.61$, write the answer to
    a 0.047×630 **b** 47×0.0063 **c** $2961 \div 0.47$

This spread will show you how to:

- Use written methods to calculate with decimal numbers
- Understand where to place the decimal point in calculations

Keywords

Place value

You already know written methods for multiplying and dividing whole numbers. You can use the same methods with decimal numbers.

- Always start with an estimate.
- Carry out the calculation.
- Use the estimate to position the decimal point in the answer.

Example

Work out $34.27 \div 2.3$

Rhian uses repeated subtraction.

$34.27 \div 2.3$
- Estimate: $30 \div 2 = 15$
- Work out $3427 \div 23$

Use only whole numbers in the calculation

$$
\begin{array}{ll}
3427 & \\
2300 & (100 \times 23) \\
\hline
1127 & 10 \times 23 = 230 \\
920 & 20 \times 23 = 460 \\
\hline
207 & (40 \times 23 = 920) \\
207 & (9 \times 23) = \begin{array}{c} 230 - 23 \\ 207 \end{array} \\
\hline
- &
\end{array}
$$

$3427 \div 23 = 100 + 40 + 9 = 149$
So, $34.27 \div 2.3 = 14.9$

Tom uses long division.

$34.27 \div 2.3$
- Estimate: $30 \div 2 = 15$
- Work out $3427 \div 23$

$$
\begin{array}{r}
149 \\
23\overline{)3427} \\
23 \\
\hline
112 \\
92 \\
\hline
207 \\
207 \\
\hline
- \\
\end{array}
$$

$3427 \div 23 = 149$
So, $34.27 \div 2.3 = 14.9$

Rhian and Tom both use a written method to calculate $3427 \div 23$, and an estimate to position the decimal point in the answer, so that the digits have the correct place values.

Example

Solve 2.93×18.5

Rhian uses grid multiplication.

2.93×18.5
- Estimate: $3 \times 20 = 60$
- Work out 293×185

×	200	90	3	
100	20000	9000	300	29300
80	16000	7200	240	23440
5	1000	450	15	1465
				54205
				111

$293 \times 185 = 54205$
$\Rightarrow 2.93 \times 18.5 = 54.205$

Tom uses long multiplication.

2.93×18.5
- Estimate: $3 \times 20 = 60$
- Work out 293×185

$$
\begin{array}{r}
293 \\
\times \ 185 \\
\hline
1465 \\
23440 \\
29300 \\
\hline
54205 \\
111 \\
\end{array}
$$

$293 \times 185 = 54205$
$\Rightarrow 2.93 \times 18.5 = 54.205$

1 Solve these, using a mental method.
   a 14×7 **b** 19×8 **c** 21×13 **d** 17×19
   e 11×28

2 Now use a written method to work out the answers for question **1**.
   Check that you get the same answers with both methods.

3 Use your answers to questions **1** and **2** to write the answers to
   a 1.4×7 **b** 19×0.8 **c** 2.1×1.3 **d** 1.7×0.019
   e 1.1×0.28

4 Solve these, using a mental method.
   a $320 \div 4$ **b** $180 \div 15$ **c** $276 \div 23$ **d** $357 \div 17$
   e $440 \div 20$

5 Now use a written method to work out the answers for question **4**.
   Check that you get the same answers with both methods.

6 Use your answers to questions **4** and **5** to write the answers to
   a $3.2 \div 4$ **b** $18 \div 15$ **c** $2.76 \div 2.3$ **d** $0.357 \div 1.7$
   e $0.44 \div 2$

7 Use a written method to work out these.
   a 4.7×5.3 **b** 1.53×2.8 **c** 21.6×4.9 **d** 33.65×3.89
   e 21.58×1.99

8 Use a calculator to check your answers to question **7**.

9 Use a written method to calculate these.
   a $34.83 \div 9$ **b** $5.425 \div 7$ **c** $7.328 \div 8$ **d** $451.8 \div 60$
   e $54.39 \div 3$

10 Use a calculator to check your answers to question **9**.

11 Use a written method to calculate these.
   a $58.65 \div 17$ **b** $66.4 \div 16$ **c** $185.76 \div 24$ **d** $7.752 \div 1.9$
   e $3.055 \div 1.3$

12 Use a calculator to check your answers to question **11**.

13 Use a written method to work out these, giving your answers correct
   to two decimal places.
   a $14.73 \div 2.8$ **b** $51.99 \div 1.8$ **c** $193.8 \div 0.14$ **d** $1013 \div 5.77$
   e $23.78 \div 0.83$

14 Use a calculator to check your answers to question **13**.

Exam review

Key objectives

- Round to a given number of significant figures
- Estimate answers to problems involving decimals
- Develop a range of strategies for mental calculation
- Understand where to position the decimal point by considering what happens if you multiply equivalent fractions

1 Work these out, rounding your answers correct to two significant figures.

   a $5.635 - 2.91$

   b $9.5 + 10.56$

   c 10.01×4.92

   d $17.28 \div 3.6$ (4)

2 Using the information that

$$97 \times 123 = 11\ 931$$

write down the value of

   a 9.7×12.3

   b $0.97 \times 123\ 000$

   c $11.931 \div 9.7$ (3)

(Edexcel Ltd., 2003)

This unit will show you how to

- Understand and use inverse operations
- Use function machines
- Set up and solve simple equations
- Solve equations involving fractions
- Understand and use reciprocals
- Solve simple inequalities, representing the solution on a number line

Before you start ...

You should be able to answer these questions.

Review

1 What value must the $\square$ represent in each case?

   a $\square + 3 = 12$ **b** $\square \times 5 = 20$

   c $\square - 11 = 19$ **d** $\square^2 = 100$

Key stage 3

2 What do each of these expressions mean?

E.g. 5x means that 'I think of a number and I multiply it by 5'.

   a $6y$ **b** p^2

   c $2m - 3$ **d** $4(h - 2)$

   e $\dfrac{t}{7}$ **f** $2x^2$

   g $(2x)^2$ **h** $10 - 3b$

Unit A1

3 Expand these brackets.

   a $3(x + 9)$ **b** $4(2x - 1)$

   c $-3(4y - 2)$ **d** $x(x - 7)$

Unit A1

4 Insert the symbol < (less than) or > (more than) between these pairs of numbers.

   a 5 3 **b** -9 1

   c -2 -5 **d** 0.9 0.85

   e $-\frac{1}{4}$ $-\frac{1}{2}$

Key stage 3

A2.1 **Working with inverse operations**

This spread will show you how to:

- Understand and use inverse operations
- Use function machines

Keywords
Inverse
Operation
Variable

- Addition, subtraction, multiplication and division are all **operations**.
- An operation has an **inverse**, for example subtraction is the inverse of addition.

An operation acts on a number or a variable.

The inverse operation changes the variable back to its original value

$$a \longrightarrow \boxed{\times 3} \longrightarrow 3a \longrightarrow \boxed{\div 3} \longrightarrow a$$

x and $2y^2$ are examples of variables.

If two or more operations act on a variable then you do their inverse operations in reverse order to get back to the original value

$$x \longrightarrow \boxed{\times 2} \longrightarrow 2x \longrightarrow \boxed{-5} \longrightarrow 2x - 5$$

$$x \longleftarrow \boxed{\div 2} \longleftarrow 2x \longleftarrow \boxed{+5} \longleftarrow 2x - 5$$

The operations are 'multiply by 2' and 'subtract 5'.

The inverse operations are 'add 5' and 'divide by 2'.

Example

Find the value of each letter.

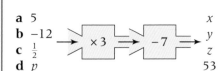

- **a** 5
- **b** −12
- **c** $\frac{1}{2}$
- **d** p

$\boxed{\times 3} \longrightarrow \boxed{-7} \longrightarrow$

- x
- y
- z
- 53

a $5 \xrightarrow{\times 3} 15 \xrightarrow{-7} 8$ so $x = 8$

b $-12 \xrightarrow{\times 3} -36 \xrightarrow{-7} -43$ so $y = -43$

c $\frac{1}{2} \xrightarrow{\times 3} 1\frac{1}{2} \xrightarrow{-7} -5\frac{1}{2}$ so $z = -5\frac{1}{2}$

d $53 \xrightarrow{+7} 60 \xrightarrow{\div 3} 20$ so $p = 20$

To find p, do the inverse operations in the reverse order.

Example

Find the starting number in each case.

a I think of a number, subtract 8 and square root it. The answer is 5.

b I think of a number, multiply by 2, subtract 4 and cube it. The answer is −216.

a $5 \xrightarrow{\text{square}} 25 \xrightarrow{+8} 25 + 8 = 33$ The starting number is 33.

b $-216 \xrightarrow{\text{cube root}} -6 \xrightarrow{+4} -2 \xrightarrow{\div 2} -1$ The starting number is −1.

Undo the operations in the reverse order.

1 Copy and complete these diagrams.

a
```
20
−8 →  ÷4  →  +5  →  ?
 x                   ?
 ?                   ?
                    15
```

b
```
 5   →  Square  →  ×1.5  →  ?
 ?                          1.5
 ?                          13.5
 ?                          y
```

c
```
 5   →  ?  →  −2  →  18
−7                   ?
 ½                   ?
 ?                   w
```

d
```
 5   →  Cube  →  ×2  →  ?
−2                      ?
 ?                      128
 ?                      2000
```

2 Explain why 'adding 10%' cannot be undone using the inverse operation 'subtracting 10%'.

3 In each case, use inverse operations to find the starting number.

  a I think of a number, double it and subtract 4. This gives me 7.

  b I think of a number, square it, multiply by 3 and get 75.

  c I think of a number, add 11, cube root it and get 2.

  d I think of a number, halve it, treble it then subtract 6. I get 54.

  e I think of a number, multiply it by $\frac{1}{4}$, square root it and get 5.

4 Is there another starting number for question **3b**?
Explain your answer.

5 a Write, in order, the operations that have acted on x.

  i $x + 3 = 17$ **ii** $2x - 1 = 17$ **iii** $\dfrac{x + 5}{2} = 24$ **iv** $3x^2 = 48$

  b Using the inverse operations, in reverse order, find the value of x in each case.

6 Look at this function machine.

Start	×2	−9	÷5	+2	$\div \frac{1}{2}$	Finish

  a What value do you end up with if (−3) enters the machine?

  b What value did you start with if these values leave the machine?

  i 6   **ii** 8.4   **iii** $\frac{1}{4}$

7 a Invent your own function machine that uses five operations to convert a value of 5 into 53.

  b Use your machine to describe what value you started with if the output is 100.

Solving one-sided equations

This spread will show you how to:
- Set up and solve simple equations

Keywords
Equation
Operation
Solve

- An **equation** is a statement with an equals sign. For example $2x - 4 = 18$ is an equation.
- To **solve** an equation, do the same **operation** to both sides. For example

$2x - 4 = 18$ Add 4 to both sides
$2x = 22$ Divide both sides by 2
$x = 11$

This equation is only true when $x = 11$.

Example

Solve these equations.

a $\dfrac{3x - 5}{2} = 8$ **b** $4(y + 3) = 40$ **c** $17x = 35$

a $\dfrac{3x - 5}{2} = 8$ Undo the operations in reverse order.

$3x - 5 = 16$ Multiply both sides by 2.
$3x = 21$ Add 5 to both sides.
$x = 7$ Divide both sides by 3.

b $4(y + 3) = 40$ Divide both sides by 4.   **or** $4(y + 3) = 40$ Expand the bracket.

$y + 3 = 10$ Subtract 3 from both sides. $4y + 12 = 40$ Subtract 12 from both sides.

$y = 7$ $4y = 28$ Divide both sides by 7.

$y = 7$

c $17x = 35$ Divide both sides by 17.

$x = \dfrac{35}{17}$ Change to a mixed number.

$= 2\dfrac{1}{17}$ Avoid decimals. $\frac{1}{17}$ is a recurring decimal and has to be rounded, so a decimal answer is not exact.

Example

The perimeter of this rectangle is 34 cm. Find x.

$3x + 1$

x

Perimeter $= x + x + 3x + 1 + 3x + 1$

$= 8x + 2$ 3x

so $8x + 2 = 34$

$8x = 32$

$x = 4$

1 Solve these one-sided equations.

a $3x - 7 = 8$ 　　　　**b** $4(x - 1) = 20$ 　　　　**c** $\dfrac{x}{2} + 8 = 13$

d $\dfrac{2x - 8}{4} = -3$ 　　**e** $2(x^2 + 9) = 68$ 　　**f** $2\left(\dfrac{x + 1}{2} - 3\right) = 8$

g $\dfrac{3x - 5}{2} = 10$ 　　**h** $x^3 - 4 = 12$ 　　**i** $\sqrt{x} - 1 = 9$

2 Copy and complete this crossword by solving the equations in the clues.

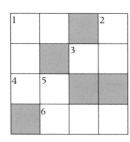

Across

1 $2(x + 2) = 30$
3 $\frac{x}{2} - 5 = 15$
4 $\frac{3x - 11}{2} = 71$
6 $3(x - 100) = 675$

Down

1 $3x + 15 = 330$
2 $x - 10 = 30$
5 $x^2 + 1 = 170$

3 The perimeter of this shape is 30 mm. Find the length of each side.

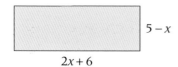

$5 - x$

$2x + 6$

4 This shape is a quadrilateral. Work out the value of x.

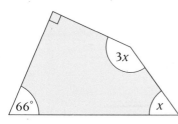

$3x$

$66°$

x

The interior angle sum of a quadrilateral is 360°.

$90 + 66 + 3x + x.$

$\dfrac{360}{160}$ 　　2.25

5 **a** Given that $y = 4x - 8$, work out the value of x when $y = 12$.

b Repeat for $y = -4$.

6 The diagram shows an isosceles triangle. Given that the equal angles are 10 less than double the third angle:

x

a Write an expression for each equal angle.

b Write an equation connecting the angles.

c Solve the equation, using your answer to find the size of each angle.

Solving double-sided equations

This spread will show you how to:
- Set up and solve simple equations

Keywords
One-sided
Unknown

When you solve an equation with **unknowns** on both sides

1 Subtract the smaller algebra term from both sides, for example

$5x - 3 = 3x + 1$ Subtract $3x$ from both sides.

$2x - 3 = 1$ This is now a **one-sided** equation.

2 Solve the one-sided equation by using inverse operations.

$2x - 3 = 1$ Add 3 to both sides.

$2x = 4$ Divide both sides by 2.

$x = 2$

Example

Solve

a $3x + 7 = 5x - 1$ **b** $5 - 6y = 2y - 3$

a $3x + 7 = 5x - 1$ Subtract $3x$ from both sides ($3x$ is smaller than $5x$).

$7 = 2x - 1$ Add 1 to both sides.

$8 = 2x$ Divide both sides by 2.

$4 = x$

$x = 4$

b $5 - 6y = 2y - 3$ Subtract $-6y$ from both sides ($-6y$ is smaller than $2y$):
Subtracting $-6y$ is the same as adding $6y$.

$5 = 8y - 3$ Add 3 to both sides.

$8 = 8y$ Divide both sides by 8.

$y = 1$

Example

The diagram shows an isosceles triangle.
Find the length of the equal sides.

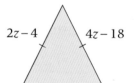

$2z - 4 = 4z - 18$ Write an equation and solve it.

$-4 = 2z - 18$ After subtracting $2z$ from both sides.

$14 = 2z$ After adding 18 to both sides.

$7 = z$

Hence, $2z - 4 = 14 - 4 = 10$ and $4z - 18 = 28 - 18 = 10$
The equal sides of the triangle are 10 units long.

1 Solve these equations.

a $2x + 4 = x + 3$

b $10 - 3x = 7x - 10$

c $8 - 3x = 5 - 2x$

d $4(7 + 2z) = 15 - 8z$

2 In each row of equations, one has a different solution from the other two. Find the odd one out.

a $2a + 7 = 4a + 1$ $6a - 2 = 2a + 6$ $10 + 2a = 7a - 5$

b $10 - 3b = 6b + 1$ $15 - 2b = 14 - b$ $3b - 14 = b$

c $2c + 2 = 4c - 1$ $8c - 7 = 6c - 4$ $1 - 4c = 2 - 8c$

d $2d + d + 8 = 3d + 2d - 4$ $5d + 7d - 3 = 10 - d$ $15 - d - d = 9 - d$

3 In each case, use the information to write an equation and solve it to find the starting number.

a I think of a number, multiply it by 8 and subtract 2. I get the same answer as when I multiply this number by 2 and add 10.

b I think of a number, multiply it by 5 and add 3. I get the same answer as when I multiply it by 2 and subtract it *from* 24.

c Taking double a number from 11 is equal to taking treble that number from 14.

4 In each case, use the information to form an equation and solve it.

a The angles in a triangle are $x°$, $x + 20°$ and $x + 40°$. Find the angles of the triangle.

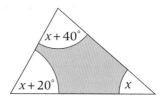

b The perimeters of these two shapes are equal. What are the dimensions of each shape?

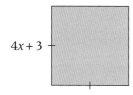
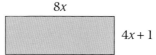

$4x + 3$ $8x$ $4x + 1$

c Abdul is 180 cm tall and Mark is $10x$ cm tall.
Alice is 164 cm tall and Miranda is $9x$ cm tall.
The difference in height between the two boys is equal to the difference in height between the two girls.
How tall are Mark and Miranda?

180 cm   164 cm $10x$ $9x$

Abdul   Alice   Mark   Miranda

Solving equations with fractions

This spread will show you how to:

- Solve equations involving fractions

Keywords

Reciprocal

- When you solve an equation involving fractions, clear the fractions first.

You can use the fact that $x \times \dfrac{1}{x} = 1$ to help you.

$\dfrac{1}{x}$ is called the **reciprocal** of x.

- Any non-zero number multiplied by its reciprocal is 1.

Example

Solve the equation $\dfrac{15}{x} = \dfrac{3}{7}$

$\dfrac{15}{x} = \dfrac{3}{7}$ Multiply both sides by the common denominator $7x$ to clear the fractions.

$\dfrac{15}{x} \times 7x = \dfrac{3}{7} \times 7x$ Cancel common factors.

$15 \times 7 = 3 \times x$, so $105 = 3x$, so $x = 35$

$7x$ is the reciprocal of $\dfrac{1}{7x}$.

When the equation has just a single fraction on each side you can **cross multiply** to clear the fractions, for example, so

$\dfrac{15}{x} \diagdown\!\!\!\!\diagup \dfrac{3}{7}$ $15 \times 7 = x \times 3 \rightarrow 105 = 3x \rightarrow x = 35$ as above

Cross multiply: LH numerator × RH denominator and vice versa.

Example

Solve these equations.

a $\dfrac{3}{x} = \dfrac{4}{9}$ **b** $\dfrac{12}{p} + 9 = 28$

a $\dfrac{3}{x} = \dfrac{4}{9}$ Simple fraction each side, so cross multiply.

$3 \times 9 = 4x$ Divide both sides by 4.

$27 = 4x$

$x = \dfrac{27}{4} = 6\tfrac{3}{4}$

b $\dfrac{12}{p} + 9 = 28$ Subtract 9 from both sides.

$\dfrac{12}{p} = 19$ Cross multiply ($19 = \tfrac{19}{1}$).

$12 \times 1 = 19p$ Divide both sides by 19.

$p = \dfrac{12}{19}$

It is tempting to use cross multiplication straightaway but you cannot cross multiply until you have a simple fraction on each side.

Example

I divide 15 by a certain number and get $\tfrac{3}{4}$. What is the number? Write an equation and solve it to find the starting number.

If x is the missing number then: $\dfrac{15}{x} = \dfrac{3}{4}$ Cross multiply

$15 \times 4 = 3x$, so $3x = 60$, so $x = 20$

1 Solve these equations, using the reciprocal function.

  a $\dfrac{7}{x} = 21$ **b** $15 = \dfrac{5}{x}$ **c** $\dfrac{4}{y} = 3$ **d** $\dfrac{7}{p} = 8$

  e $\dfrac{10}{x} = -2$ **f** $11 = \dfrac{5}{y}$ **g** $-3 = \dfrac{7}{y}$ **h** $-\dfrac{3}{x} = -9$

2 Solve these equations.

  a $\dfrac{5}{x} + 9 = 10$ **b** $\dfrac{10}{p} + 7 = 8$ **c** $\dfrac{x}{4} + 3 = 10$ **d** $-2 = 1 + \dfrac{3}{x}$

3 Solve these equations.

  a $\dfrac{16}{x} + 4 = 2$ **b** $\dfrac{12}{2y} - 3 = 5$ **c** $\dfrac{6}{3p} - 1 = 10$ **d** $\dfrac{15}{2x} + 4 = -2$

4 Solve these equations.

  a $\dfrac{x+1}{3} = \dfrac{x-1}{4}$ **b** $\dfrac{2y-1}{3} = \dfrac{y}{2}$ **c** $\dfrac{5}{w+5} = \dfrac{15}{w+7}$ **d** $\dfrac{3}{x-1} = \dfrac{9}{2x-1}$

5 Write an equation and solve it to find the starting number in each case.

  a I think of a number, divide it into 16 and I get 10. What is my number?

  b I think of a number, add 4, divide it into 12 and get 7. What is my number?

  c I think of a number, take 3 and divide it into 11. This gives me the same answer as when I take the same number and divide it into 8. What is my number?

6 Use the formula $P = \dfrac{180}{n+1}$ to find

  a the value of P when $n = 4$

  b the value of n when $P = 12$

  c the value of n when $P = 2\frac{2}{9}$.

7 The means of each set of expressions are equal. Use this information to find x and, hence, the value of each expression.

To find the mean add all the expressions and divide by the number of expressions.

Set 1

$2x-1$		$3x+2$
	$5x+4$	$7x$
$6x-4$		$10-2x$

Set 2

$3x-7$		$5x+8$
	$13-x$	
$2(3x+1)$		$12+4x$

Inequalities

This spread will show you how to:
- Solve simple inequalities, representing the solution on a number line

Keywords

Inequality
Number line
Reverse
Solve

- An **inequality** tells you about two quantities that are unequal.
- You can **solve** an inequality using inverse operations, for example

$3x + 2 > 5$ Subtract 2 from both sides.

$3x > 3$ Divide both sides by 3.

$x > 1$

$x < 3, 2x > 5,$
$x + 1 > 0$ are
inequalities.

- If you multiply or divide by a negative number, an inequality becomes false.
  For example:

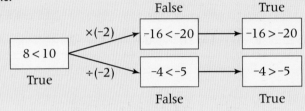

When you multiply or divide an inequality by a negative number you must **reverse** the inequality sign, for example

$7 - 2x \leqslant 3$ Subtract 7 from both sides.

$-2x \leqslant -4$ Multiply by –1 and **reverse** the inequality sign.

$2x \geqslant 4$ so $x \geqslant 2$

You can show the solution of an inequality on a **number line**, for example

This shows that x can have a value equal to or greater than 1.

The full circle shows that x can have the value 1.

This shows that x can only have values less than −1.

The open circle shows that x cannot have the value −1.

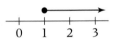

Solve these inequalities and represent the solutions on a number line.

a $5x - 7 \geqslant 3x + 13$ **b** $16 > -4x$

a $5x - 7 \geqslant 3x + 13$ Subtract 3x from both sides.

$2x - 7 \geqslant 13$ Add 7 to both sides.

$2x \geqslant 20$

$x \geqslant 10$

b $16 > -4x$ Divide both sides by 4.

$4 > -x$ Multiply by −1 and reverse sign.

$-4 < x,$

so $x > -4$

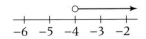

1 Use an inequality to represent each of these. For example, I think of a number and it is more than 11: $x > 11$

   a I think of a number and it is 3 or below.

   b I think of a number and it is between 2 and 8 inclusive.

   c I think of a number and it is over −5 but below 12.

2 What inequalities are represented on these number lines?

   a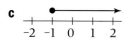
                b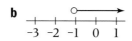

   c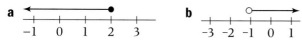
                d

   e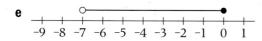

3 Is this statement true or false? The inequality $3 \geqslant x$ is represented on this number line.

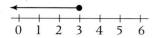

4 Solve these inequalities and represent the solutions on a number line.

   a $3x \leqslant 21$
         b $2x - 5 > 17$
         c $\dfrac{p}{2} + 6 \leqslant -2$

   d $28 < 7x + 49$
         e $5y + 3 \leqslant 2y + 5$
         f $-3y > 9$

   g $4(x + 2) \leqslant 16$
         h $-6x < 30$
         i $\dfrac{x}{-5} \geqslant -2$

   j $4p - 3 \leqslant 3(p - 2)$
         k $3(x - 2) < 5(x + 6)$
         l $6x - 4 \geqslant -2x$

5 **a** The area of this rectangle exceeds its perimeter. Write an inequality and solve it to find the range of values of x.

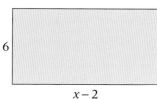

   b Given that x is an integer, find the smallest possible value that x can take.

Exam review

Key objectives

- Set up simple equations
- Solve linear equations in one unknown, with integer or fractional coefficients
- Solve simple linear inequalities in one variable, and represent the solution set on a number line

1 a Solve the inequality.

$$3x - 2 \leqslant 5x + 4$$ (2)

b Represent the solution set of the following inequality on a number line.

$$-6 < 2x \leqslant 4$$ (2)

2 a Solve

$$20y - 16 = 18y - 9$$ (3)

b Solve

$$\frac{40 - x}{3} = 4 + x$$ (3)

(Edexcel Ltd., 2004)

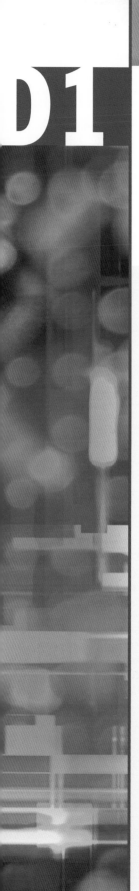

This unit will show you how to

- Design a questionnaire, recognising bias and taking steps to avoid it
- Design and use data collection sheets and two-way tables
- Understand the concept of random sampling
- Find an average and a measure of spread for a data set
- Find the mean of two combined data sets

Before you start ...

You should be able to answer these questions.

Review

1 Jon wanted to find out who used the vending machine at his school. In his survey Jon asked everyone at his school. This is called a census.

   a Explain what a census is.

   b How is a census different from a sample?

Key stage 3

2 Nicola and Maddy wanted to find out how many people catch flu in December. Nicola asked 100 people in the town centre one morning. Maddy looked up flu statistics on the internet.

   a Who collected primary data and who collected secondary data?

   b Explain the difference between primary and secondary data.

Key stage 3

3 Lee drew this tally table to collect data on hours of TV people watch:

TV hours	Frequency
3 – 4	
4 – 5	
more than 5	

Key stage 3

   a Explain why it would be difficult to write a tally mark for

     i 4 hours   **ii** 2 hours.

   b How could the difficulties discussed in part **a** be overcome?

This spread will show you how to:

● Design a questionnaire, recognising bias and taking steps to avoid it

Keywords
Biased
Data
Questionnaire
Survey

Organisations carry out statistical surveys to collect data that help them plan for the future.

You can use a **questionnaire** and collect **data** in a **survey**.

You have to choose suitable questions for a questionnaire.

Suitable questions
● can be answered yes or no
● ask for facts.

Unsuitable questions
● may be vague
● are leading or **biased**
● could be embarrassing.

Questions must have responses that
● cover all possible answers
● do not overlap or have gaps.

> *Do you have an MP3 player?*
> *How many CDs do you own?*
> *none* ☐ *1–10* ☐ *11–20* ☐ *over 20* ☐
>
> *Do you listen to a lot of music?*
> *Do you agree that Coldplay is the best band in the world?*
> *How cool are you?*
>
> *Where do you buy your CDs?*
> *internet* ☐ *store* ☐ *other* ☐
> *How much do you spend on CDs each month?*
> *£0 to £14.99* ☐ *£15 to £29.99* ☐
> *£30 and up* ☐

Example

A recently restyled breakfast radio show conducted this survey.

1 What is your opinion of the new breakfast show?
        Fantastic ☐   Good ☐
2 How long do you listen to the show?
        10 min ☐   1 hour ☐   1–2 hours ☐

  a Comment on these questions.
  b Write two questions for the survey to find out if listeners like the new show and for how long they listen.

Examiner's tip
The techniques described in this unit will be useful for your statistical coursework task.

a Question **1** does not include all possible responses, for example if you do not like the new show or don't think it's an improvement.
In question **2** there are gaps, between 10 min and 1 hour, and an overlap: 1 hour appears twice.

b 1 What is your opinion of the new breakfast show compared to the old one?
        Much better ☐   Better ☐
        Neither better nor worse ☐   Not as good ☐
  2 How much of the breakfast show do you listen to?
        all of it ☐   over an hour ☐
        half an hour to 1 hour ☐   less than half an hour ☐

1 Katy is doing a survey to find out how often people go to the cinema and how much they spend. She writes this question:

> How many times a month do you go to the cinema?

a What is wrong with this question?

b Write an improved question to find out how often people visit the cinema.

c Write a question to find out how much people spend when they go to the cinema.

> In this exercise you can find out if any questions you write 'work' by testing them out on groups of people in your class. The larger the sample the more reliable the results.

2 Sally put this question in a questionnaire:

> Do you agree that tennis is the best sport?

a i What is wrong with this question?

ii Write a better question to find out the favourite sport. Include some response boxes.

Sally also wants to find out how often people play sport.

b Design a question for Sally to use. Include some response boxes.

3 James wants to know which flavour crisps he should stock in the school tuck shop. He asks this question:

> Do you prefer plain or ketchup flavoured crisps?

a What is wrong with this question?

b Think about what crisps you and your friends like and design a better question for James to use. You should include some response boxes.

c James also put this question in his questionnaire:

> How many times have you visited the tuck shop?
> Once ☐   Lots of times ☐

i Write two things that are wrong with this question.

ii Design a better question for James to use. Include some response boxes.

4 Merlin wants to find out how far people would travel to see their favourite band perform. He writes this question:

> How far would you travel to see your favourite band?
> less than 1 mile ☐   5–10 miles ☐   any distance ☐

a i What is wrong with this question?

ii Design a better question for Merlin to use. Include some response boxes.

Merlin also wants to find out how much people would pay for a ticket to see their favourite band.

b Design a question for Merlin to use. Include some response boxes.

Collecting data – choosing a sample

This spread will show you how to:

- Design a questionnaire, recognising bias and taking steps to avoid it
- Understand the concept of random sampling

Keywords
Bias
Population
Random
Sample

It may be time-consuming, too costly, too long or too impractical to collect data from everyone. In these cases you should ask a representative **sample**.

You must choose the sample so that it is not biased. For example, a survey of preferred music using a sample of friends is biased as friends are more likely to have similar opinions.

- A sample should represent a whole **population**.

One way of avoiding **bias** is to use a **random** sample.

The population is the group of people or items being surveyed

- In a random sample each member of the population has the same chance of being included.
- Methods for choosing a random sample include:
  - taking names out of a hat
  - giving everyone a number and using a calculator or random number tables to pick numbers.

The larger the sample the more reliable the results

Example

James carries out a survey to find out if people in his town enjoy sport.

He stands outside a football ground and surveys people's opinions as they go in to watch a match.

Write two reasons why this is not a good sample to use.

People who watch football usually enjoy sport.
More men than women go to watch football so the survey could be gender biased.

Example

A train company carried out a survey about a local rail service.

They telephoned 100 people from a page of the telephone directory to answer a questionnaire on the rail service.

Write three reasons why this sample could be unrepresentative.

Only people who have a land-line telephone (and are not ex-directory) can be included in the sample.
Only people on one page of the telephone directory are included in the sample.
Some people may not be at home when they are phoned.

1 Katy is doing a survey to find out how often people go to the cinema and how much they spend. She stands outside a cinema and asks people as they go in.

Write a reason why this sample could be biased.

2 Sally wants to find out how often people play sport. Sally belongs to an athletics club. She asked members in her athletics club.

How could this sample be biased?

3 James wants to know which flavour crisps he should stock in the school tuck shop.

a He asks his mum, dad, auntie and uncle.

Explain why this is not a good sample to use.

b He asks only Year 11 at his school.

Explain why this sample could be biased.

c Describe how James could take a sample of 50 people. (There are 1000 people in his school.)

4 Merlin wants to find out how far people would travel to see their favourite band perform.

a He asks all his friends.

Write a reason why this sample could be biased.

b He goes into town one Saturday morning and asks anyone listening to music on a MP3 player.

How could this sample be biased?

5 Jenny carries out a survey to find out the most popular band. She asks 10 of her friends – all girls.

How could this sample be biased?

6 Wayne carries out a survey to find out the most popular car colour. He stands on a street corner and notes the colour of the first 15 cars that pass by.

Write a reason why this sample could be biased.

7 Lisa wants to find out how people travel to work.

a She asks people at a bus stop one morning.

How could this sample be biased?

b She opens the telephone directory at a random page and phones everyone on that page.

How could this sample be biased?

Designing a data collection sheet – two-way tables

This spread will show you how to:

- Design and use data collection sheets and two-way tables

Keywords
Data
Two-way table

- You can use a data collection sheet to collect **data** from a questionnaire or experiment.
- You can use a **two-way table** to collate the two sets of results.

Example

Two questions on a questionnaire are:

'Are you male or female?'   and   'How old are you?'

Design a two-way table to collect this data.

	Under 10	10–19	20–29	30–40	40+	Total
Male						
Female						
Total						

- You can use data in a two-way table to find other results.

Example

The table gives information about Key Stage 4 students at a school.

	Boys	Girls
Year 10	68	117
Year 11	89	126

a Work out the percentage of Key Stage 4 students in Year 10 who are boys.

b Work out the percentage of Key Stage 4 students who are girls.

Find the totals in the table.

	Boys	Girls	Total
Year 10	68	117	**185**
Year 11	89	126	**215**
Total	**157**	**243**	**400**

There are 400 students at Key Stage 4 (68 + 117 + 89 + 126).

a There are 68 Year 10 boys: $\dfrac{68}{400} \times 100 = 17\%$

b There are 243 (117 + 126) girls altogether: $\dfrac{243}{400} \times 100 = 60.75\%$

1 Katy is doing a survey to find out how often people go to the cinema and how much they spend.

Design a suitable data collection sheet in the form of a two-way table that she could use.

2 Sally wants to find out how often people play sport.

She wants to divide the results into those from males and those from females.

Design a suitable data collection sheet in the form of a two-way table that she could use.

3 James wants to know which flavour crisps he should stock in the school tuck shop.
He also wants to know which year groups prefer which flavours.

Design a suitable data collection sheet in the form of a two-way table that he could use.

4 Merlin wants to find out how far people would travel to see their favourite band perform and how much they would spend on a ticket to watch them.

Design a suitable data collection sheet in the form of a two-way table that he could use.

5 Jenny carries out a survey to find out people's favourite band and how many of that band's CDs they own.

Design a suitable data collection sheet in the form of a two-way table that she could use.

6 Wayne carries out a survey to find the most popular car colour and the most popular make of car.

Design a suitable data collection sheet in the form of a two-way table that he could use.

7 Lisa wants to find out how people travel to work and how long it usually takes them.

Design a suitable data collection sheet in the form of a two-way table that she could use.

8 The table gives information about the number of students in Years 7–9 that attended a school disco.

	Year 7	Year 8	Year 9
Boys	42	58	96
Girls	78	104	122

a How many students attended the disco?

b Work out the percentage of students that were

  i Year 8 girls   **ii** in Year 7   **iii** boys.

Averages and spread

This spread will show you how to:
- Find an average and a measure of spread for a data set

You can summarise data using an **average** and a **measure of spread**.

- An average is a single value.
  There are three types of average:
  - the **mode** is the value that occurs most often
  - the **median** is the middle value when the data are arranged in order
  - the **mean** is calculated by adding all the values and dividing by the number of values.

- Spread is a measure of how widely dispersed the data are.
  Two measures of spread are:
  - the range
  - the **interquartile range** (IQR).

  If there are one or more extreme values the IQR is a better measure of spread than the range.

An extreme value is a value well outside the range of the rest of the data

- Range = highest value − lowest value
- IQR = upper quartile − lower quartile

Lower quartile
$= \frac{1}{4}(n + 1)$th value

Upper quartile
$= \frac{3}{4}(n + 1)$th value

Example

Louise collected data on the number of times her friends went swimming in one month.

| 4 | 7 | 22 | 1 | 6 | 2 | 1 | 5 | 6 | 6 | 4 |

Work out the: **a** range **b** mode **c** mean
d median **e** interquartile range.

In order the data are: 1 1 2 4 4 5 6 6 6 7 22

a Range = 22 − 1 = 21

b Mode = 6

c Mean = 5.8
$(4 + 7 + 22 + 1 + 6 + 2 + 1 + 5 + 6 + 6 + 4) \div 11 = 64 \div 11 = 5.8$

The mean does not have to be an integer even if all the data values are integers.

d There are 11 values.
Median $= \frac{11 + 1}{2} = $ 6th value = 5

If there are n values
Median value
$= (\frac{n + 1}{2})$th value.

e Interquartile range = upper quartile − lower quartile

Lower quartile $= (\frac{11 + 1}{4})$th value   Upper quartile $= (\frac{3(11 + 1)}{4})$th value
$= (\frac{12}{4})$th value $= $ 9th value
$= $ 3rd value $= 6$
$= 2$

IQR = 6 − 2 = 4

1 For these sets of numbers work out the
  i range   **ii** mode   **iii** mean
  iv median   **v** interquartile range

  a 5, 9, 7, 8, 2, 3, 6, 6, 7, 6, 5

  b 45, 63, 72, 63, 63, 24, 54, 73, 99, 65, 63, 72, 39, 44, 63

  c 97, 95, 96, 98, 92, 95, 96, 97, 99, 91, 96

  d 13, 76, 22, 54, 37, 22, 21, 19, 59, 37, 84

  e 89, 87, 64, 88, 82, 88, 85, 83, 81, 89, 90

  f 53, 74, 29, 32, 67, 53, 99, 62, 34, 28, 27, 27, 27, 64, 27

  g 101, 106, 108, 102, 108, 105, 106, 109, 103, 105, 107, 104, 104, 105, 105

2 For the set of numbers in question **1e**, explain why the interquartile range is a better measure of spread to use than the range.

3 For the set of numbers in question **1f**, explain why the mode is not the best average to use.

4 **a** Subtract 100 from each of the numbers in question **1g** and write down the set of numbers you get.

  b For your set of numbers in **a**, work out the

  i range   **ii** mode   **iii** mean

  iv median   **v** interquartile range

  c Compare your answers for the measures of spread in part **b i** and **v** and **1g i** and **v**.
  What do you notice?

  d Add 100 to your answers for the measures of average in part **b ii**, **iii** and **iv**.
  Compare these answers to the answers you got in question **1g**.
  What do you notice?

  e Give a reason for what you noticed in parts **c** and **d**.

5 A scientist takes two sets of measurements from her experiment.
  Her results are:

Set A:	0	99	99	100	100	100	100	100	101	101	200
Set B:	0	0	99	99	100	100	100	101	101	200	200

  a For each set of measurements, work out the

  i range   **ii** mode   **iii** mean

  iv median   **v** interquartile range

  b Discuss whay you notice about the measurements and your answers to part **a**.

Mean of two combined data sets

This spread will show you how to:

● Find the mean of two combined data sets

Keywords

Mean

● **Mean** = $\dfrac{\text{Sum of all values}}{\text{Number of values}}$

You can combine two data sets to form one larger data set by finding the mean of all the data.

Example

32 students, 12 boys and 20 girls, in class 8Z sat a maths test.
The boys' mean mark was 63%.
The girls' mean mark is 78%.
Work out the mean mark for class 8Z.

Boys: Total sum of marks $\quad 63 \times 12 = 756$
Girls: Total sum of marks $\quad 78 \times 20 = 1560$
Total sum of marks for boys and girls $\quad 756 + 1560 = 2316$
Mean mark of all students
$\quad 2316 \div 32 = 72.375\% = 72\%$ to nearest whole mark.

Example

50 students answered a survey question about time spent on the internet one evening.
30 of the students were boys and 20 were girls.

The mean time spent on the internet by all 50 students was 18 minutes.
The mean time spent on the internet by the 30 boys was 24 minutes.

Work out the mean time spent on the internet by the 20 girls.

Total time spent on internet by all 50 students:
$\quad 50 \times 18 = 900$ minutes

Total time spent on the internet by the 30 boys:
$\quad 30 \times 24 = 720$ minutes

Total time spent on the internet by the 20 girls:
$\quad 900 - 720 = 180$ minutes

Mean time spent on internet by the 20 girls:
$\quad 180 \div 20 = 9$ minutes

Example

There are 13 boys and 16 girls in a class.

In a test, the mean mark for the boys was p.

In the same test, the mean mark for the girls was q.
Find an expression for the mean mark of all 29 students.

$\text{Mean} = \dfrac{13p + 16q}{29}$

1 An athletics club has 100 members, 60 boys and 40 girls.

The mean time the boys spent training one day was 86 minutes.
The mean time the girls spent training one day was 72 minutes.

Work out the mean time spent training on one day for all 100 members of the athletics club.

2 There are 120 students in Year 11 at St Edmunds school.
75 are girls and 45 are boys.

The mean time spent on homework each week for boys is 5.2 hours.
The mean time spent on homework each week for girls is 8.6 hours.

Work out the mean time spent on homework for all 120 students in Year 11 at St Edmunds school.

3 Of the students in Year 13 at St Edmunds school, 60 boys and 20 girls have passed the driving test.

The mean number of driving lessons that all 80 students had before passing the test was 19.75.
The mean number of driving lessons for the boys was 12.

Work out the mean number of driving lessons for the girls.

4 The mean mark in a statistics test for class 10Z was 84%.

There are 32 students in the class, 12 of whom are girls.
The mean mark in the test for these girls was 93%.

Work out the mean mark in the statistics test for the boys.

5 Thirty boys and girls were asked how many times they had visited the cinema in the past year.

The average number of times was 5.4.
The average number of times for the 12 boys that were asked was 2.5.

Work out the average number of times the girls in the group visited the cinema in the past year.

6 A fitness test was taken by 25 girls and 52 boys.

The average fitness score for the girls was 6.4, and the average fitness score for the boys was 9.2.

Work out the average fitness score for the whole group.

Exam review

Key objectives

- Design an experiment or survey
- Design and use two-way tables for discrete and grouped data
- Calculate mean, range and median of small data sets with discrete data

1 Thirty boys and twenty girls were asked how many hours of television they watch a day.

The average number of hours watched by the boys was a.

The average number of hours watched by the girls was b.

Write an expression for the mean number of hours of television watched a day for the total fifty children. (3)

2 Bill is carrying out a survey into the flavours of crisps that people like best.

He also wants to divide the results into those from males and those from females.

Design a suitable data collection sheet, in the form of a two-way table, that he could use to collect this information. (3)

(Edexcel Ltd., (Spec.))

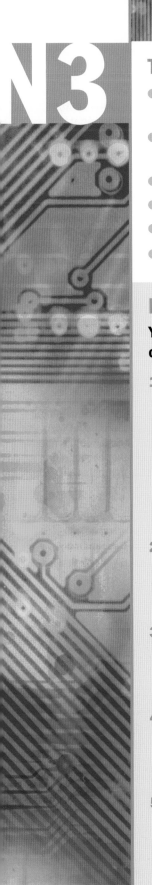

This unit will show you how to

- Compare and order fractions by rewriting them with a common denominator
- Understand the terms equivalent and improper fractions and mixed numbers
- Add, subtract, multiply and divide with fractions
- Understand unit fractions and use them as multiplicative inverses
- Distinguish between recurring and terminating decimals
- Convert fractions to decimals and percentages and vice versa

Before you start ...

You should be able to answer these questions.

Review

1 Sketch five copies of the diagram shown.

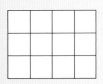

Shade your diagrams to represent each of these fractions.

a $\frac{1}{2}$ **b** $\frac{2}{3}$ **c** $\frac{3}{4}$ **d** $\frac{5}{6}$ **e** $\frac{7}{12}$

Key stage 3

2 Cancel each of these fractions down to their simplest terms.

a $\frac{2}{4}$ **b** $\frac{15}{20}$ **c** $\frac{8}{10}$ **d** $\frac{95}{100}$ **e** $\frac{6}{8}$

Key stage 3

3 Write a decimal equivalent to each of these fractions.

a $\frac{3}{4}$ **b** $\frac{2}{5}$ **c** $\frac{7}{10}$ **d** $\frac{9}{20}$ **e** $\frac{17}{100}$

Key stage 3

4 Write a fraction equivalent to each of these decimals.

a 0.5   **b** 0.25   **c** 0.3   **d** 0.8   **e** 0.45

Key stage 3

5 Write a percentage equivalent to each of these fractions.

a $\frac{1}{2}$ **b** $\frac{1}{4}$ **c** $\frac{1}{10}$ **d** $\frac{1}{5}$ **e** $\frac{1}{20}$

Key stage 3

This spread will show you how to:

● Compare and order fractions by rewriting them with a common denominator

Keywords
Ascending
Common
denominator
Denominator
Descending
Least common
multiple (LCM)
Numerator

Which fraction is bigger $\frac{2}{5}$ or $\frac{5}{12}$?

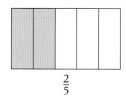

$\frac{2}{5}$

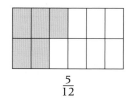

$\frac{5}{12}$

$$\frac{2}{5} \overset{\times 12}{\underset{\times 12}{=}} \frac{24}{60}$$

$$\frac{5}{12} \overset{\times 5}{\underset{\times 5}{=}} \frac{25}{60}$$

So $\frac{5}{12}$ is bigger than $\frac{2}{5}$.

● In the fraction $\frac{2}{5}$
  – The top number, 2, is the **numerator**.
  – The bottom number, 5, is the **denominator**.
  – The **common denominator** of $\frac{2}{5}$ and $\frac{5}{12}$ is 60, the **LCM** of 5 and 12 (the original denominators).

The **least common multiple** (**LCM**) is the lowest number that two (or more) numbers will divide into exactly.

To compare two or more fractions:

● Find the common denominator.

● Work out the equivalent fractions.

● Write the fractions in ascending or descending order.

Ascending – going up
Descending – going down

Example

Write these in ascending order. $\frac{7}{8}$ $\frac{5}{6}$ $\frac{3}{4}$

$8 = 2^3$
$6 = 2 \times 3$
$4 = 2^2$

LCM of 8, 6 and 4 $= 2^3 \times 3$
$\qquad\qquad\qquad = 24$

$\frac{7}{8} = \frac{21}{24}$ Multiply numerator and denominator by 3.

$\frac{5}{6} = \frac{20}{24}$ Multiply numerator and denominator by 4.

$\frac{3}{4} = \frac{18}{24}$ Multiply numerator and denominator by 6.

$\frac{18}{24} < \frac{20}{24} < \frac{21}{24}$

In ascending order the fractions are: $\frac{3}{4}, \frac{5}{6}, \frac{7}{8}$

1. Write the least common multiple of each pair of numbers.

   a 2 and 4 **b** 2 and 5 **c** 3 and 8 **d** 4 and 6

   e 4 and 10 **f** 7 and 5 **g** 15 and 20 **h** 20 and 30

2. The diagram shows that $\frac{1}{2} = \frac{2}{4}$

   Draw diagrams to show that

   a $\frac{1}{2} = \frac{3}{6}$ **b** $\frac{2}{3} = \frac{4}{6}$

   c $\frac{3}{5} = \frac{9}{15}$ **d** $\frac{3}{4} = \frac{15}{20}$

3. Write each fraction as an equivalent fraction with a denominator of 60.

   a $\frac{1}{3}$ **b** $\frac{1}{4}$ **c** $\frac{2}{3}$ **d** $\frac{2}{5}$

4. Write each fraction as an equivalent fraction with a denominator of 24.

   a $\frac{3}{4}$ **b** $\frac{1}{3}$ **c** $\frac{3}{8}$ **d** $\frac{5}{12}$

5. Rewrite each fraction with the denominator shown.

   a $\frac{2}{3} = \frac{}{30}$ **b** $\frac{3}{7} = \frac{}{42}$ **c** $\frac{7}{9} = \frac{}{45}$ **d** $\frac{5}{8} = \frac{}{40}$

6. Write the fractions in each pair as equivalent fractions with a common denominator. Say which fraction in each pair is larger.

   a $\frac{1}{5}$ and $\frac{3}{10}$ **b** $\frac{2}{3}$ and $\frac{3}{4}$ **c** $\frac{2}{5}$ and $\frac{1}{3}$ **d** $\frac{7}{10}$ and $\frac{2}{3}$

7. Draw diagrams (like the ones in question **2**) to illustrate your answers to question **6**.

8. Find the least common multiple of each set of numbers.

   a 2, 3 and 5 **b** 3, 4 and 6 **c** 2, 3 and 8 **d** 2, 4 and 7

9. Rewrite each set of fractions with a common denominator.

   a $\frac{1}{2}, \frac{2}{3}$ and $\frac{3}{4}$ **b** $\frac{1}{5}, \frac{3}{4}$ and $\frac{7}{20}$ **c** $\frac{1}{8}, \frac{7}{12}$ and $\frac{2}{3}$ **d** $\frac{2}{3}, \frac{3}{4}$ and $\frac{2}{7}$

10. Write each set of fractions in ascending order. Show your working.

    a $\frac{2}{3}, \frac{1}{5}$ and $\frac{2}{15}$ **b** $\frac{1}{4}, \frac{2}{5}$ and $\frac{7}{20}$ **c** $\frac{3}{7}, \frac{3}{8}$ and $\frac{5}{14}$ **d** $\frac{2}{3}, \frac{5}{6}$ and $\frac{2}{7}$

11. Write each set of fractions in descending order. Show your working.

    a $\frac{2}{5}, \frac{1}{2}, \frac{3}{10}$ and $\frac{1}{4}$ **b** $\frac{1}{4}, \frac{3}{20}, \frac{4}{5}$ and $\frac{1}{10}$ **c** $\frac{2}{5}, \frac{3}{8}, \frac{3}{4}$ and $\frac{17}{40}$ **d** $\frac{5}{6}, \frac{11}{24}, \frac{7}{12}$ and $\frac{5}{8}$

This spread will show you how to:

- Understand the terms equivalent and improper fractions and mixed numbers
- Add and subtract fractions

Keywords

Common
  denominator
Improper fraction
Mixed number

- You can only add and subtract fractions if they have common denominators.

$$\frac{2}{8} + \frac{5}{8} = \frac{7}{8}$$

- If the fractions have different denominators, change them to equivalent fractions with the same denominator, then add.

$$\frac{11}{12} - \frac{1}{3} = \frac{11}{12} - \frac{4}{12} = \frac{7}{12}$$

- If your answer is an improper fraction, change it to a mixed number.

$$\frac{3}{4} + \frac{2}{5} = \frac{15}{20} + \frac{8}{20} = \frac{23}{20} = 1\frac{3}{20}$$

- Cancel any common factors in the numerator and denominator.

$$\frac{4}{5} - \frac{3}{10} = \frac{8}{10} - \frac{3}{10} = \frac{5}{10} = \frac{1}{2}$$

$\frac{23}{20}$ is an **improper fraction** – its numerator is larger than its denominator.

$1\frac{3}{20}$ is a **mixed number** – a whole number and a fraction.

Example

Calculate   **a** $\frac{1}{2} + \frac{1}{3}$ **b** $\frac{3}{4} - \frac{1}{5}$ **c** $\frac{1}{4} + \frac{5}{6}$

a $\frac{1}{2} + \frac{1}{3} = \frac{3}{6} + \frac{2}{6} = \frac{3+2}{6} = \frac{5}{6}$ The lowest common denominator is 6

b $\frac{3}{4} - \frac{1}{5} = \frac{15}{20} - \frac{4}{20} = \frac{15-4}{20} = \frac{11}{20}$ The lowest common denominator is 20

c $\frac{1}{4} + \frac{5}{6} = \frac{3}{12} + \frac{10}{12} = \frac{3+10}{12} = \frac{13}{12} = 1\frac{1}{12}$ The lowest common denominator is 12

A common mistake is to simply add or subtract the numerators and denominators.

Example

Both of these students calculated incorrectly.
Say what they did wrong and find the correct answer.

a $\frac{2}{3} + \frac{3}{4}$

Jodi wrote

$\frac{2}{3} + \frac{3}{4} = \frac{2+3}{3+4} = \frac{5}{7}$ ✗

b $\frac{7}{8} - \frac{2}{3}$

Abdul wrote

$\frac{7}{8} - \frac{2}{3} = \frac{7-2}{8-3} = \frac{5}{5} = 1$ ✗

a The answer $\frac{5}{7}$ cannot be correct. Both $\frac{2}{3}$ and $\frac{3}{4}$ are bigger than $\frac{1}{2}$, so the answer must be greater than 1.

a $\frac{2}{3} + \frac{3}{4} = \frac{8}{12} + \frac{9}{12}$

     $= \frac{17}{12}$

     $= 1\frac{5}{12}$

b $\frac{7}{8} - \frac{2}{3} = \frac{21}{24} - \frac{16}{24}$

     $= \frac{5}{24}$

Jodi added the numerators and denominators instead of finding the common denominator first.

Abdul subtracted the numerators and denominators instead of finding the common denominator first.

b $\frac{7}{8}$ is smaller than 1, so when $\frac{2}{3}$ is taken away the answer must be less than 1.

1 The diagram shows that $\frac{1}{4}+\frac{1}{4}=\frac{1+1}{4}=\frac{2}{4}=\frac{1}{2}$

Draw diagrams to show that

a $\frac{1}{3}+\frac{2}{3}=\frac{2+1}{3}=\frac{3}{3}=1$ **b** $\frac{1}{5}+\frac{3}{5}=\frac{1+3}{5}=\frac{4}{5}$

2 Work out **a** $\frac{1}{5}+\frac{2}{5}$ **b** $\frac{1}{4}+\frac{3}{4}$ **c** $\frac{2}{7}+\frac{3}{7}$ **d** $\frac{3}{8}+\frac{5}{8}$

3 Draw diagrams (like the ones from question **1**) to illustrate your answers to question **2**.

4 Work out **a** $\frac{2}{3}-\frac{1}{3}$ **b** $\frac{4}{5}-\frac{1}{5}$ **c** $\frac{5}{6}-\frac{1}{6}$ **d** $\frac{9}{10}-\frac{3}{10}$

5 The diagram shows that $\frac{1}{3}+\frac{1}{4}=\frac{4}{12}+\frac{3}{12}=\frac{7}{12}$

Draw diagrams to show that

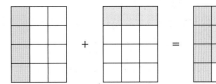

a $\frac{1}{3}+\frac{1}{2}=\frac{2}{6}+\frac{3}{6}=\frac{5}{6}$

b $\frac{3}{5}+\frac{3}{10}=\frac{6}{10}+\frac{3}{10}=\frac{9}{10}$

6 Calculate **a** $\frac{1}{5}+\frac{1}{10}$ **b** $\frac{2}{3}+\frac{1}{6}$ **c** $\frac{2}{5}+\frac{3}{20}$ **d** $\frac{1}{8}+\frac{1}{4}$

7 Draw diagrams (like the ones from question **5**) to illustrate your answers to question **6**.

8 Work out **a** $\frac{3}{8}-\frac{1}{4}$ **b** $\frac{5}{6}-\frac{1}{3}$ **c** $\frac{4}{5}-\frac{1}{20}$ **d** $\frac{3}{8}-\frac{1}{16}$

9 Work out **a** $\frac{4}{7}-\frac{1}{3}$ **b** $\frac{4}{5}-\frac{2}{3}$ **c** $\frac{8}{9}-\frac{5}{6}$ **d** $\frac{3}{5}-\frac{1}{4}$

10 The diagram shows what happens when you add fractions with a total that is greater than 1.

In this example, $\frac{3}{4}+\frac{4}{5}=\frac{15}{20}+\frac{16}{20}=\frac{15+16}{20}=\frac{31}{20}=1\frac{11}{20}$

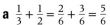

Draw diagrams to find the answers to **a** $\frac{3}{5}+\frac{1}{2}$ **b** $\frac{5}{6}+\frac{3}{4}$

11 Calculate **a** $\frac{2}{3}+\frac{1}{2}$ **b** $\frac{4}{5}+\frac{1}{2}$ **c** $\frac{1}{3}+\frac{4}{5}$ **d** $\frac{4}{7}+\frac{1}{2}$

12 Change these improper fractions into mixed numbers.

a $\frac{5}{4}$ **b** $\frac{9}{5}$ **c** $\frac{13}{8}$ **d** $\frac{17}{4}$

13 Change these mixed numbers to improper fractions.

a $1\frac{3}{4}$ **b** $1\frac{7}{16}$ **c** $1\frac{5}{9}$ **d** $2\frac{4}{7}$

14 Calculate

a $2\frac{3}{5}+1\frac{1}{3}$ **b** $2\frac{3}{4}+1\frac{5}{6}$ **c** $4\frac{3}{7}+3\frac{1}{2}$ **d** $5\frac{4}{9}+2\frac{3}{7}$

e $3\frac{3}{5}-2\frac{1}{4}$ **f** $2\frac{1}{2}-1\frac{3}{4}$ **g** $3\frac{3}{4}-1\frac{4}{5}$ **h** $7\frac{3}{7}-2\frac{1}{2}$

This spread will show you how to:

- Understand unit fractions and use them as multiplicative inverses
- Multiply and divide fractions

Keywords
Multiplicative inverse
Reciprocal
Unit fraction

The diagram shows the multiplication

$$\frac{2}{3} \times \frac{3}{4} = \frac{6}{12} = \frac{1}{2}$$

You get the same result if you multiply the numerators together and multiply the denominators together.

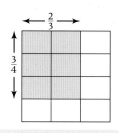

- To multiply fractions, multiply the numerators and then the denominators, then cancel any common factors.

Example

Find **a** $\frac{2}{3} \times \frac{4}{5}$ **b** $\frac{4}{9} \times \frac{3}{5}$ **c** $\frac{7}{4} \times \frac{5}{2}$

a $\frac{2}{3} \times \frac{4}{5} = \frac{2 \times 4}{3 \times 5}$ **b** $\frac{4}{9} \times \frac{3}{5} = \frac{4 \times 3}{9 \times 5}$ **c** $\frac{7}{4} \times \frac{5}{2} = \frac{7 \times 5}{4 \times 2}$

$= \frac{8}{15}$ $= \frac{12}{45} = \frac{4}{15}$ $= \frac{35}{8} = 4\frac{3}{8}$

- Multiplying by a **unit fraction** is the same as dividing by its denominator. For example, multiplying by $\frac{1}{5}$ is the same as dividing by 5.

$$10 \times \frac{1}{5} = \frac{10}{5} = 2 \text{ and } 10 \div 5 = 2$$

- Dividing by a unit fraction is the same as multiplying by its denominator.

$$10 \div \frac{1}{2} = 10 \times 2 = 20$$

You combine these two ideas when you divide by a fraction.

For example,

$10 \div \frac{5}{2} = 10 \div 5 \div \frac{1}{2}$ ÷5 is the same as $\times\frac{1}{5}$

$= 10 \times \frac{1}{5} \times 2$ $÷\frac{1}{2}$ is the same as ×2

$= 10 \times \frac{2}{5}$

$= \frac{20}{5} = 4$

A unit fraction has a numerator of 1.

The **multiplicative inverse** of an integer is its **reciprocal**.
For example, the reciprocal of 5 is $\frac{1}{5}$.

- To divide by a fraction, multiply by its multiplicative inverse.

Example

Find **a** $\frac{3}{7} \div 5$ **b** $\frac{5}{12} \div \frac{3}{4}$ **c** $3\frac{1}{2} \div 2\frac{1}{5}$

a $\frac{3}{7} \div 5 = \frac{3}{7} \times \frac{1}{5}$ **b** $\frac{5}{12} \div \frac{3}{4} = \frac{5}{12} \times \frac{4}{3}$ **c** $3\frac{1}{2} \div 2\frac{1}{5} = \frac{7}{2} \div \frac{11}{5}$

$= \frac{3}{35}$ $= \frac{20}{36} = \frac{5}{9}$ $= \frac{7}{2} \times \frac{5}{11}$

$= \frac{35}{22} = 1\frac{13}{22}$

The **multiplicative inverse** of a fraction is the original fraction 'turned upside down'. The inverse of $\frac{3}{5}$ is $\frac{5}{3}$.

Change mixed numbers to improper fractions first.

Exercise N3.3

1 Write the reciprocal (multiplicative inverse) of

   a 4   **b** 6   **c** 10   **d** 12

2 Write the multiplicative inverse of

   a $\frac{1}{5}$ **b** $\frac{1}{9}$ **c** $\frac{1}{2}$ **d** $\frac{1}{3}$

3 Rewrite each of these divisions as multiplications.

   a $8 \div 5$ **b** $6 \div 4$ **c** $9 \div 5$ **d** $17 \div 3$

   > For example, you can write $7 \div 5$ as $7 \times \frac{1}{5}$.

4 Copy and complete these sentences.

   a Dividing a number by 4 is the same as multiplying the number by ____

   b Multiplying a number by $\frac{1}{2}$ is the same as dividing the number by ____

   c Dividing a number by $\frac{1}{3}$ is the same as multiplying the number by ____

5 Draw diagrams, like the one on page 78, to show that

   a $\frac{1}{4} \times \frac{1}{2} = \frac{1}{8}$ **b** $\frac{3}{4} \times \frac{1}{3} = \frac{3}{12} = \frac{1}{4}$ **c** $\frac{2}{5} \times \frac{2}{3} = \frac{4}{15}$ **d** $\frac{1}{4} \times \frac{3}{5} = \frac{3}{20}$

6 Calculate these, giving your answers in their simplest form.

   a $\frac{3}{4} \times \frac{1}{5}$ **b** $\frac{2}{3} \times \frac{2}{9}$ **c** $\frac{2}{7} \times \frac{1}{4}$ **d** $\frac{5}{16} \times \frac{4}{5}$

   e $\frac{5}{9} \times \frac{4}{7}$ **f** $\frac{7}{8} \times \frac{2}{21}$ **g** $\frac{4}{5} \times \frac{3}{13}$ **h** $\frac{8}{35} \times \frac{7}{24}$

7 Calculate, giving your answers as fractions in their simplest terms.

   a $3 \div 4$ **b** $6 \div 8$ **c** $4 \div 5$ **d** $8 \div 10$

   e $3 \div 7$ **f** $9 \div 5$ **g** $24 \div 7$ **h** $22 \div 8$

   > For example, $13 \div 4 = 13 \times \frac{1}{4}$ $= \frac{13}{4} = 3\frac{1}{4}$.

8 Calculate

   a $\frac{5}{8} \div 4$ **b** $\frac{3}{4} \div 6$ **c** $\frac{2}{3} \div 7$ **d** $\frac{3}{16} \div 9$

   e $4 \div \frac{1}{5}$ **f** $6 \div \frac{2}{3}$ **g** $5 \div \frac{2}{5}$ **h** $11 \div \frac{3}{7}$

9 Calculate

   a $\frac{1}{8} \div \frac{3}{8}$ **b** $\frac{1}{5} \div \frac{4}{5}$ **c** $\frac{1}{14} \div \frac{3}{7}$ **d** $\frac{1}{10} \div \frac{2}{5}$

   e $\frac{2}{3} \div \frac{3}{4}$ **f** $\frac{5}{8} \div \frac{7}{9}$ **g** $\frac{1}{12} \div \frac{3}{8}$ **h** $\frac{2}{7} \div \frac{3}{4}$

10 Calculate

   a $1\frac{1}{2} \times \frac{3}{4}$ **b** $2\frac{3}{4} \times \frac{2}{5}$ **c** $1\frac{1}{5} \times 2\frac{1}{2}$ **d** $1\frac{2}{3} \times 1\frac{4}{5}$

   e $2\frac{7}{8} \div \frac{3}{5}$ **f** $4\frac{1}{4} \div 3\frac{1}{2}$ **g** $5\frac{3}{8} \div 2\frac{3}{4}$ **h** $9\frac{1}{3} \div 2\frac{1}{4}$

This spread will show you how to:

- Distinguish between recurring and terminating decimals
- Convert fractions to decimals and percentages and vice versa

Keywords
Denominator
Numerator
Recurring
Terminating

- To convert a fraction to a decimal divide the numerator by the denominator.

Example

Write these fractions as decimals.   **a** $\frac{5}{8}$ **b** $\frac{5}{9}$ **c** $\frac{1}{7}$

a $\frac{5}{8} = 5 \div 8 = 8\overline{)5.000}$ $\begin{array}{r} 0.625 \\ \hline \end{array}$

0.625 is a
terminating
decimal

b $\frac{5}{9} = 5 \div 9 = 9\overline{)5.000 \dots}$ $\begin{array}{r} 0.555\dots \\ \hline \end{array}$

$0.555 \dots = 0.\dot{5}$ is a
recurring decimal.
The dot over the 5
shows the
recurring digit.

c $\frac{1}{7} = 1 \div 7 = 7\overline{)1.000000000 \dots}$ $\begin{array}{r} 0.142857142\dots \\ \hline \end{array}$

$0.\dot{1}4285\dot{7}$ is a
recurring decimal.
The dots show the
recurring group of
digits.

To decide if a fraction will be a terminating or a recurring decimal, look at the denominator.

- If the only factors of the denominator are 2 and/or 5 or combinations of 2 and 5 then the fraction will be a terminating decimal.
- If the denominator has any factors other than 2 and/or 5 then the fraction will be a recurring decimal.

Example

Say whether these fractions are terminating or recurring decimals.

a $\frac{9}{20}$ **b** $\frac{4}{30}$ **c** $\frac{7}{16}$ **d** $\frac{11}{13}$

a denominator $= 20 = 2 \times 2 \times 5 \rightarrow \frac{9}{20}$ is a terminating decimal

b $\frac{4}{30} = \frac{2}{15}$ denominator $= 15 = 3 \times 5 \rightarrow \frac{2}{15}$ is a recurring decimal

Simplify the
fraction.

c denominator $= 16 = 2^4 \rightarrow \frac{7}{16}$ is a terminating decimal

d denominator $= 13 \rightarrow \frac{11}{13}$ is a recurring decimal

To convert a fraction to a percentage

- write it as a decimal
- multiply the decimal by 100%.

$100\% = \frac{100}{100} = 1$
so you are
multiplying by 1
which does not
change the value
of the decimal.

Example

Write as percentages   **a** $\frac{5}{8}$ **b** $\frac{5}{9}$ **c** $\frac{1}{7}$

a $\frac{5}{8} = 0.625 = 0.625 \times 100\% = 62.5\%$

b $\frac{5}{9} = 0.\dot{5} = 0.\dot{5} \times 100\% = 55.5\dot{5}\% = 55.6\%$ to 1 dp

c $\frac{1}{7} = 0.\dot{1}4285\dot{7} = 0.\dot{1}4285\dot{7} \times 100\% = 14.\dot{2}8571\dot{4}\% = 14.3\%$ to 1 dp

1 Write a decimal equivalent for each of these fractions.

   a $\frac{1}{2}$ 0.5   **b** $\frac{3}{4}$ 0.75   **c** $\frac{2}{5}$ = 0.25

   d $\frac{1}{10}$ 0.1   **e** $\frac{1}{5}$ 0.20   **f** $\frac{1}{4}$ 0.4

You should be able to do all of these mentally.

2 Convert each of the decimals from question **1** to a percentage.

3 Convert each fraction to a percentage, using a written method.
   Show your working.

   a $\frac{5}{8}$ **b** $\frac{4}{5}$ **c** $\frac{7}{8}$

   d $\frac{3}{5}$ **e** $\frac{3}{8}$ **f** $\frac{1}{8}$

You should practise using a written method here, even if you can do these mentally.

4 Use a calculator to check your answers to question **3**.

5 Use a calculator to convert these fractions to decimals.

   a $\frac{1}{16}$ **b** $\frac{7}{25}$ **c** $\frac{7}{125}$

   d $\frac{3}{40}$ **e** $\frac{7}{16}$ **f** $\frac{1}{32}$

6 Convert the decimal answers from question **5** to percentages.

7 Use a written method to convert each fraction to a decimal.
   Use the 'dot' notation to represent recurring decimals.

   a $\frac{1}{3}$ **b** $\frac{1}{6}$ **c** $\frac{2}{3}$

   d $\frac{1}{7}$ **e** $\frac{1}{9}$ **f** $\frac{5}{6}$

DID YOU KNOW?
Blaise Pascal's mechanical calculator, invented in 1645, used dials and gears to add and subtract positive numbers and was based on the decimal system.

8 Convert your decimal answers from question **7** to percentages.

9 Use a calculator to check your answers to questions **7** and **8**.

10 Use an appropriate method to convert these fractions to decimals.

   a $\frac{3}{7}$ **b** $\frac{3}{16}$ **c** $\frac{17}{80}$

   d $\frac{5}{9}$ **e** $\frac{4}{25}$ **f** $\frac{5}{7}$

11 State whether each of these fractions will give a recurring decimal or
   a terminating decimal. Explain your answers.

   a $\frac{1}{25}$ **b** $\frac{3}{20}$ **c** $\frac{4}{11}$

   d $\frac{1}{126}$ **e** $\frac{1}{125}$ **f** $\frac{1}{128}$

12 Shula says, 'I used my calculator to change $\frac{1}{13}$ to a decimal, and I got
   the answer 0.07692308. There is no repeating pattern, so the decimal
   does not recur.' Explain why Shula is wrong.

This spread will show you how to:

- Distinguish between recurring and terminating decimals
- Convert fractions to decimals and percentages and vice versa

Keywords
Denominator
Recurring
Terminating

- To convert a percentage to a fraction, divide the percentage by 100.
  For example,
  $$43.7\% = \tfrac{43.7}{100} = 0.437$$

- To convert a **terminating** decimal to a fraction:
  1 Write the decimal as a fraction with **denominator** 10, 100, 1000, ..., according to the number of decimal places. For example, $0.45 = \tfrac{45}{100}$
  2 Simplify the fraction.
  $$\tfrac{45}{100} = \tfrac{9}{20}$$

2 decimal places
so denominator is
100.

Convert these to fractions   **a** 0.306   **b** 45%   **c** 32.5%   **d** 0.52

a $0.306 = \tfrac{306}{1000} = \tfrac{153}{500}$

b $45\% = \tfrac{45}{100} = \tfrac{9}{20}$

c $32.5\% = 0.325 = \tfrac{325}{1000} = \tfrac{65}{200} = \tfrac{13}{40}$

d $0.52 = \tfrac{52}{100} = \tfrac{26}{50} = \tfrac{13}{25}$

The next example shows how to convert **recurring** decimals to fractions. You multiply by a power of 10, then subtract one lot of the original decimal to produce a fraction.

Write as fractions   **a** $0.\dot{4}$ **b** $0.\dot{5}\dot{6}$

a $\qquad 10 \times 0.\dot{4} = 4.\dot{4}$ one recurring figure so multiply decimal by 10

$\qquad 10 \times 0.\dot{4} - 1 \times 0.\dot{4} = 4.\dot{4} - 0.\dot{4}$ subtract one lot of the original decimal.

$\qquad\qquad 9 \times 0.\dot{4} = 4$

$\qquad\qquad\quad 0.\dot{4} = \tfrac{4}{9}$

b $\qquad 100 \times 0.\dot{5}\dot{6} = 56.\dot{5}\dot{6}$ two recurring figures so multiply decimal by 100

$\quad 100 \times 0.\dot{5}\dot{6} - 1 \times 0.\dot{5}\dot{6} = 56.\dot{5}\dot{6} - 0.\dot{5}\dot{6}$ subtract one lot of the original decimal.

$\qquad\qquad 99 \times 0.\dot{5}\dot{6} = 56$

$\qquad\qquad\quad 0.\dot{5}\dot{6} = \tfrac{56}{99}$

1 Convert these percentages to decimals.

   a 43%   **b** 86%   **c** 94%   **d** 45.5%   **e** 3.75%   **f** 105%

2 Convert these decimals to fractions. You should be able to do these mentally.

   a 0.5   **b** 0.25   **c** 0.2   **d** 0.125   **e** 0.75   **f** 0.9

3 Convert these decimals to fractions.

   a 0.51   **b** 0.43   **c** 0.413   **d** 0.719   **e** 0.91   **f** 0.871

4 Convert these percentages to fractions.

   a 49%   **b** 53%   **c** 73%   **d** 81%   **e** 37%   **f** 19%

5 Convert these decimals to fractions. Give your answers in their simplest form.

   a 0.32   **b** 0.55   **c** 0.44   **d** 0.155   **e** 0.64   **f** 0.265

6 Convert these percentages to fractions. Give your answers in their simplest form.

   a 55%   **b** 62%   **c** 84%   **d** 65%   **e** 72%   **f** 18.5%

7 Which of these fractions will make recurring decimals?
   Explain your answer.

   $\frac{22}{25}$ $\frac{17}{20}$ $\frac{8}{11}$ $\frac{2}{5}$

8 Write each of these recurring decimals using 'dot' notation.

   a 0.111 …   **b** 0.555 …   **c** 0.75555 …   **d** 0.346346346 …

   e 0.7656565 …

9 Use the method shown on page 82 to convert these recurring decimals to fractions.

   a $0.\dot{2}$ **b** $0.\dot{6}$ **c** $0.\dot{2}\dot{5}$ **d** $0.\dot{2}\dot{7}$ **e** $0.5\dot{4}\dot{5}$ **f** $0.\dot{6}0\dot{5}$

10 Convert these percentages to fractions.

   a $52.\dot{2}$%   **b** $5.\dot{5}$%   **c** $45.\dot{2}\dot{7}$%   **d** $8.3\dot{5}$%   **e** $66.\dot{6}$%   **f** $8.\dot{2}0\dot{5}$%

11 Find a fraction equal to the recurring decimal $0.\dot{0}12345678\dot{9}$, giving your answer in its simplest form. Show your working.

Key objectives

- Order fractions by rewriting them with a common denominator
- Add and subtract fractions by writing them with a common denominator
- Distinguish between fractions with denominators that have only prime factors of 2 and 5 (which are represented by terminating decimals), and other fractions (which are represented by recurring decimals)
- Convert a recurring decimal to a fraction and vice-versa
- Multiply and divide a given fraction by an integer, by a unit fraction and by a general fraction

1 a Convert the following to fractions in their simplest form.

    i 60%   **ii** 0.52   (2)

  b Work out the following, giving your answers as fractions in their simplest form.

    i $0.36 + 0.61$ **ii** $33\% - 20\%$ **iii** $\frac{5}{7} \div \frac{10}{9}$ (2)

2 Write these numbers in order of size.
Start with the smallest number.

  a 0.56, 0.067, 0.6, 0.65, 0.605

  b 5, −6, −10, 2, −4

  c $\frac{1}{2}$, $\frac{2}{3}$, $\frac{2}{5}$, $\frac{3}{4}$ (4)

(Edexcel Ltd., 2003)

This unit will show you how to

- Understand and use lines, alternate angles and corresponding angles
- Calculate and use the interior and exterior angles of polygons
- Recognise the parts of a circle, using correct vocabulary to describe them
- Calculate angles in a circle by using the circle theorems
- Apply the tangent theorems to circle problems

Before you start ...

You should be able to answer these questions.

Review

1 The diagrams show angles on a straight line.
Work out the missing angles.

Key stage 3

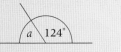

2 The diagrams show angles at a point.
Work out the missing angles.

Key stage 3

3 Work out the missing angle in these triangles.

Key stage 3

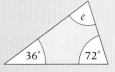 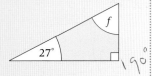

4 Work out the missing angle in these quadrilaterals.

Key stage 3

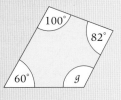 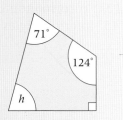

This spread will show you how to:

- Use parallel lines, alternate angles, corresponding angles and interior angles

Keywords

Alternate
Corresponding
Interior
Parallel
Supplementary
Vertically
  opposite

- Angles are formed when two lines cross, as at this crossroads.

$$a = c \quad \text{and} \quad b = d$$

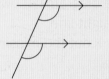

- **Vertically opposite** angles are equal.

Parallel lines are lines that never cross.

You need to remember these angle facts for parallel lines.

- **Alternate** angles are equal.

These are sometimes called Z angles.

- **Corresponding** angles are equal.

These are sometimes called F angles.

- **Interior** angles are **supplementary**. $a + b = 180°$

Supplementary angles add up to 180°.

When you work out angles, you should always say which angle fact you are using.

Example

Find the missing angles. Give reasons for your answers.

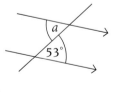

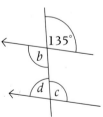

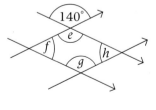

$a = 53°$ alternate angles

$b = 135°$ vertically opposite
$c = 135°$ corresponding
$d = 45°$ interior angles

$e = 140°$ vertically opposite
$f = 40°$ interior angles
$g = 140°$ interior angles
$h = 40°$ interior angles

1 Copy the diagrams.

   i Use colour to show alternate angles in each diagram.

   ii Use another colour to show corresponding angles in each diagram.

You may be able to find more than one pair in each diagram.

  a   **b**   **c**

2 Find the missing angles in each diagram.
Write down which angle fact you are using each time.

  a 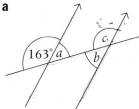   **b** 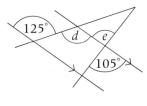   **c**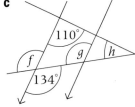

3 **a** Angles x, y and z are on a straight line.
Write down the value $x + y + z$.

   b Use alternate angles to work out the
two missing angles in the triangle.

   c Use your answers to **a** and **b** to show
that angles in a triangle add up to 180°.

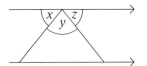

4 **a** Work out the missing angles in this
diagram.

   b Describe the quadrilateral formed
between the pairs of parallel lines.

5 Work out the missing angles.

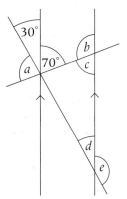

Angles in polygons

This spread will show you how to:

● Calculate and use the interior and exterior angles of polygons

Keywords

Exterior angle
Interior angle
Polygon
Vertex

A **polygon** is a closed shape with three or more straight sides.

> ● A regular polygon has all sides the same length and all interior angles equal.

The **interior angles** are inside the polygon.

The **exterior angles** are made by extending each side in the same direction.
Exterior angles are outside the polygon.

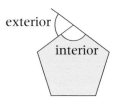

A square is a regular quadrilateral.

> ● For any polygon the exterior angle sum = 360°.
> ● For a regular polygon with *n* sides, each exterior angle = 360° ÷ *n*.

You can divide any polygon into triangles by drawing diagonals from a **vertex** (corner).

The number of triangles is always two less than the number of sides.

A pentagon divides into 3 triangles.
Angles in a triangle add up to 180°.
So interior angle sum of a pentagon = 3 × 180 = 540°

> ● For a polygon with *n* sides the interior angle sum = $(n - 2) \times 180°$.

Example

In a regular octagon find

a an interior angle   **b** an exterior angle.

a An octagon has 8 sides and divides into 6 triangles.
  Interior angle sum is 6 × 180° = 1080°
  Each interior angle is 1080° ÷ 8 = 135°.

b Each exterior angle is 360° ÷ 8 = 45°.

Example

An irregular hexagon has angles 108°, 92°, 120°, 134°, 115° and *x*.

Find the size of angle *x*.

Any hexagon has 6 sides and divides into 4 triangles.

Sum of interior angles is 4 × 180° = 720°

Sum of the 5 given angles is
108° + 92° + 120° + 134° + 115° = 569°

So *x* = 720° − 569° = 151°

1 a Copy the table of regular polygons.

b Draw each polygon and divide it into triangles by drawing diagonals from a vertex.

c Complete the table.

Shape	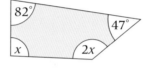					
Number of sides						
Number of triangles the shape splits into						
Sum of the interior angles in the shape						
Size of one interior angle						
Size of one exterior angle						

2 Find the missing angles in these quadrilaterals.

a

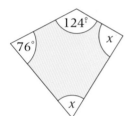

b

c

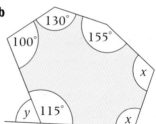

3 Find the missing angles in these irregular polygons.

a

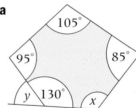

b

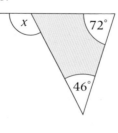

4 a Find the exterior angle marked x in each triangle.

i   **ii**   **iii**

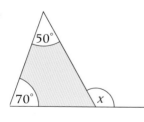
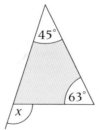

b Use your answers to part **a** to help you copy and complete this statement.

The exterior angle of a triangle is equal to_____

5 Use a diagram to explain why the interior angle sum of a quadrilateral is 360°.

This spread will show you how to:
- Recognise the parts of a circle, using correct vocabulary to describe them
- Calculate angles in a circle by using the circle theorems

Keywords
Arc
Chord
Diameter
Segment

A straight line joining two points on the circumference is a **chord**.

A chord divides the circle into two **segments**.

An **arc** is the part of the circumference that joins two points.

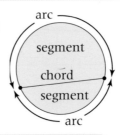

The **circumference** is the distance around the outside of a circle.

The **diameter** is the longest chord.

- The angle at the centre of a circle is double the angle at the circumference from the same arc.

- Angles from the same arc in the same segment are equal.

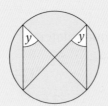

Example

Find the missing angles. Give a reason for each answer.

a

$\mathbf{a} = 70°$ angle at the centre is double the angle at the circumference

b

$\mathbf{b} = 360° - 104° = 256°$ angles at a point

c

$\mathbf{c} = 110°$ angle at the centre is double the angle at the circumference

The angle at the centre is a reflex angle.

d, e

$\mathbf{d} = 30°$ angles on same arc are equal
$\mathbf{e} = 65°$ angles on same arc are equal

f

$\mathbf{f} = 104°$ angle at the centre is double the angle at the circumference

1 Find the missing angles.
Give a reason for each answer.

a

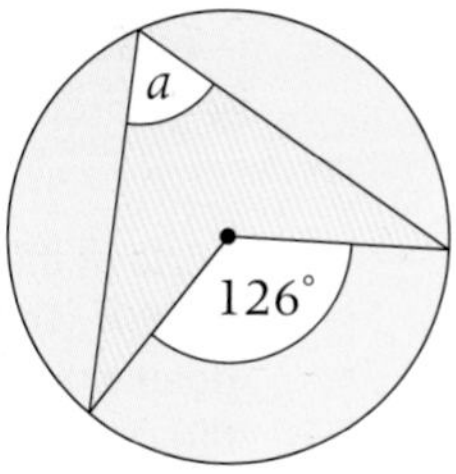

b

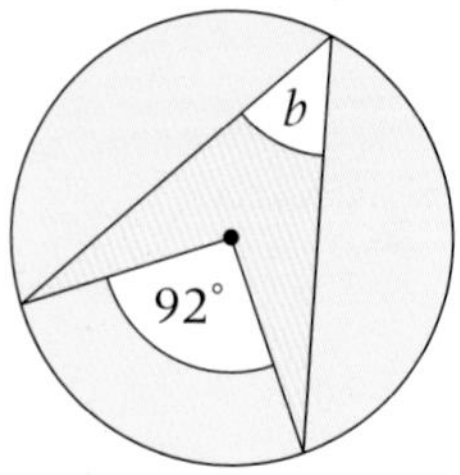

c 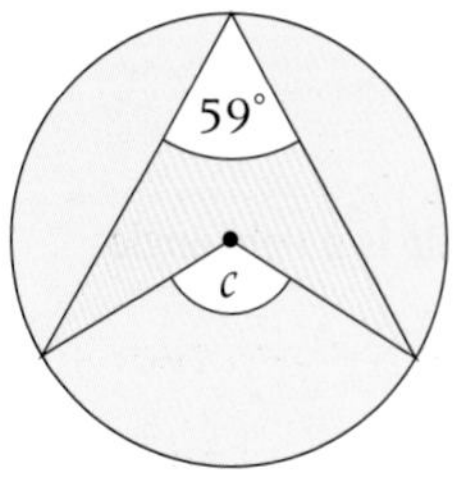
59°
c

d
32°
d

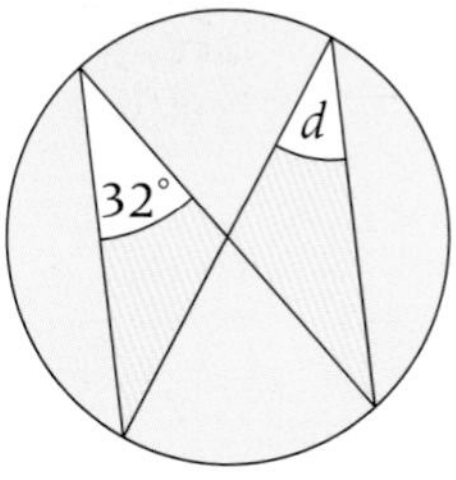

e
e
78°

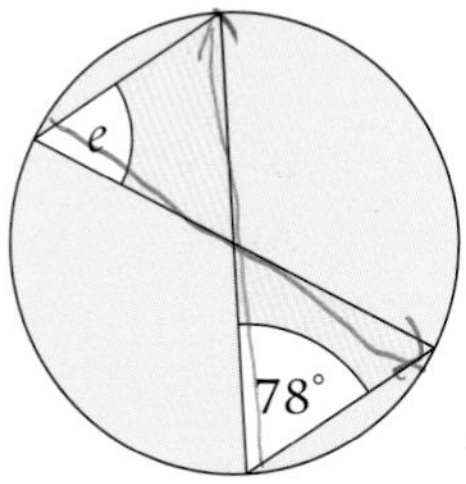

f 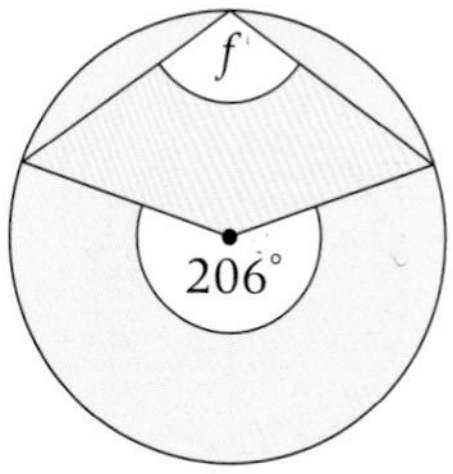
f
206°

g
122°
g

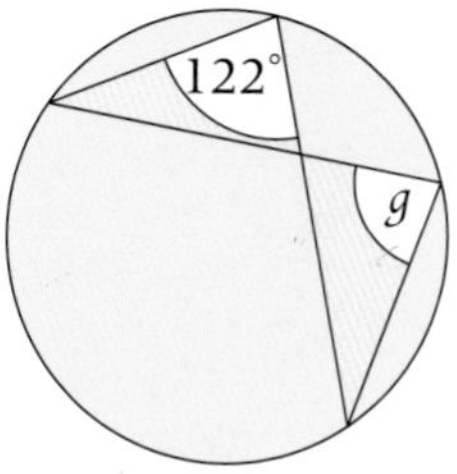

h
21°
h

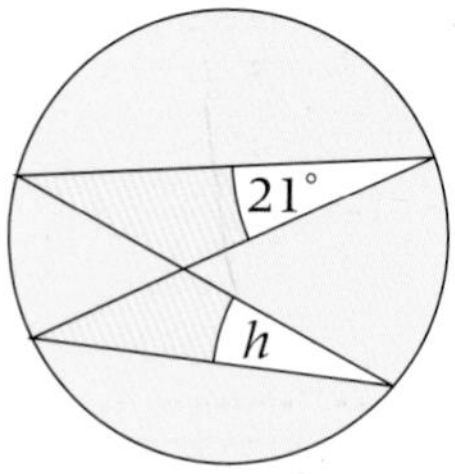

i
i
280°

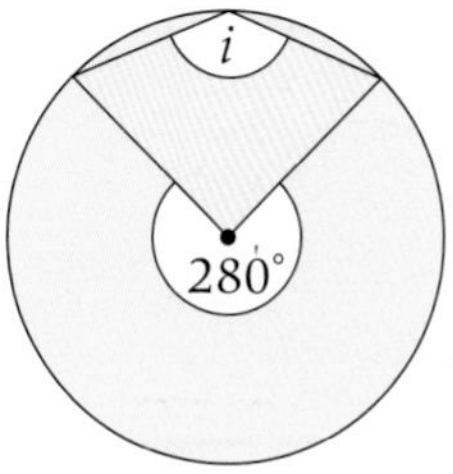

j 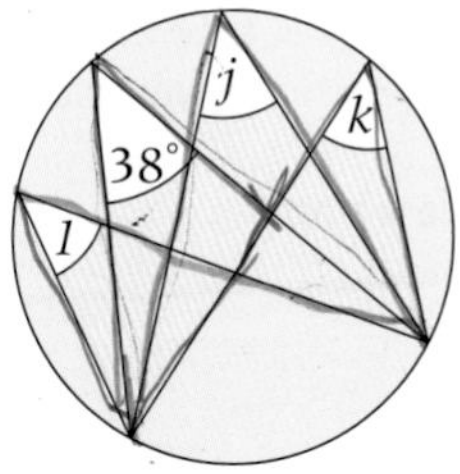
j
38°
k 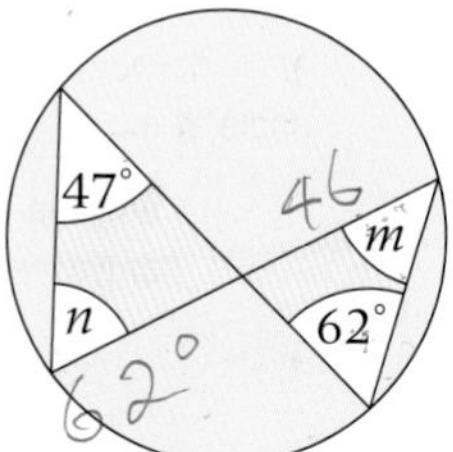
l

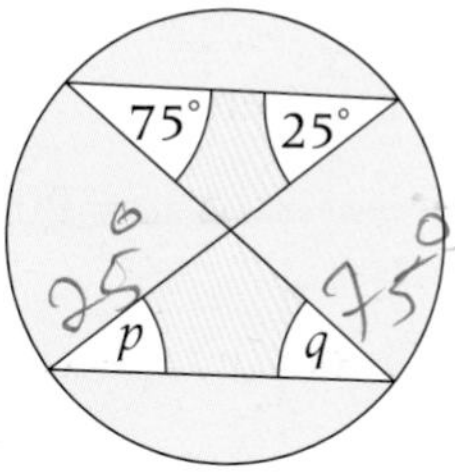

k
47°
46
m 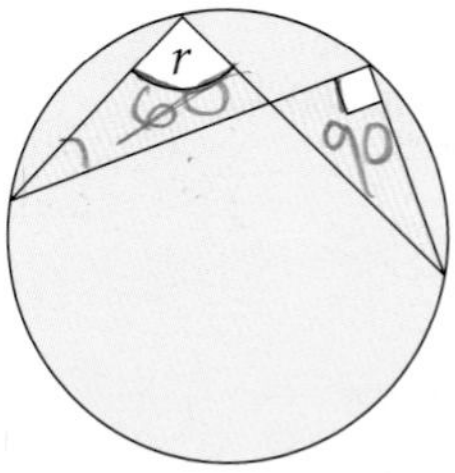
n
62°
62°

l
75° 25°
25
75
p 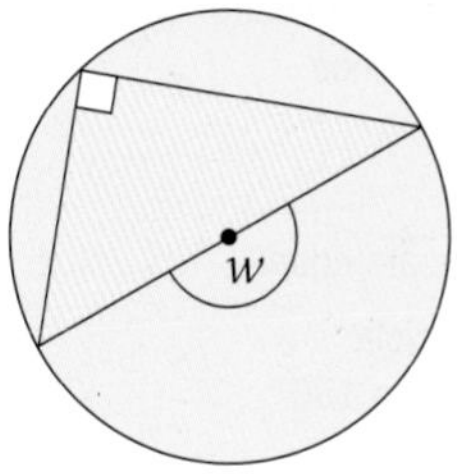 q

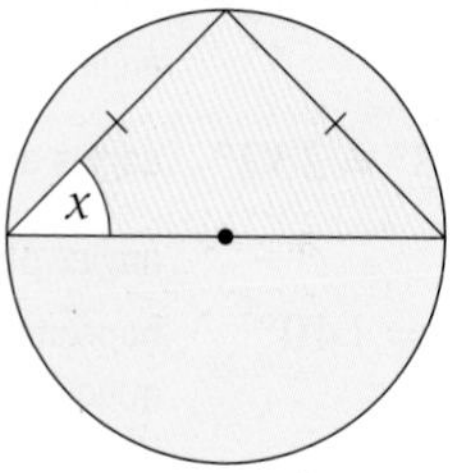

m
r
90

n 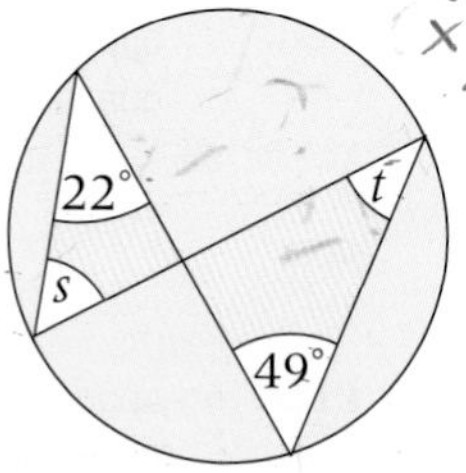
22°
t
s
49°

o
22
+ 4a
71

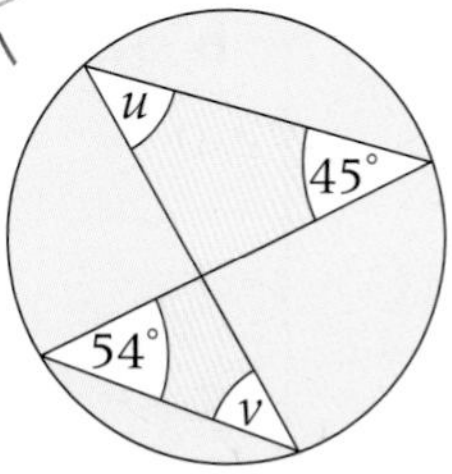

u
45°
54°
v

p
w

q
x

r
y
104

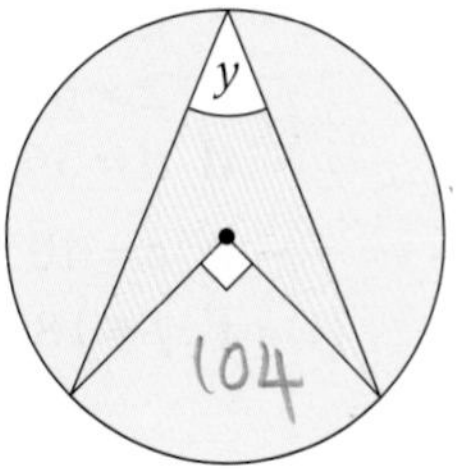

S2.4　More circle theorems

This spread will show you how to:

- Calculate angles in a circle by using the circle theorems

Keywords
Circumference
Cyclic
  quadrilateral
Semicircle

- An angle in a **semicircle** is a right-angle.

Angle at the centre
is 180°. This is
double the angle at
the circumference.
So the angle at the
circumference is
$\frac{1}{2}$ of 180° = 90°

A **cyclic quadrilateral** has all four vertices on the **circumference** of a circle.

- Opposite angles in a cyclic quadrilateral add up to 180°.

$a + b = 180°$
$x + y = 180°$

Example

Find the missing angles.

Give a reason for each answer.

You will need to
use the angle facts
from S2.3.

a

a $a = 90°$ 　　angle in a semicircle is a right angle

b

b $82° + b = 180$ 　　opposite angles in a cyclic
　　$b = 98°$ 　　quadrilateral add up to 180°

c,d

c $c = 180° - 64° = 116°$ 　　opposite angles in cyclic
The obtuse angle at 　　quadrilateral add up to 180°
the centre is 　　angle at the centre is double the
$2 \times 64° = 128°$ 　　angle at the circumference
d $d = 360° - 128° = 232°$ 　　angles at a point

e,f

e $e = 40°$ 　　angles on same arc are equal
f $f = 180° - 40° = 140°$ 　　opposite angles in cyclic
　　quadrilateral add up to 180°

1 Find the missing angles.
Give a reason for each answer.

You may need to use angle facts from page 90.

a
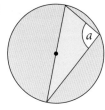

b
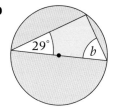

c
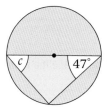

d
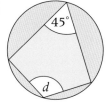

e
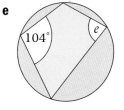

f
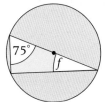

g
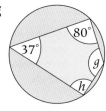

h
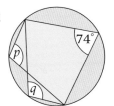

i
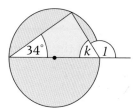

j
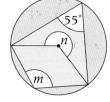

k
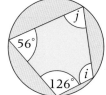

l
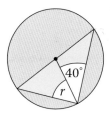

m
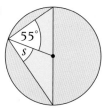

n
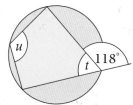

o
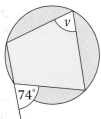

p
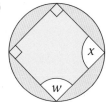

q
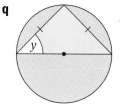

r
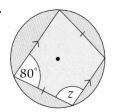

Tangents to circles

This spread will show you how to:

- Apply the tangent theorems to circle problems

Keywords
Radius
Right angle
Tangent

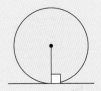

- The angle between the **tangent** and the **radius** at a point is a **right angle**.

From a point outside a circle you can draw two tangents to the circle.

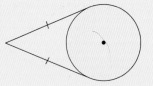

- Two tangents drawn from a point to a circle are equal.

Example

Find the missing angles and the missing length.

Give a reason for each answer.

You may need to use angle facts from pages 90 and 92.

a

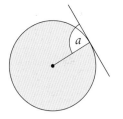

$a = 90°$

angle between tangent and radius at a point is a right angle

b

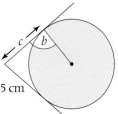

5 cm

$b = 90°$

$c = 5$ cm

angle between tangent and radius at a point is a right angle

two tangents drawn from a point to a circle are equal

c

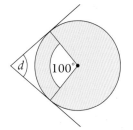

The given angles are d and 100°. The two other angles are 90°

$d + 100° + 90° + 90° = 360°$

$d = 80°$

angles in a quadrilateral add up to 360°

angle between radius and tangent is a right angle

1 Find the missing angles and the missing lengths.
Give a reason for each answer.

a

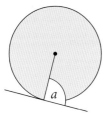

b

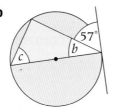

c

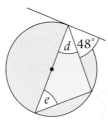

d

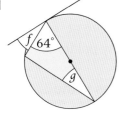

e

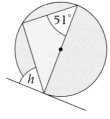

f

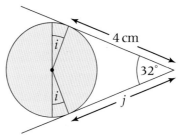

g

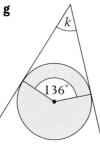

h

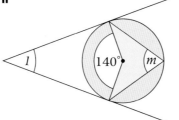

i

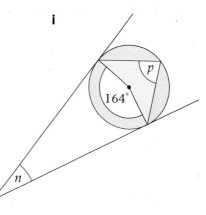

j

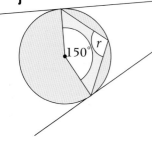

k

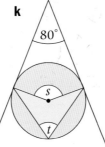

l

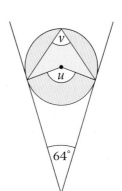

Exam review

Key objectives

- Use parallel lines, alternate angles and corresponding angles
- Calculate and use the sums of the interior and exterior angles of polygons
- Recall the definition of a circle and the meaning of related terms, including chord, tangent, arc, sector and segment
- Understand that the tangent at any point on a circle is perpendicular to the radius at that point
- Prove and use the facts that the angle subtended by an arc at the centre of a circle is twice the angle subtended at any point on the circumference, the angle subtended at the circumference by a semicircle is a right angle, that angles in the same segment are equal, and that opposite angles of a cyclic quadrilateral sum to 180°

1 Work out the size of an exterior angle of a regular hexagon.

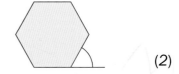

(2)

2 In the diagram, A, B and C are points on the circle, centre O. Angle BCE = 63°. FE is a tangent to the circle at point C.

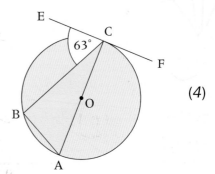

(4)

   a Calculate the size of angle ACB. Give reasons for your answer.

   b Calculate the size of angle BAC. Give reasons for your answer.

(Edexcel Ltd., 2003)

This unit will show you how to

- Generate common sequences and describe how number patterns are formed
- Use position-to-term rules to write sequences
- Find and describe the *n*th term of a sequence, using this to find other terms
- Explain how the formula for the *n*th term of a sequence works
- Use quadratic expressions to describe the *n*th term of a sequence

Before you start ...

You should be able to answer these questions.

Review

1 Complete the next two values in each pattern.

    a 2, 4, 6, 8, ...   **b** 100, 94, 88, 82, 76, ...

    c 1, 2, 4, 7, 11, ... **d** 10, 7, 4, 1, ...

    e 3, 6, 12, 24, ... **f** $1, \frac{1}{2}, \frac{1}{3}, \frac{1}{4}, \frac{1}{5}, \ldots$

Key stage 3

2 Given that $n = 3$, put these expressions in ascending order.

$2n + 7$ $4(n - 1)$

$2n^2$ $\frac{9}{n} + 15$

$15 - n$

Key stage 3

3 This pattern has been shown in three different ways.

Key stage 3

1	4	9	16
1×1	2×2	3×3	4×4
□	⊞		

It has a special name.
What is it called?
Describe how it got this name.

Number patterns

This spread will show you how to:

- Generate common sequences and describe how number patterns are formed

Keywords

Sequence
Term

- A **sequence** is a set of numbers that follow a pattern, for example

  5, 9, 13, 17, 21, ... are the first five **terms** of a sequence that goes up in 4s

  3, 6, 12, 24, 48, ... are the first five terms of a sequence that doubles

  1, 4, 9, 16, 25, ... is the sequence of square numbers
  (1×1, 2×2, 3×3 and so on)
  1, 8, 27, 64, 125, ... are the cube numbers
  ($1 \times 1 \times 1$, $2 \times 2 \times 2$, $3 \times 3 \times 3$ and so on)

Examiner's tip
The techniques learned in this unit will be useful for your Using and Applying Mathematics coursework task.

You can find sequences in patterns, for example,

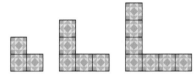

Number of tiles is:
3, 5, 7, ...
The next diagram would need 9 tiles.

Example

Write the next two terms in each sequence.

a 4, 5, 7, 10, 14, 19, ...   **b** 0.6, 0.7, 0.8, 0.9, ...

a 4, 5, 7, 10, 14, 19, ... The terms increase by 1, then 2, then 3.
so the next two terms are $19 + 6 = 25$ and $25 + 7 = 32$.

b 0.6, 0.7, 0.8, 0.9, ... The terms increase by 0.1. The next two terms
are $0.9 + 0.1 = 1.0$ and $1.0 + 0.1 = 1.1$.

You may be tempted to continue 0.7, 0.8, 0.9, 0.10, ... but 0.10 is less than 0.9!

Example

Name each of these sequences.

a 1, 3, 5, 7, 9, ...   **b** 25, 36, 49, 64, ...

a These are the odd numbers.   **b** These are the square numbers, starting at 5×5 (or 5^2).

Example

How many tiles would be in the tenth diagram in this pattern?

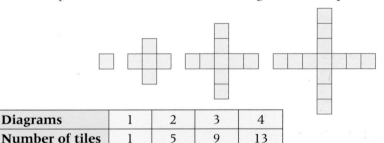

Diagrams	1	2	3	4
Number of tiles	1	5	9	13

A table is a good way of organising results so that you can see what is happening.

The number of tiles increases by 4 each time. If you continue this, you get 1, 5, 9, 13, 17, 21, 25, 29, 33, **37**.
The tenth diagram would have 37 tiles.

1 Copy each sequence and add the next two terms.

   a 4, 9, 14, 19, 24, ___, ___   **b** 100, 93, 86, 79, 72, ___, ___

   c 1, 2, 4, 7, 11, ___, ___   **d** 9, 99, 999, 9999, 99 999, ___, ___

   e 1, 1, 2, 3, 5, 8, ___, ___   **f** 54, 27, 13.5, 6.75, ___, ___

2 Copy and fill in the missing numbers in each sequence.

   a 4, ___, 10, ___, 16, ...   **b** 4, ___, ___, 32, 64, ...

   c 95, ___, ___, ___, 87, ...   **d** 1, ___, 27, 64, ___, ...

3 Write the first five terms of each of these well-known number patterns.

   a Multiples of 3   **b** Powers of 2

   c Prime numbers   **d** Square numbers over 100

4 The triangular numbers form a sequence.

   a Copy the table and use the diagram to complete it.

Diagrams	1	2	3	4	5
Number of dots	1	3			

   b Hence, write the first 10 triangular numbers.

   c Why do you think the square numbers (1, 4, 9, 16, 25, ...) got their name?

   d How did the cube numbers get their name? Write the first five cube numbers.

5 Find the tenth term in each of these number patterns.

   a (1×2), (2×3), (3×4), (4×5), ...

   b $\frac{1}{2}, \frac{2}{3}, \frac{3}{4}, \frac{4}{5}, ...$

   c (5×2), (5×4), (5×8), (5×16), ...

   d (1×1), (4×8), (9×27), (16×64), ...

6 Look at this number pattern.

   a Write the next two lines in the pattern.

   b What is $66\ 666\ 666\ 667^2$?

   c What is $\sqrt{4\ 444\ 444\ 444\ 888\ 888\ 889}$?

$$7^2 = 49$$
$$67^2 = 4489$$
$$667^2 = 444\ 889$$

This spread will show you how to:

- Use position-to-term rules to write sequences
- Find and describe the nth term of a sequence, using this to find other terms

Keywords

nth term
Position
Position-to-term
  rule
Sequence
Term

- You can use a **position-to-term rule** to write a **sequence**.

For example, this is a sequence using the position-to-term rule

   Multiply by 5 and add 2

The first 5 terms are:

Position	1	2	3	4	5
Term	$1 \times 5 + 2 = 7$	$2 \times 5 + 2 = 12$	$3 \times 5 + 2 = 17$	$4 \times 5 + 2 = 22$	$5 \times 5 + 2 = 27$

You can write a position-to-term rule in algebraic notation.
For example, T_n represents the **nth term** of the sequence

$$T_n = 3n - 2$$

To find T_1, the first term, put $n = 1$ in the rule.

Hence, $T_1 = 3 \times 1 - 2 = 1$
$T_2 = 3 \times 2 - 2 = 4$
$T_3 = 3 \times 3 - 2 = 7$
$T_4 = 3 \times 4 - 2 = 10$
The sequence $T_n = 3n - 2$ is 1, 4, 7, 10, ...

The rule connects the **position** of a **term** in a sequence to its value.

Example

Find the first five terms of the sequences with these position-to-term rules.

a $T_n = 5n + 7$

b $T_n = 3n^2$

a $T_1 = 5 \times 1 + 7 = 12$
$T_2 = 5 \times 2 + 7 = 17$
$T_3 = 5 \times 3 + 7 = 22$
$T_4 = 5 \times 4 + 7 = 27$
$T_5 = 5 \times 5 + 7 = 32$, ...
Hence, the sequence with $T_n = 5n + 7$ is 12, 17, 22, 27, 32, ...

b $T_1 = 3 \times 1^2 = 3$
$T_2 = 3 \times 2^2 = 12$
$T_3 = 3 \times 3^2 = 27$
$T_4 = 3 \times 4^2 = 48$
$T_5 = 3 \times 5^2 = 75$, ...
Hence, the sequence with $T_n = 3n^2$ is 3, 12, 27, 48, 75, ...

You may notice a pattern that allows you to generate the terms quickly, for example this sequence goes up in 5s.

Remember 'BIDMAS' ... the power comes before multiplication, so square before multiplying by 3.

1 Write the first five terms of each sequence.

  a $T_n = 8n + 2$ **b** $T_n = 5n - 4$ **c** $T_n = 7n$

  d $T_n = 10 - 2n$ **e** $T_n = n^2 - 3$ **f** $T_n = 2n^2$

2 Generate the first five terms of each sequence.

  a $T_n = 9 - n$ **b** $T_n = (n + 1)(n + 2)$ **c** $T_n = \frac{1}{n}$

  d $T_n = (-n)^3$ **e** $T_n = (2n - 1)(n + 1)(n - 3)$ **f** $T_n = n^4$

3 Here are four position-to-term rules.

    $\boxed{T(n) = n^2}$ $\boxed{T_n = 18 + 2n}$

        $\boxed{T_n = n^3}$ $\boxed{T_n = n(n + 5)}$

Write rules that give

a two sequences with an identical first term

b two sequences whose third term is 24

c two sequences whose tenth term is greater than 100.

4 **a** Generate all the terms and the formula for T_n, the nth term of the sequence described by the flow chart.

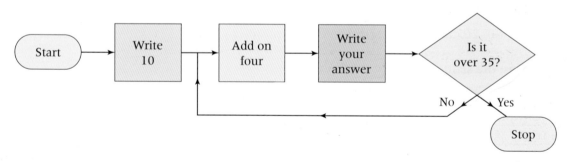

b Design your own flow chart for the sequence 2, 5, 10, 17, 26, 37, 50.

5 Write the algebraic position-to-term rule for

  a **i** Two sequences with identical 1st terms

     ii Two sequences with identical 5th terms

  b If two sequences have identical 1st terms and identical 5th terms, do they have to be identical sequences? Discuss your answer.

This spread will show you how to:

● Find and describe the *n*th term of a sequence, using this to find other terms

Keywords

Formula
Linear
*n*th term

● In a **linear** sequence the terms go up or down by the same amount.

> 12, 15, 18, 21, 24, ... goes up in 3s, so it is linear
> 100, 95, 90, 85, 80, ... goes down in 5s, so it is linear

You can find the **nth term** of a linear sequence by comparing the sequence to the times table to which it is related. For example the sequence 10, 17, 24, 31, 38, ... is connected to the 7× table, hence:

Position	1	2	3	4	5
Multiples of 7	7	14	21	28	35
Sequence term	10	17	24	31	38

The terms go up in 7s.

By comparing the sequence to the 7× table, you can see that you have to add 3 to the multiples of 7 to get the terms of the sequence. Hence *n*th term T_n = Position × 7 + 3

$$T_n = 7n + 3$$

You can use the *n*th term **formula** to find any term of the sequence quickly, for example the 100th term of the above sequence is

$$T_{100} = 7 \times 100 + 3 \quad \text{so } T_{100} = 703$$

Example

Find the *n*th term and, hence, the 100th term of

a 2, 8, 14, 20, 26, ...   **b** 25, 20, 15, 10, 5, ...

a Compare the sequence to the multiples of 6.

Position	1	2	3	4	5
Multiples of 6	6	12	18	24	30
Term	2	8	14	20	26

This sequence goes up in 6s, so it must be connected to the 6 times table.

First term = 2 = 6 − 4
Subtract 4 from the multiple of 6 to get each term.
$$T_n = 6n - 4$$
Hence, $T_{100} = 6 \times 100 - 4 = 596$

b Compare this sequence to the multiples of −5.

Position	1	2	3	4	5
Multiples of 5	−5	−10	−15	−20	−25
Term	25	20	15	10	5

If a sequence goes down use negative multiples.

First term = 25 = 30 − 5
Second term = 20 = 30 − 10
Add 30 to the multiple of −5 to get each term.
$$T_n = -5n + 30 \text{ or } T_n = 30 - 5n$$
Hence, $T_{100} = 30 - 5 \times 100 = -470$

On a number line, to get from −5 to 25, you need to add 30.

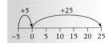

1 Find a formula for the nth term, T_n, of each sequence.

   a 4, 9, 14, 19, 24, ...   **b** 1, 3, 5, 7, 9, ...

   c 10, 12, 14, 16, 18, ...   **d** 1, 1.5, 2, 2.5, 3, ...

   e −4, −2, 0, 2, 4, ...   **f** 1, 2, 3, 4, 5, ...

   g The multiples of 13   **h** Counting up in 10s, starting from 4

   i 10, 8, 6, 4, 2, ...   **j** 100, 95, 90, 85, 80, ...

   k 50, $49\frac{3}{4}$, $49\frac{1}{2}$, $49\frac{1}{4}$, ...

   l Counting down in multiples of 4 from 75

2 Write five linear sequences of your own that have a third term of 15. Find a formula for the nth term of each one, and use this to find the 100th term in each case.

3 Are these true or false?

   a The 50th term of 2, 5, 8, 11, 14, ... is more than 150.

   b The 50th term of 5, 9, 13, 17, 21, ... is even.

   c The 100th term of 1000, 990, 980, 970, 960, ... is negative.

> Find the position-to-term rule for each sequence.

4 Here are some terms of a sequence. In each case find the formula for the nth term, T_n.

   a The 5th term is 20, the 6th term is 28 and the 7th term is 36.

   b The 100th term is 302, 101st term is 305 and the 102nd term is 308.

   c The 153rd term is 260, the 154th term is 262 and the 155th term is 264.

5 How could you find the nth term formula for these fractional sequences? See if you can work out what it would be in each case.

   a $\frac{3}{7}$, $\frac{5}{10}$, $\frac{7}{13}$, $\frac{9}{16}$, $\frac{11}{18}$ ...

   b $\frac{10}{30}$, $\frac{12}{27}$, $\frac{14}{24}$, $\frac{16}{21}$, $\frac{18}{18}$ ...

   c $\frac{7}{1}$, $\frac{8}{4}$, $\frac{9}{9}$, $\frac{10}{16}$, $\frac{11}{25}$ ...

   d $\frac{1}{11}$, $\frac{8}{9}$, $\frac{27}{7}$, $\frac{64}{5}$, $\frac{125}{3}$, ...

> Find separate formulae for the numerator and denominator.

6 a What is the nth term formula for $\frac{1}{1}$, $\frac{1}{2}$, $\frac{1}{3}$, $\frac{1}{4}$, $\frac{1}{5}$, ...?

   b Plot this sequence on a graph, with the term number on the x-axis and the term on the y-axis.

   c What happens if you continue this sequence indefinitely?

   d Investigate sequences of your own that behave in a similar way.

Describing patterns

This spread will show you how to:

● Explain how the formula for the *n*th term of a sequence works

Keywords
Formula
*n*th term
Pattern

Examiner's tip
The situations described on this page are particularly relevant to coursework tasks.

● You can describe a **pattern** using a **formula**.

This pattern is made of triangles.

   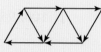

Number of triangles (*n*)	1	2	3	4
Number of arrows	3	5	7	9

+2   +2   +2

Number of arrows = 2 × number of triangles +1
so **nth term** of pattern = 2*n* + 1
This works because each extra triangle needs two arrows and you need one arrow at the start.

This pattern is made from cubes.
The outside faces of the cubes in each block are painted.

How many faces are painted each time?

Block number (*n*)	1	2	3	4
Number of painted faces	6	10	14	18
4× table	4	8	12	16

The terms go up in 4s.

Each cube in the pattern has 4 painted faces and the 2 end faces are painted.

In the *n*th block,

number of painted faces = 4 × block number + 2
so, *n*th term = 4*n* + 2

Example

Find a formula connecting the number of edges and the number of hexagons and explain why it works.

   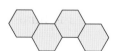

Make a table.

Number of hexagons (*n*)	1	2	3	4
Number of edges	6	11	16	21
5× table	5	10	15	20

The terms go up in 5s.

Number of edges = 5 × number of hexagons + 1
$$E = 5n + 1$$

Each hexagon needs five edges to join it to the previous one, and there is one edge to start the whole pattern off.

1 For each pattern, there is a formula. Explain why each formula works.

a

$E = 3s + 1$ where E = number of edges, s = number of squares

b

$W = B + 4$ where W = number of white tiles, B = number of coloured tiles

c

$M = L(L + 1)$ where M = number of nails and L = width of rectangle

2 Derive your own formulae for these patterns and explain why they work.

a Connect the number of coloured tiles (B) and the number of white tiles (W).

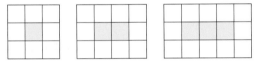

b i Connect the number of white circles (W) with the number of coloured circles (B).

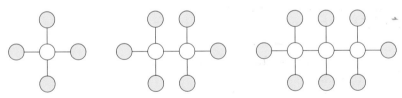

ii Use the same pattern to connect the number of white circles (W) with the number of lines (L).

c Find a formula for

i the number of presents (P) swapped at a party with n people, if all the guests give one another a present

ii the number of handshakes (H) at a party with n people, if all the guests shake hands with each other.

> Find the first 3 or 4 terms and make a table.

3 a A square jigsaw puzzle has n^2 pieces. It has C corner pieces, E edge pieces and M middle pieces. Write a formula for C, E and M in terms of n and show that your formula accounts for all the pieces.

b Extend to a rectangular puzzle with m pieces in its length and n pieces in its width.

This spread will show you how to:

- Use quadratic expressions to describe the *n*th term of a sequence

In a quadratic sequence the difference between terms increases.
For example, the square numbers:

Term   1,   4,   9,   16,   25,   36...

Difference   3   5   7   9   11

However, if you work out the differences a second time they are **constant**.

Constant means 'staying the same'.

Term   1,   4,   9,   16,   25,   36,

1st difference   +3   +5   +7   +9   +11

2nd difference   +2   +2   +2   +2

- **If a sequence has a constant second difference it is a quadratic sequence.**

So this sequence is a quadratic sequence.

3,   6,   11,   18,   27...

1st difference   +3   +5   +7   +9

2nd difference   +2   +2   +2

Compare the sequence to the square numbers.

Position	1	2	3	4	5
Square numbers	1	4	9	16	25
Term	3	6	11	18	27

Notice that you need to add 2 to the square numbers to get the sequence

$$T_n = n^2 + 2$$

This is the position-to-term rule for the sequence.

Find a formula for each of these quadratic sequences.

a 0, 3, 8, 15, 24, ...   **b** 2, 8, 18, 32, 50, ...

a Make a table.

Position	1	2	3	4	5
Square numbers	1	4	9	16	25
Term	0	3	8	15	24

The terms are 1 less than the square numbers, so $T_n = n^2 - 1$

Learn the square numbers.

b Make a table.

Position	1	2	3	4	5
Square numbers	1	4	9	16	25
Term	2	8	18	32	50

The terms are double the square numbers so $T_n = 2n^2$

Examiner's tip
These are difficult and unlikely to be examined on. However, you may encounter quadratic sequences in your coursework.

1 Generate the first five terms of each of these quadratic sequences.

a $T_n = n^2 + 5$ **b** $T_n = n^2 - 2$ **c** $T_n = 3n^2$

d $T_n = n^2 + 2n$ **e** $T_n = 2n^2 - 1$ **f** $T_n = 3n^2 + 1$

2 Find the formula for T_n, the nth term of these sequences.

a 4, 7, 12, 19, 28, ... **b** −2, 1, 6, 13, 22, ...

c 10, 40, 90, 160, 250, ... **d** 3, 12, 27, 48, 75, ...

e 2, 6, 12, 20, 30, ... **f** $\frac{1}{2}$, 2, $4\frac{1}{2}$, 8, $12\frac{1}{2}$, ...

3 The value of the second difference can tell us something about the formula for the nth term of a quadratic sequence.

a Copy and complete this table.

nth term	First five terms	First differences	Second differences
$2n^2$	2, 8, 18, 32, 50, ...	6, 10, 14, 18, ...	4, 4, 4, ...
$3n^2$	3, 12, __, __, __, __, ...		
$4n^2$	4, __, __, __, __, __, ...		
$5n^2$	5, __, __, __, __, __, ...		
$10n^2$	10, __, __, __, __, __, ...		

b What do you notice about the second difference and how it is connected to the quadratic formula for each sequence?

c Use your findings to help you to find the nth term for these sequences.

i 6, 24, 54, 96, 150, ... **ii** 1, 7, 17, 31, 49, ...

iii 4, 14, 30, 52, 80, ...

4 a Using differences, find a formula connecting the height of each rectangle (h) with the number of tiles (T).

b Explain why your formula works.

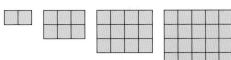

Start by making a table.

Rectangle height (h)	1	2	3	4
Number of tiles	2			
Square numbers	1			

5 a Write a formula for the area, A, of each rectangle, connecting it to the pattern number, n.

b Explain **i** why your formula works

ii why your formula is quadratic.

A3

Exam review

Key objectives

- Generate common integer sequences
- Generate terms of a sequence using term-to-term and position-to-term definitions of the sequence
- Use linear expressions to describe the nth term of an arithmetic sequence

1 Here are some terms of a sequence.

3rd term is 5, 4th term is 7, 5th term is 9, 6th term is 11.

a Find the formula for the nth term, T_n. (2)

b Hence find the 1st and 2nd term of the sequence. (2)

2 Here are some patterns made with sticks. (2)

The graph shows the number of sticks m used in pattern number n.

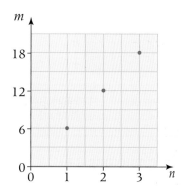

Write down a formula for m in terms of n.

(Edexcel Ltd., 2003)

D2

Displaying and interpreting data

This unit will show you how to

- Draw and use scatter diagrams
- Understand the concept of correlation
- Draw and use lines of best fit
- Draw stem-and-leaf diagrams, using them to find the median and range of data sets
- Draw and use back-to-back stem-and-leaf diagrams
- Draw box plots

Before you start ...

You should be able to answer these questions.

Review

1 Draw a grid with both axes from −4 to +5. Draw these points on the grid. — Key stage 3

A (2, 4) ***B*** (1, −3)
C (5, −3) ***D*** (−2, 2)
E (−4, −1) ***F*** (0, 3)
G (4, 0)

2 If £1 = €1.40, — Key stage 3

a how many euros you would get for
i £2 **ii** £4.50?

b how many pounds you would get for
i €4 **ii** €7?

3 Work out — Key stage 3

a 50% of 80 **b** 50% of 120
c 50% of 160 **d** 25% of 60
e 25% of 200 **f** 25% of 120
g 75% of 100 **h** 75% of 200
i 75% of 160 **j** 75% of 840

This spread will show you how to:

● Draw and use scatter diagrams

● Understand the concept of correlation

Keywords

Correlation

Scatter diagram

● A **scatter diagram** shows you the relationship between two numerical variables.

If there is a close relationship, or **correlation**, between the variables, the points will lie roughly on a straight line.

Examiner's tip
The diagrams described in this unit may be useful for your statistical coursework task.

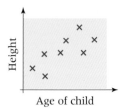

If the line slopes upwards there is **positive correlation**

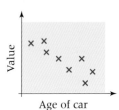

If the line slopes downwards there is **negative correlation**

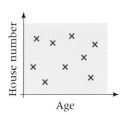

If the points are widely scattered, there is **no correlation**.

Example

Sally recorded the circumference and weights of eight pumpkins.

Circumference, cm	142	124	136	140	139	128	132	135
Weight, kg	28	21	25	26	26	23	23	24

a Draw a scatter diagram of these data.

b Describe the correlation shown.

c Describe the relationship between the circumference and weight of these pumpkins.

a

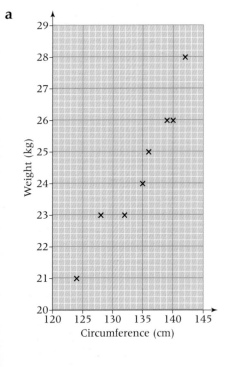

b There is positive correlation.

c The correlation shown suggests that larger pumpkins weigh more.

1 Kevin recorded the heights and weights of eight boys.

	Adam	Ben	Carl	Don	Evan	Fred	Gavin	Henry
Height, cm	157	136	142	140	138	152	156	160
Weight, kg	47	32	36	37	33	39	42	51

a Draw a scatter diagram of these data.

b Describe the correlation shown.

c Describe the relationship between the height and weight of these boys.

> You will need to use the graphs you draw in this exercise for Exercise D2.2.

2 Iain recorded the percentages achieved in Statistics and Mathematics exams by 10 students.

Statistics %	78	82	74	75	93	70	66	62	77	89
Mathematics %	70	76	61	70	89	65	59	58	73	82

a Draw a scatter diagram of these data.

b Describe the correlation shown.

c Describe the relationship between the percentages achieved in Statistics and Mathematics for these students.

3 Louise recorded information about the average number of minutes per day spent playing computer games and the reaction times of nine students.

Minutes per day spent playing computer games	40	60	75	40	35	20	80	50	45
Reaction time, seconds	5.2	4.3	3.9	5.5	6.0	7.2	3.6	4.8	5.0

a Draw a scatter diagram of these data.

b Describe the correlation shown.

c Describe the relationship between the average number of minutes per day spent playing computer games and reaction time.

4 Bob caught nine fish during one angling session.
He recorded information about the weights and lengths of the fish he caught in this table.

Weight, g	500	560	750	625	610	680	600	650	580
Length, cm	30	32	50	44	39	48	40	45	36

a Draw a scatter diagram of these data.

b Describe the correlation shown.

c Describe the relationship between the weights and lengths of the fish.

This spread will show you how to:
● Draw and use lines of best fit

Keywords
Correlation
Line of best fit
Prediction
Scatter diagram

● You can use a **scatter diagram** that shows **correlation** to predict other results.

To predict results you first draw a **line of best fit**.

● A line of best fit should follow the trend of the plotted points

The line of best fit can slope downwards. It doesn't have to start at (0, 0).

Example

a Draw a line of best fit for the data Sally collected on the circumference and weight of eight pumpkins in D2.1.
b Use the line to predict
   i the weight of a pumpkin, circumference 137 cm
   ii the circumference of a pumpkin that weighs 22 kg.

a

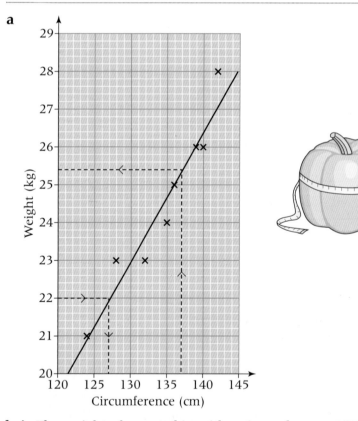

Draw a vertical line from 137 cm up to the graph, then across to the vertical axis

b i The weight of a pumpkin with a circumference 137 cm is 25 kg.
  ii The circumference of a pumpkin weighing 22 kg is 127 cm.

It would not be sensible to use this graph to predict the weight of a pumpkin that is 120 cm around as this is outside the range of the collected data plotted in the graph.

1 Look at the graph you drew in Exercise D2.1 question **1**.

  a Draw a line of best fit for the data Kevin collected on the height and weight of eight boys.

  b Use the line to predict

    i the weight of a boy 145 cm tall

    ii the height of a boy who weighs 45 kg.

2 Look at the graph you drew in Exercise D2.1 question **2**.

  a Draw a line of best fit for the data Iain collected on Statistics and Mathematics results.

  b Use the line to predict

    i the percentage in Mathematics for Imogen who achieved 80% in Statistics

    ii the percentage in Statistics for Katherine who achieved 80% in Mathematics.

  c Give a reason why it would not be sensible to use this graph to predict the percentage achieved in the Mathematics exam for a person who achieves 46% in the Statistics exam.

3 Look at the graph you drew in Exercise D2.1 question **3**.

  a Draw a line of best fit for the data Louise collected on time spent playing computer games and reaction time.

  b Use the line to predict

    i the reaction time of Jim, who spends on average 70 minutes per day playing computer games

    ii the time per day spent playing computer games for Ellis, who has a reaction time 6.4 seconds.

  c Give a reason why it would not be sensible to use this graph to predict the reaction time of Tom who spends on average 180 minutes per day playing computer games.

4 Look at the graph you drew in Exercise D2.3 question **4**.

  a Draw a line of best fit for the data Bob collected on the weight and length of fish.

  b Use the line to predict

    i the length of a fish that weights 700 g

    ii the weight of a fish that is 42 cm long.

Stem-and-leaf-diagrams

This spread will show you how to:

- Draw stem-and-leaf diagrams, using them to find the median and range of data sets

You can represent small amounts of data on a **stem-and-leaf** diagram.

A stem-and-leaf diagram shows all the data and the overall trend.

- **You can use a stem-and-leaf diagram to find statistics such as the median.**

When you draw a stem-and-leaf diagram choose the stem according to the data.

For example, for heights given to nearest cm, data will be in the 100s:
165, 154, 178, ... so in the stem you use

```
15
16
17
```

Example

These are the percentages for a Statistics test taken by 23 students.

```
58   54   78   66   67   40   45   38   58   73   51   49
47   53   41   36   59   64   52   43   39   80   37
```

a Draw a stem-and-leaf diagram for these data.

b Describe the trend.

c For these data find the

  i range

  ii median

  iii lower quartile and upper quartile

  iv interquartile range.

a
```
3 | 8 6 9 7              3 | 6 7 8 9           Now put the stem and
4 | 0 5 9 7 1 3          4 | 0 1 3 5 7 9       leaves in order
5 | 8 4 8 1 3 9 2        5 | 1 2 3 4 8 8 9
6 | 6 7 4               6 | 4 6 7
7 | 8 3                 7 | 3 8
8 | 0                   8 | 0
```

↑ Use the units digit as the leaves

↑ Use the tens digit as the stem

Key: 7 | 3 stands for 73% ← Always give a key

b Most students scored less than 60%.

c
  i Range = 44%   highest value – lowest value: $80 - 36 = 44$

  ii Median = 52%   data is in order so count up to find 12th value: $\frac{1}{2}(23 + 1) = 12$

  iii Lower quartile (LQ) 41%   lower quartile is the 6th value in the 1st quarter of data: $\frac{1}{4}(23 + 1) = 6$
       Upper quartile (UQ) 64%   then count back 6 values to find upper quartile

  iv Interquartile range (IQR) 23%   IQR = UQ – LQ $64 - 41 = 23$

For these data sets

a Draw a stem-and-leaf diagram to show the information.
(Remember to include a key.)

b Describe the trend.

c Find the

   i range   **ii** median   **iii** lower quartile and upper quartile

   iv interquartile range.

1 Percentage achieved in a Statistics test

   78 82 74 45 69 75 93 54 61 70 48 66
   62 51 77 59 51 89 81 52 63 71 59

2 Minutes per day spent playing computer games

   40 26 75 84 33 39 28 66 67 71 80
   37 52 47 63 49 41 44 58 69 43
   73 55 59 43 61 38 29 30 77 60

3 Time taken, in minutes, to solve a crossword puzzle

   12 24 21 16 8 9 3 31 18 27 35
   41 26 12 17 6 5 19 29 32 37 40
   15 22 10 33 11 7 20 27 29 34

4 Weight in kg of Year 10 boys

   47 51 63 39 42 57 36 37 49 32 60 54
   56 45 52 48 61 58 56 59 70 66 56

5 Height in cm of a group of Year 10 girls

   153 147 160 146 162 158 159 171 149 152 150 163
   172 167 165 155 155 157 154 168 150 172 152

6 IQ scores of a Year 10 tutor group in a comprehensive school

   101 112 125 109 98 107 108 117 121 116 94
   91 105 106 114 118 126 131 92 88 129
   89 116 103 108 127 110 117 104 119 133

7 Time taken to the nearest minute to change a flat tyre

   10 17 3 19 22 27 16 9 6 30 23 21
   12 21 9 25 23 18 33 8 32 15 11

Interpreting stem-and-leaf diagrams

This spread will show you how to:

- Draw and use back-to-back stem-and-leaf diagrams

Keywords
Average
Interquartile
    range (IQR)
Median
Measure of
    spread
Stem-and-leaf
    diagram

You can represent two data sets of the same variable on a back-to-back **stem-and-leaf** diagram.
The data sets will share a common stem.

- You can compare the two data sets using an **average** and a **measure of spread**.

The range measures the spread of all the data.

Range = highest value − lowest value

The **interquartile range** (**IQR**) measures the spread of the middle half of the data.

IQR = upper quartile − lower quartile

Example

Draw a back-to-back stem-and-leaf diagram to show the test results of a group of girls and a group of boys.

Girls   52   34   58   46   41   57   47   35   49   47   54

Boys   61   43   47   56   59   39   58   69   52   46   54

Compare the performances of the boys and the girls.

```
       Girls              Boys
         5  4 | 3 | 9
  9 7 7 6 1 | 4 | 3  6  7
     8 7 4 2 | 5 | 2  4  6  8  9
             | 6 | 1  9          Key: 1│4│3 stands for 41% for girls, 43% for boys
```

There are 11 girls.   There are 11 boys.

$$\text{Median} = \frac{(11 + 1)}{2}\text{th result}$$ $$\text{Median} = \frac{(11 + 1)}{2}\text{th result}$$
$$= \text{6th result}$$ $$= \text{6th result}$$
$$= 47$$ $$= 54$$

Range = 58 − 34   Range = 69 − 39
$\quad= 24$ $\quad= 40$

$$\text{Lower quartile} = \frac{(11 + 1)}{4}\text{th value}$$ $$\text{Lower quartile} = \frac{(11 + 1)}{4}\text{th value}$$
$$= \text{3rd result}$$ $$= \text{3rd result}$$
$$= 41$$ $$= 46$$

Upper quartile = 54   Upper quartile = 59
$\quad\quad\text{IQR} = 54 − 41$ $\quad\quad\text{IQR} = 54 − 41$
$\quad\quad= 13$ $\quad\quad= 13$

1 The boys' average is 7 marks higher than the girls' average.
2 The IQR is the same for the boys and the girls.
3 The range of the boys' marks is greater than that of the girls.

For these data sets

a Draw a back-to-back stem-and-leaf diagram. (Remember to include a key.)

b For both data sets in each question work out the

   i median   **ii** interquartile range   **iii** range.

c Use your answers to part **b** to make comparisons between the two data sets.

1 Percentages achieved in two tests

Test A: 38   37   62   45   42   55   56   61   49   52
          47   58   43   51   44   56   41   44   53

Test B: 65   72   57   79   66   48   53   54   41   75
          56   63   69   72   53   44   57   61   70

2 Height to the nearest cm of a sample of men and a sample of women

Men: 176   183   184   172   168   175   183   159   169   174
          160   180   178   167   182   188   171   178   158   165
          177   169   167

Women: 157   148   151   167   174   165   169   158   153   155
          161   158   155   172   156   162   166   149   178   154
          152   150   162

3 IQ scores of two classes, X and Y

X: 105   123   131   117   118   104   98   96   103   112   110
      117   126   129   123   109   108   115   99   89   121   134
      106   105   122   124   116   110   118   115   130

Y: 118   119   104   121   126   118   109   97   114   129   130
      107   116   87   93   128   121   118   113   103   102   114
      107   131   99   106   124   132   119   126   114

4 Reaction times to the nearest tenth of a second by a sample of boys and a sample of girls

Boys: 4.2   5.7   3.2   3.8   6.4   3.8   6.1   5.9   5.3   5.6   3.6   4.4
        5.2   3.2   3.8   5.8   4.7   4.5   6.2   6.8   7.1   6.6   7.2

Girls: 4.4   4.6   5.2   4.3   6.7   7.2   8.0   4.0   8.2   7.7   6.3   7.6
        4.8   5.9   5.2   7.4   7.3   6.2   6.5   5.6   6.6   6.3   5.5

5 Times taken to the nearest minute for a sample of children to complete two jigsaw puzzles

Puzzle P: 13   17   19   10   8   22   31   11   24   27
             6   37   18   12   29   14   8   17   9

Puzzle Z: 28   21   29   15   12   9   32   17   18   11
            19   16   8   33   24   14   25   17   23

This spread will show you how to:

● Draw box plots

Keywords

Box plot
Box and whisker
Interquartile range
  (IQR)
Lower quartile
Median
Upper quartile

● You can represent data sets on a **box plot**.

● To draw a box plot you need the **median**, the **upper** and **lower quartiles**, and the highest and lowest values.

Example

A group of 15 students took a science test.

These are their results.

52, 62, 71, 46, 41, 49, 36, 57, 60, 80, 41, 40, 79, 39, 64

a Find the median, the range and the upper and lower quartiles.
b Draw a box plot to represent these results.
c Find the interquartile range.

A box plot is sometimes called a **box and whisker diagram**.

a First arrange the data in order.

36, 39, 40, 41, 41, 46, 49, 52, 57, 60, 62, 64, 71, 79, 80

$$\text{Median} = \frac{(15 + 1)}{2}\text{th value}$$
$$= \text{8th value}$$
$$= 52$$

$$\text{Lower quartile} = \frac{(15 + 1)}{4}\text{th value}$$
$$= \text{4th value}$$
$$= 41$$

$$\text{Range} = 80 - 36$$
$$= 44$$

$$\text{Upper quartile} = \frac{3}{4}(15 + 1)\text{th value}$$
$$= \text{12th value}$$
$$= 64$$

b Lower quartile   Median   Upper quartile

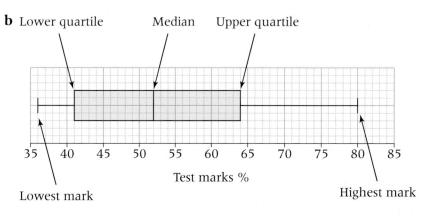

Lowest mark

Highest mark

Use graph paper and remember to scale and label the axis.

c Interquartile range = 64 − 41 = 23

The data sets for questions **1** to **7** were used in Exercise D2.3 to draw stem-and-leaf diagrams.

Use the values you found for the median, lower quartile and upper quartile (or use the data to find them again) and draw a box plot to show the information for each of these data sets.

1 Percentage achieved in a Statistics test

| 78 | 82 | 74 | 45 | 69 | 75 | 93 | 54 | 61 | 70 | 48 | 66 |
| 62 | 51 | 77 | 59 | 51 | 89 | 81 | 52 | 63 | 71 | 59 | |

2 Minutes per day spent playing computer games

40	26	75	84	33	39	28	66	67	71	80
37	52	47	63	49	41	44	58	69	43	
73	55	59	43	61	38	29	30	77	60	

3 Time taken, in minutes, to solve a crossword puzzle

12	24	21	16	8	9	3	31	18	27	35
41	26	12	17	6	5	19	29	32	37	40
15	22	10	33	11	7	20	27	29	34	

4 Weight in kg of Year 10 boys

| 47 | 51 | 63 | 39 | 42 | 57 | 36 | 37 | 49 | 32 | 60 | 54 |
| 56 | 45 | 52 | 48 | 61 | 58 | 56 | 59 | 70 | 66 | 56 | |

5 Height in cm of a group of Year 10 girls

153	147	160	146	162	158	159	171	149
152	150	163	172	167	165	155	155	157
154	168	150	172	152				

6 IQ scores of a Year 10 tutor group in a comprehensive school

101	112	125	109	98	107	108	117	121
116	94	91	105	106	114	118	126	131
92	88	129	89	116	103	108	127	110
117	104	119	133					

7 Time taken to the nearest minute to change a flat tyre

| 10 | 17 | 3 | 19 | 22 | 27 | 16 | 9 | 6 | 30 | 23 | 21 |
| 12 | 21 | 9 | 25 | 23 | 18 | 33 | 8 | 32 | 15 | 11 | |

D2

Exam review

Key objectives

- Draw and produce, using paper and ICT, diagrams for continuous data, including scatter graphs and stem-and-leaf diagrams
- Draw and produce, using paper and ICT, cumulative frequency tables and diagrams, box plots and histograms for grouped continuous data
- Draw lines of best fit by eye, understanding what these represent
- Distinguish between positive, negative and zero correlation using lines of best fit

1 Here are the results, to the nearest minute, of the times of runners in a 10 kilometre race.

60 54 69 55 60 72 52 53 61 59

61 52 55 58 49 61 58 70 52 50

a Draw a stem-and-leaf diagram for this data. (4)

b Work out the

  i median **ii** interquartile range **iii** range. (3)

2 The table shows the number of pages and the weight, in grams, for each of 10 books.

Number of pages	80	130	100	140	115	90	160	140	105	170
Weight (g)	160	270	180	290	230	180	320	270	210	300

a Copy and complete the scatter graph to show the information in the table. The first six points in the table have been plotted for you. (1)

b For these books, describe the relationship between the number of pages and the weight of a book. (1)

c Draw a line of best fit on the scatter diagram. (1)

d Use your line of best fit to estimate:

  i the number of pages in a book of weight 280g

  ii the weight of a book with 120 pages. (2)

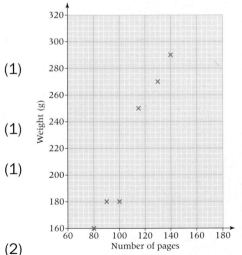

(Edexcel Ltd., 2004)

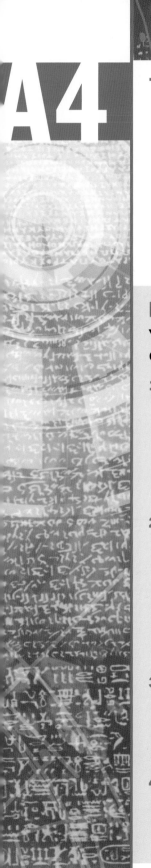

This unit will show you how to

- Recognise that equations of the form $y = mx + c$ have straight line graphs
- Plot straight line graphs, given a linear equation
- Recognise and understand the form of equations corresponding to horizontal, vertical and diagonal line graphs
- Find the gradient and y-axis intercept of straight line graphs
- Find the equation of a line joining several points

Before you start ...

You should be able to answer these questions.

Review

1 Draw a set of axes labelled from -5 to $+10$ and plot these coordinates.

Key stage 3

A (3, 4)	B (9, 0)
C (0, 3)	D (−1, 5)
E (−3, −4)	F (5, −2)
G $(2\frac{1}{2}, -\frac{1}{4})$	H (−0.3, −0.7)

2 Plot these coordinate groups on one set of axes, joining them to produce a line in each case.

Key stage 3

  i   (3, 1), (3, 2), (3, 3)
  ii (1, −2), (2, −2), (3, −2)
  iii (1, 3), (2, 5), (3, 7)
  iv (1, 4), (2, 3), (3, 2)

3 Convert these top heavy fractions to mixed numbers and vice versa.

Key stage 3

  a $\frac{13}{2}$ **b** $\frac{21}{4}$ **c** $\frac{33}{7}$ **d** $-\frac{100}{11}$
  e $2\frac{1}{3}$ **f** $7\frac{1}{2}$ **g** $5\frac{3}{4}$ **h** $-4\frac{3}{8}$

4 Find the value of the unknown in each equation.

Unit A2

  a $9 = 3n + 3$ **b** $6 = 1 + 5m$
  c $10p - 4 = 1$

This spread will show you how to:

- Recognise that linear equations have straight-line graphs
- Plot straight-line graphs, given a linear equation

Keywords
Axes
Linear
Plot

- The graphs of equations such as $y = 2x + 1$ (with no x^2 or higher powers) are straight lines. For example,

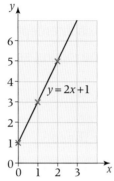

$y = 2x + 1$ is a **linear** equation

The graph of a linear equation is a straight line.

- You need three points to plot a straight line.

For example, to draw the graph of $y = 3x - 2$, first make a table of x and y values.

x	1	2	3
y	1	4	7

when $x = 1$, $y = (3 \times 1) - 2 = 1$

Draw x- and y-**axes** and **plot** the three points.

For an equation such as '$2x + 3y = 12$' it is a little harder to find the points.

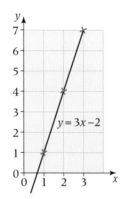

Two points are enough to fix a straight line but, as a check, work out a third point.

$2x + 3y = 12$ is an **implicit** equation – you can't calculate the value of y directly from the equation.

Example

Plot the graph of $2x + 3y = 12$.

x	**0**	6	**3**
y	4	**0**	2

Draw the x and y axes and plot the points.

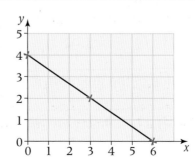

When the equation is in implicit form, as here, putting $x = 0$ and $y = 0$ gives you two points. For the third, choose an x-value between the first and second values.

1 Which of these equations have straight line graphs?

$y = 2x + 3$	$y = 7 - 3x$	$y = x^2 + 1$	$y = 5x$
$y = 7$	$y = 2x - x^3$	$2x + 7y = 8$	$x = -2$

2 **a** For each equation, copy and fill in the table of values.

i $y = 3x + 2$

x	0	1	2
y			8

ii $y = 2x - 4$

x	1	2	3
y	−2		

iii $2x + 5y = 10$

x	0		1
y		0	

b Draw x- and y-axes from −8 to 8 and plot the graphs.

3 I have a mobile phone. Each month, I pay £5 line rental. For every hour I then spend on the phone, I pay £7.

a Copy and complete this table of values to show my total phone bill for different lengths of time spent on the phone.

x **(hours on phone)**	1	2	3	4	5
y **(total bill £)**	12				

b Plot a graph to show hours against total bill.

c Use your graph to find the approximate cost if I spend 3 hours and 15 minutes on the phone one month.

d What is the equation of the graph you have drawn?

4 **a** The point (2, 5) lies on the graph $y = 2x + 1$. Does (3, 8) lie on this graph? Explain your answer.

b The point (2, 5) lies on the graph $y = 2x + 1$. Name another point that lies on this line.

5 **a** Plot the graphs $y = 3x - 1$ and $y = 3x + 2$ on the same axes.
   b Explain why there is no point that lies on both of these graphs.
   c Name the equation of another line that would have no points in common with these two.

6 Would you prefer to get £3 per week pocket money, with 20p for every chore done (such as washing up) or £5 per week pocket money with 15p for every chore done? Use line graphs to report on your favoured option and to find how many chores you would need to do to receive the same amount under both options.

This spread will show you how to:

● Recognise and understand the form of equations corresponding to horizontal, vertical and diagonal line graphs

A straight line can be **diagonal**, **vertical** or **horizontal**.

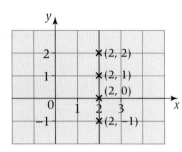

The x-coordinate of every point on this vertical line is 2.

The y-coordinate can have any value.

The equation of the line is x = 2

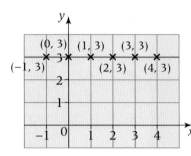

The y-coordinate of every point on this horizontal line is 3.

The x-coordinate can have any value.

The equation of the line is y = 3

● Horizontal lines have equations of the form x = c.
● Vertical lines have equations of the form y = c.

c stands for a number.

Give three points that would lie on each of these lines
a $y = 5$ **b** $x = -2$.

a $y = 5$
Since the y-coordinate is 5, possible points are (1, **5**), (2, **5**) and (17, **5**)
b $x = -2$
Since the x-coordinate is −2, possible points are (**−2**, 1), (**−2**, 2) and (**−2**, 11)

Where do the graphs $x = 4$ and $y = -1$ intersect?

When lines **intersect** they cross.

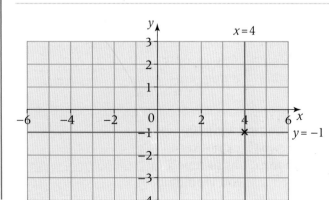

All points on the line x = 4 have x-coordinate 4.

All points on the line y = −1 have y-coordinate −1.

The lines intersect at (4, −1).

1 **a** Copy and complete the table, deciding if each equation is that of a horizontal, vertical or diagonal line, or none of these.

$$x = 9 \qquad y = 2x - 1 \qquad x = -0.5 \qquad y = 7 \qquad y = x^2 + x$$

Horizontal	Vertical	Diagonal	None of these

   b Add an equation of your own to each column in the table.

2 Match each line with its equation.

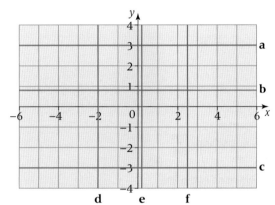

$$y = -3 \qquad x = -2 \qquad y = 3 \qquad x = 2.5 \qquad x = \tfrac{1}{4} \qquad y = \tfrac{3}{4}$$

3 On one set of axes labelled from −6 to +6, plot these graphs.

   a $x = 5$ **b** $y = 2$ **c** $x = 1.6$ **d** $y = -3$ **e** $y = 1$ **f** $x = -1\tfrac{1}{4}$

4 Where do these graphs intersect? Plot the graphs if you need to.

   a $x = 5$ and $y = 2$ **b** $x = 4$ and $y = -3$

   c $x = -2$ and $y = 9$ **d** $y = -4$ and $x = -2$

5 **a** Give the equations of four lines which, when plotted, form the sides of a rectangle.

   b Repeat part **a** for a square.

   c Repeat part **a** for an isosceles right-angled triangle.

6 **a** Which point with integer coordinates fits these clues?

   Above $y = -1$, below $y = 3$, below $y = 2x + 1$, above $y = 2 - x$ and left of $x = 2$.

   b Write your own clues to describe the point (3, 4).

Gradients and intercepts

This spread will show you how to:

● Find the gradient and *y*-axis intercept of straight line graphs

Keywords
Coefficient
Constant
Gradient
Rise
Run
y-axis intercept

● The **gradient** of a straight line tells you how steep it is.

To work out the gradient find how many units the line **rises** for each unit it **runs** across the page.

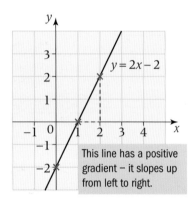

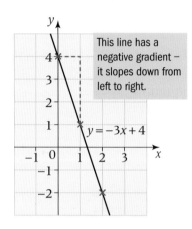

This line has a positive gradient – it slopes up from left to right.

This line has a negative gradient – it slopes down from left to right.

For the line $y = 2x - 2$, **gradient** $= \dfrac{\text{rise}}{\text{run}} = \dfrac{2}{1} = 2$ For the line $y = -3x + 4$, **gradient** $= \dfrac{\text{rise}}{\text{run}} = \dfrac{3}{-1} = -3$

● The gradient is the **coefficient** of *x* (the number of *x*s) in the equation of the line.

The **intercept** is the distance from the origin to where the line cuts the *y*-axis.

The line $y = 2x - 2$ cuts the *y*-axis at $(0, -2)$.
The *y*-intercept is -2.
The line $y = -3x + 4$ cuts the *y*-axis at $(0, 4)$.
The *y*-intercept is 4.

● The intercept is the **constant** term (the number) in the equation of the line.

Example

Find the gradient and intercept of the lines
a $y = 3x - 4$ **b** $y = \frac{1}{2}x + 5$

a gradient = 3, intercept = -4 **b** gradient = $\frac{1}{2}$, intercept = 5

If the equation is not in the form $y = ...$, rearrange it first, for example

$$3x + 2y = 12 \implies 2y = -3x + 12 \implies y = -\tfrac{3}{2}x + 6$$
Now you can see that the gradient is $-\frac{3}{2}$ and the intercept is 6.

Example

Find the gradient and intercept of the lines
a $x + y = 4$ **b** $2x - 5y = 10$

a $x + y = 4$
$y = 4 - x$
$y = -x + 4$
gradient = -1, intercept = 4

b $2x - 5y = 10$
$-5y = 2x + 10$ Divide both sides by -5
$y = \frac{2}{5}x - 2$
gradient = $\frac{2}{5}$, intercept = -2

1 Write the gradient and intercept of each line in the diagram.

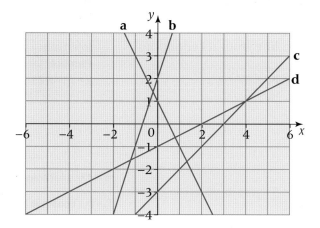

2 On a pair of axes labelled from −8 to +8, draw lines with these key characteristics.

a positive gradient

c gradient = $\frac{1}{4}$

b y-intercept = 3

d negative gradient and y-intercept = −2

e y-intercept = 5 and gradient = 3

f y-intercept = 1 and gradient = $\frac{2}{3}$

3 a Plot the line $y = 3x - 2$ on a pair of axes.

b Write its gradient and intercept.

c What do you notice about the connection between the gradient and intercept and the equation of the line?

d Repeat for the line $y = 2 - \frac{1}{2}x$.

4 The points (2, 1) and (5, 3) are shown.

a What is the gradient of the line joining these two points?

b Explain how you could have found the gradient without counting squares and only using the coordinates given.

c What is the gradient of a line joining (2, 9) to (31, 67)?

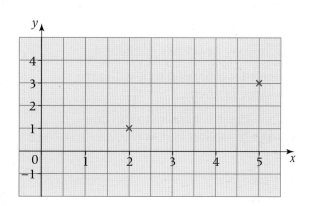

DID YOU KNOW?

To warn drivers of steep sections of road, gradient is shown on road signs as a percentage or ratio (see N4). This road has a gradient of one metre up for every seven metres along.

This spread will show you how to:

- Recognise that equations of the form $y = mx + c$ have straight-line graphs

Keywords

Coefficient
Constant
Gradient
Intercept
Parallel
$y = mx + c$

You can find the **gradient** and **intercept** of a line from its equation.

- You can write the equation of any straight line in the form **$y = mx + c$**.

For example

$x + y = 5 \implies y = -x + 5$

$4x - y = 6 \implies 4x = 6 + y \implies y = 4x - 6$

- The gradient of the line $y = mx + c$ is m.
- The y-axis intercept of the line $y = mx + c$ is c.

m is the **coefficient** of x.

c is the **constant** (number).

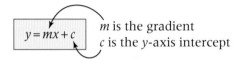

$y = mx + c$ m is the gradient
                c is the y-axis intercept

Example

What are the gradient and intercept of each of these lines?

a $y = 2x + 5$ **b** $y = 1 - 3x$

a Compare the equation to $y = mx + c$.

$m = 2$ and $c = 5$

Hence the gradient = 2 and the intercept = 5.

b In this equation $m = -3$ and $c = 1$.
Hence the gradient = -3 and the intercept = 1.

- **Parallel** lines have the same gradient.

Example

Find the equation of a line parallel to $y = 4x + 5$.

The line $y = 4x + 5$ has gradient 4.

A line that is parallel to it will have the same gradient.

So, $y = 4x + 1$ is parallel to the line $y = 4x + 5$.

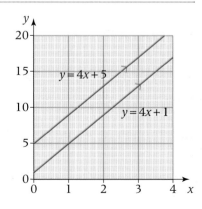

Many other equations are possible.

1 Match each line with its equation.

$$y = 4x - 2$$

$$y = 3x + 1$$

$$y = x$$

$$y = 2 - 4x$$

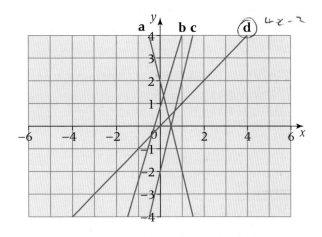

2 Copy and complete the table.

Equation	Gradient	Direction	Intercept
$y = 4x + 3$			
$y = 3x + 4$			
$y = 9x - 2$			
$y = 4x - 5$			
$2y = 8x + 6$			

3 For each graph

   i write its gradient and intercept

   ii write the equation of the line.

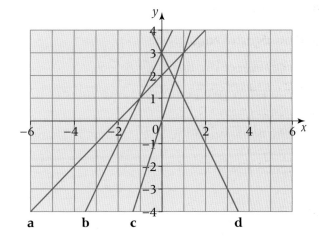

4 Write the equation of a straight line that is

   a parallel to $y = 3x + 3$

   b parallel to $y = 7 - 2x$ and cuts the y-axis at $(0, 3)$

   c a mirror image, in the y-axis, of $y = 3x + 1$.

5 **a** A straight line passes through $(0, 4)$ and $(2, 10)$.
       What is its equation?

   b A straight line is parallel to $2y = 6x - 9$ and goes through $(1, 7)$.
       What is its equation?

This spread will show you how to:

- Find the equation of a line joining several points

Keywords

Equation
Gradient
Intercept
Parallel
$y = mx + c$

- If you know the **gradient** of a line and the y-axis **intercept** you can write the equation of the line.

Example

What is the equation of a line with gradient 9 passing through (0, 5)?

Gradient = 9 and intercept = 5,
so the equation of the line is $y = 9x + 5$.

Remember,
$y = mx + c$

- If you know the gradient and a point on the line you can find the equation of the line.

Example

What is the equation of a line with gradient 8 that passes through the point (2, 7)?

Gradient = 8, so equation is $y = 8x + c$.
The line goes through (2, 7) so,

$7 = 8 \times 2 + c$ Put $x = 2$ and $y = 7$ in the equation $y = 8x + c$
$7 = 16 + c$
$c = -9$

The equation of the line is $y = 8x - 9$.

- If you know two points on a line you can find the equation of the line.

Example

Find the equation of the line joining (1, 2) and (4, 3).

Gradient $= \frac{\text{rise}}{\text{run}} = \frac{1}{3}$

Equation of line is $y = \frac{1}{3}x + c$

The line goes through (1, 2) so substitute 1 for x and 2 for y in $y = \frac{1}{3}x + c$.

$2 = \frac{1}{3} \times 1 + c$
$2 = \frac{1}{3} + c \implies c = \frac{5}{3}$

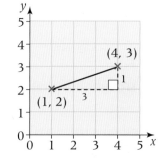

The equation is $y = \frac{1}{3}x + \frac{5}{3} \implies 3y = x + 5$

Check using the point (4, 3):
When $x = 4$, $3y = 4 + 5 = 9 \rightarrow y = 3$

1 Copy and complete the table.

Gradient	Intercept	Equation
3	5	
5	−2	
−2	7	
$\frac{1}{2}$	9	
$-\frac{1}{4}$	−3	
0	4	
1	0	

2 a Which of these lines will pass through the point (2, 8)?

$$y = 4x \qquad y = 2x + 3 \qquad y = 12 - 2x \qquad y = 7x - 5 \qquad y = 5x - 1$$

b Write the equations of two lines that pass through (1, 4).

3 Find the equations of the nine lines described in the table.

a Gradient of 7 and intercepts y-axis at (0, 5)	**b** Gradient of $\frac{1}{2}$ and passes through (0, 3)	**c** Parallel to a line with gradient 4 and passing through (3, 8)
d Gradient of 3 and passing through (4, 7)	**e** Gradient of −2 and cutting through (4, −3)	**f** Parallel to $y = \frac{1}{4}x - 1$ and passing through (0, −2)
g Passing through (0, 1) and (1, 5)	**h** Passing through (0, 2) and (5, 7)	**i** Passing through the midpoint of (1, 7) and (3, 13) with a gradient of 8

4 a What is the gradient of the line joining (0, 5) to (12, 41)?

b What is the equation of the line joining (0, 5) to (12, 41)?

c Repeat **a** and **b** for the points (3, 10) and (5, 6).

5 Where does the line $2y = 9x - 5$ cross

a the y-axis

b the x-axis

c the line $4y = x + 24$?

6 Find the equation of this line in the form $ax + by = c$, where a, b, c are values.

Key objectives

- Plot graphs of functions in which y is given explicitly in terms of x, or implicitly
- Understand that the form $y = mx + c$ represents a straight-line and that m is the gradient of the line and c is the value of y-intercept
- Find the equation of a given straight line in the form $y = mx + c$

1 a Plot the graphs of $y = 2x - 1$ and $y = 2x + 2$ on the same axes. (2)

   b Will these graphs ever intersect? Explain your answer. (1)

2

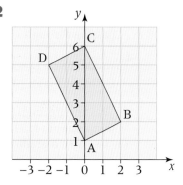

ABCD is a rectangle.

A is the point (0, 1).

C is the point (0, 6).

The equation of the straight line through A and B is $y = 2x + 1$.

Find the equation of the straight line through D and C. (2)

(Edexcel Ltd., 2004)

This unit will show you how to

- Understand and use the probability scale
- Calculate and explain the probability of an outcome of an event
- Calculate the probability of mutually exclusive events
- Use two-way tables to calculate probabilities
- Calculate theoretical probabilities and relative frequencies
- Understand the concepts of experimental and theoretical probability

Before you start ...

You should be able to answer these questions.

Review

1 Work out.

Unit N3

   a $1 - \frac{2}{5}$ **b** $1 - \frac{4}{7}$

   c $1 - \frac{3}{8}$ **d** $1 - \frac{4}{11}$

2 Work out.

Unit N3

   a $\frac{2}{3} + \frac{1}{6}$ **b** $\frac{1}{5} + \frac{1}{4}$

   c $\frac{1}{3} + \frac{5}{8}$ **d** $\frac{1}{4} + \frac{3}{8}$

3 Calculate.

Unit N3

   a $\frac{1}{4} \times \frac{2}{3}$ **b** $\frac{1}{5} \times \frac{3}{4}$

   c $\frac{1}{3} \times \frac{2}{5}$ **d** $\frac{5}{6} \times \frac{1}{2}$

4 Change these fractions to decimals.

Unit N3

   a $\frac{7}{10}$ **b** $\frac{3}{4}$

   c $\frac{3}{8}$ **d** $\frac{2}{5}$

   e $\frac{1}{3}$ **f** $\frac{1}{16}$

This spread will show you how to:

- Understand and use the probability scale
- Calculate and explain the probability of an outcome of an event

Keywords
Event
Outcome
Probability

- **Probability** is a measure of how likely an **outcome** of an **event** is.
- Probability is measured on a scale from 0 to 1.

0 ——————————————————— 1

Outcome: Outcome:
cannot happen   certain to happen

For all other outcomes the probability is a fraction between 0 and 1.

- Probability of an outcome = $\dfrac{\text{number of favourable outcomes}}{\text{total number of outcomes}}$

Throwing a dice is an event.
Getting a 6 is an outcome.

You can write probability as a fraction, a decimal or a percentage.

Example

There are 30 students in Class 10Z, 18 girls and 12 boys.
All of the students are aged 14 or 15 and all own a mobile phone.
A student is chosen at random from Class 10Z. What is the probability that the student chosen

a is aged 10   **b** owns a mobile phone   **c** is a girl   **d** is a boy?

a This outcome cannot happen. All the students are 14 or 15, so
P(aged 10) = 0.
b This outcome is certain to happen. All the students own a mobile phone, so P(owns mobile) = 1.
c There are 18 girls and a total of 30 students, so P(girl) = $\frac{18}{30}$.
d There are 12 boys and a total of 30 students, so P(boy) = $\frac{12}{30}$.

The probability that a girl is chosen or that a boy is chosen is certain to happen as the students are all either girls or boys.

Notice that $\frac{18}{30}$ (P(girl)) + $\frac{12}{30}$ (P(boy)) = 1 (P(certainty))

- Sum of probabilities of all possible outcomes = 1

P(aged 10) is a short way of writing 'the probability that a student is aged 10'.

All possible outcomes are accounted for.

Example

A spinner has circles, squares and triangles on its face.
The table gives the probabilities of landing on circle, square and triangle.
Work out the value of x.

Outcome	Circle	Square	Triangle
Probability	0.25	0.625	x

Total probability = 1 $0.25 + 0.625 + x = 1$
$0.875 + x = 1$
So, $x = 0.125$

1 The probability that a girl chosen at random in Class 10Y has a
cat is 0.47.
What is the probability that a girl chosen at random in Class 10Y
does not have a cat?

2 The probability that a boy chosen at random in Class 10X does not
wear glasses is 0.35.
What is the probability that a boy chosen at random in Class 10X
does wear glasses?

3 A dish contains 5 orange, 3 lemon and 10 strawberry sweets.
One sweet is chosen at random.
What is the probability that the sweet is

    a orange **b** banana **c** strawberry **d** lemon **e** not lemon?

4 A bag contains 7 green, 8 blue and 10 red marbles.
Nine of the marbles are large, 16 are small.
One marble is chosen at random.
What is the probability that the marble is

    a green **b** not green **c** large **d** small **e** red **f** not blue?

5 Four girls compete to become head girl.
The probabilities of being chosen are shown in the table.

Amy	Beth	Cathy	Debs
0.14	0.32	0.27	x

Work out the value of x.

6 Five boys want to be captain of the cricket team.
The probabilities of their being chosen are shown in the table.

Gerry	Harry	Iain	Jim	Ken
0	0.24	x	0.15	0.21

    a Explain what the probability that Gerry is chosen is 0 means.

    b Work out the value of x.

7 Four teams are left in a football competition.
The probabilities of their winning the competition are shown in
the table.

City	United	Rovers	Rangers
0.18	0.22	x	$2x$

    a Explain the chances of Rangers winning compared to the chances
of Rovers winning.

    b Work out the value of x.

This spread will show you how to:

- Calculate the probability of mutually exclusive outcomes

Keywords
Mutually
exclusive
Outcome

- Two or more **outcomes** are **mutually exclusive** if they cannot happen at the same time.

For example,

When you roll a dice, the outcomes 'an even number' and 'a 3' can both happen, but not at the same time.

When you flip a coin, the outcomes 'head' and 'tail' can both happen, but not at the same time.

So the outcomes are mutually exclusive.

- When two or more outcomes are mutually exclusive the addition rule (sometimes called the OR rule) is used to find their probability.

- For two outcomes, A and B, the probability of A or B happening is the sum of P(A) and P(B).

$P(A \text{ or } B) = P(A) + P(B)$

- Probability of an event **not** happening = 1 − Probability an event happens
- For an event A, $P(\text{not } A) = 1 - P(A)$ or $P(A) = 1 - P(\text{not } A)$

Example

A spinner has 12 equal sides: five green, four blue, two red and one white. The spinner is spun.

a What is the probability that the spinner lands on

   i green or white   **ii** green or blue   **iii** blue or white

   iv blue or red or white   **v** not green?

b Why are the answers to parts **iv** and **v** the same?

a **i** $P(\text{green}) = \frac{5}{12}$ $P(\text{white}) = \frac{1}{12}$ $P(\text{green or white}) = \frac{5}{12} + \frac{1}{12} = \frac{6}{12}$

  ii $P(\text{green}) = \frac{5}{12}$ $P(\text{blue}) = \frac{4}{12}$ $P(\text{green or blue}) = \frac{5}{12} + \frac{4}{12} = \frac{9}{12}$

  iii $P(\text{blue}) = \frac{4}{12}$ $P(\text{white}) = \frac{1}{12}$ $P(\text{blue or white}) = \frac{4}{12} + \frac{1}{12} = \frac{5}{12}$

  iv $P(\text{blue}) = \frac{4}{12}$ $P(\text{red}) = \frac{2}{12}$ $P(\text{white}) = \frac{1}{12}$

     $P(\text{blue or red or white}) = \frac{4}{12} + \frac{2}{12} + \frac{1}{12} = \frac{7}{12}$

  v $P(\text{not green}) = 1 - P(\text{green}) = 1 - \frac{5}{12} = \frac{7}{12}$

b The possible outcomes are green, blue, red and white.
Part **v**, P(not green), is the same as part **iv**, P(blue or red or white).

1 A spinner has 20 equal sides: 7 have circles, 5 have pentagons, 4 have squares, 3 have triangles, 1 has a rectangle.
The spinner is spun.
What is the probability that the spinner lands on

   a a circle   **b** not a circle   **c** a rectangle

   d not a rectangle   **e** a circle or a square   **f** a pentagon or a square

   g a triangle or a square or a rectangle

   h a circle or a pentagon or a triangle   **i** not a square

2 A nine sided dice has one digit (1, 2, 3, 4, 5, 6, 7, 8, 9) on each of its sides.
The dice is rolled.
What is the probability that the dice lands

   a on the number 7   **b** not on the number 7

   c on an odd number   **d** on a multiple of 3

   e on a multiple of 5   **f** not on a multiple of 5

   g on a multiple of 4   **h** not on a multiple of 4

   i on a prime number?

3 The table shows information about the type of pet owned by students in Class 10A. No student owns more than one pet.

Pet	Cat	Dog	Hamster	Fish	No pet
Number of students	7	8	2	4	9

   a How many students are there in the class?

   b One student is chosen at random from the class.
     Work out the probability that the student chosen will own

     i a cat   **ii** a cat or a dog   **iii** a dog or a fish

     iv a cat or a hamster   **v** no pets   **vi** a pet.

4 The table shows information about the number of driving lessons students in Class 12C had in February one year.

Number of driving lessons	0	1	2	3	More than 3
Number of students	5	2	8	12	3

One student is chosen at random from the class.
Work out the probability that the student chosen had

   a 1 driving lesson   **b** 1 or 2 driving lessons

   c 2 or more driving lessons   **d** 2 or fewer driving lessons.

D3.3 Probability and expectation

This spread will show you how to:

- Use two-way tables to calculate probabilities

- Probabilities can be found from information given in a **two-way table**.

Example

Each of the students in Class 10Z went abroad last year.
The two-way table shows some information about the countries visited.

	Europe	America	Rest of the world	Total
Girls	13			18
Boys		2		12
Total	22	5		

a Complete the table.

b One student is chosen at random from Class 10Z. Write down the probability that the student

   i visited Europe   ii is a boy who visited America.

a

	Europe	America	Rest of the world	Total
Girls	13	3	2	18
Boys	9	2	1	12
Total	22	5	3	30

Use the totals to find the missing numbers.
$3 + 2 = 5$
$13 + 9 = 22$

b i Probability that the student visited Europe $= \frac{22}{30}$
  ii Probability that the student is a boy who visited America $= \frac{2}{30}$

You can use the probability of a particular outcome happening to calculate the **expected number** of times it will occur.

- Expected number = Total number of outcomes × Probability of a particular
    outcome happening

Example

The foreign countries visited by the students in Class 10Z is typical of the foreign countries visited by the students in Year 10 at the same school.
There are 240 students in Year 10 at the school.
How many students would you expect to visit

a Europe   b America   c America or the Rest of the world?

a P(Europe) $= \frac{22}{30}$ so expected number: $240 \times \frac{22}{30} = 176$
b P(America) $= \frac{5}{30}$ so expected number: $240 \times \frac{5}{30} = 40$
c P(America or Rest of the world) = P(America) + P(Rest of the world)
                    $= \frac{5}{30} + \frac{3}{30} = \frac{8}{30}$

Expected number: $240 \times \frac{8}{30} = 64$

There are 240 students, so total number of outcomes is 240.

1 The two-way table shows where each of the 32 students in Class 10Y spent their Easter holiday.

	France	Spain	UK	Total
Girls	3			18
Boys		8		
Total	6	12		32

a Copy and complete the table.

b One student is chosen at random from Class 10Y. Find the probability that the student

   i visited France

   ii did not visit France

   iii is a boy

   iv is a boy who visited Spain

   v visited France or Spain

   vi is a girl who did not visit France.

2 The two-way table shows some information about the favourite activity of 50 students.

	Orienteering	Paintballing	Quadbiking	Total
Girls	11		4	23
Boys		8		
Total	16			50

a Copy and complete the table.

b One student is chosen at random. Find the probability that the student's favourite activity is

   i paintballing

   ii not paintballing

   iii paintballing or orienteering

   iv quadbiking or paintballing.

c These students are a sample chosen from a larger group of 400 students. How many of the larger group would you expect

   i to prefer orienteering

   ii to be girls?

3 The two-way table shows some information about the preferred subject of the 120 students in Year 10.

	Science	Humanities	Other subjects	Total
Girls	29		32	
Boys		3		
Total	56	21		120

a Copy and complete the table.

b One student is chosen at random. Find the probability that the student

   i is a girl

   ii prefers humanities

   iii prefers humanities or science

   iv is a boy who prefers science.

c This year group is typical of the whole school. There are 600 students in the school. How many of the whole school would you expect to prefer

   i science

   ii humanities?

This spread will show you how to:

- Calculate theoretical probabilities and relative frequencies
- Understand the concepts of experimental and theoretical probability

Keywords
Bias
Experimental
Fair
Theoretical

- You can calculate the **theoretical** probability when an event is **fair** or unbiased.

For example:
A fair coin is thrown.
Probability of a fair coin landing on heads = $\frac{1}{2}$

one head, two outcomes

An unbiased dice is rolled.
Probability of an unbiased dice landing on 2 = $\frac{1}{6}$

one '2', six outcomes

- Theoretical probability is based on equally likely outcomes.
- Theoretical probabilities are calculated on the assumption that the number of favourable outcomes and the number of possible outcomes are as expected.

Example

A fair spinner with 8 sides, three yellow, two red, two blue and one white, is spun.
Work out the probability that the spinner lands on

a yellow **b** red or blue **c** not white.

a P(yellow) = $\frac{3}{8}$ **b** P(red or blue) = $\frac{4}{8}$ **c** P(not white) = $1 - \frac{1}{8} = \frac{7}{8}$

- You use **experimental** probability when the event is **unfair** or **biased** or when the theoretical outcome is known.

Example

A biased coin is thrown 200 times. It lands on heads 140 times.

a Estimate the probability of this coin landing on heads on the next throw.
b Estimate the number of heads you expect to get when the coin is thrown 500 times.

a P(Head) = $\frac{140}{200}$ **b** Estimated expected number = $500 \times \frac{140}{200} = 350$

Example

Tom carries out a survey about the number of people in his village who are left-handed.
He asks 40 people and five of them are left-handed.

a Estimate the probability that a person in the village is left-handed.
b 192 people live in his village. Estimate how many are left-handed.

a P(left-handed) = $\frac{5}{40}$ **b** Estimated number = $192 \times \frac{5}{40} = 24$

- The closer experimental probability is to theoretical probability the less likely it is that there is **bias**.

1 The probability that a biased dice will land on a six is 0.35.
The dice is rolled 400 times.
Estimate the number of times the dice will land on a six.

2 The probability that a biased coin will land on tails is 0.24.
The coin is thrown 500 times.
Estimate the number of times the coin will land on tails.

3 A biased coin is thrown 80 times. It lands on heads 50 times.
Estimate the probability that this coin will land on heads on the next throw.

4 A biased four-sided dice is rolled 120 times. The table shows the outcomes.

Score	1	2	3	4
Frequency	22	34	44	20

  a Explain why it is twice as likely that the dice will land on 3 as on 1.

  b The dice is rolled once more. Estimate the probability that it lands on 2.

  c The dice is rolled a further 300 times.
  How many of those times would you expect the dice to land on 4?

5 A biased dice is rolled 100 times. The table shows the outcomes.

Score	1	2	3	4	5	6
Frequency	7	20	32	17	13	11

  a The dice is rolled once more. Estimate the probability that the dice will land on

  i 6 **ii** 3 **iii** 1 or 2 **iv** 4 or more **v** a number that is not 5.

  b The dice is going to be rolled a further 400 times.
  How many times would you expect the dice to land on

  i 2 **ii** 4 or 5?

6 There are 240 students in Year 10 at Endeavour School.

  a In a survey of 30 students from Year 10, 13 owned an MP3 player.
  How many of the whole of Year 10 would you expect to own an MP3 player?

  b In a survey of 40 students from Year 10, 19 said they liked peaches.
  How many of the whole of Year 10 would you expect to like peaches?

  c In a survey of 48 students from Year 10, 5 were left-handed.
  How many of the whole of Year 10 would you expect to be left-handed?

This spread will show you how to:

● Calculate theoretical probabilities and relative frequencies

● Experimental probability is also known as **relative frequency**.
● Relative frequency is the proportion of successful trials in an experiment.
● The more trials that are carried out the more reliable the estimate of probability.

Example

Rachael has a mixed colours packet of seeds.
She plants 10 seeds each week for 7 weeks.
The table shows the number of purple flowers
that grew in each group of seeds.

Week	1	2	3	4	5	6	7
Number of purple flowers	4	6	5	7	4	6	5

a Work out the relative frequency of a purple flower.
b Find the best estimate of the probability of getting a purple flower.

a

Week	1	2	3	4	5	6	7
Number of purple flowers	4	6	5	7	4	6	5
Relative frequency	$\frac{4}{10}$	$\frac{10}{20}$	$\frac{15}{30}$	$\frac{22}{40}$	$\frac{26}{50}$	$\frac{32}{60}$	$\frac{37}{70}$

For each successive week, find the total number of purple flowers and the total number of seeds planted.

b The best estimate includes all the results. P(purple flower) = $\frac{37}{70}$

You can compare the relative frequency of an outcome with the theoretical probability. If they are quite different, the experiment may be biased.

Example

Dan suspects that a particular dice has a bias towards the number 3.
Dan rolls the dice 30 times and gets these results

4	3	3	3	6	5	1	3	2	5
1	3	4	5	3	3	6	2	5	4
6	3	1	3	6	5	3	2	4	3

a What is the relative frequency of rolling a 3?
b Is the dice biased toward 3? Explain your answer.

a The number 3 is rolled 11 times. Relative frequency = $\frac{11}{30}$

b Theoretical probability of rolling a '3' is P(3) = $\frac{1}{6}$
In 30 rolls expected number of 3s = $30 \times \frac{1}{6} = 5$
5 is not close to 11 so the dice does appear to be biased towards 3.

1 Jim carries out an experiment.
He throws a coin 320 times.
The coin lands on tails 114 times.
Is the coin fair? Explain your answer.

2 A spinner has 10 equal sides, 5 black and 5 red.
Dave carries out an experiment.
He spins the spinner 280 times.
The spinner lands on black 133 times.
Is the spinner fair? Explain your answer.

3 Jane carries out an experiment.
She rolls a dice 200 times.
The table shows the outcomes.

Outcome	1	2	3	4	5	6
Frequency	32	34	35	31	35	33

Is the dice fair? Explain your answer.

4 Tom has a four-sided dice.
He rolls the dice 100 times.
The table shows the outcomes.

Outcome	1	2	3	4
Frequency	18	44	19	19

Is the dice fair? Explain your answer.

5 Clara suspects that a coin is biased. She flips the coin and notes how
many heads she gets in each group of 10 flips.
Clara flips the coin 100 times in total.
The table shows her results.

Group of 10 flips	1	2	3	4	5	6	7	8	9	10
Number of heads	4	3	4	2	5	4	4	3	3	2
Relative frequency										

a Copy the table and complete for the relative frequency.

b Write the best estimate of the probability of the coin landing
on heads.

c Is the coin biased? Explain your answer.

D3

Exam review

Key objectives

- Understand and use the probability scale
- Understand and use estimates or measures of probability from theoretical models, or from relative frequency
- Identify different mutually exclusive outcomes and know that the sum of the probabilities of all these outcomes is 1

1 A 6-sided dice is rolled 100 times.
The results are recorded in the table.

Score	1	2	3	4	5	6
Frequency	18	10	17	20	12	23

a Work out the probability of rolling

   i a two   **ii** a six   **iii** a two or a six.   (3)

b Do you think the dice is fair? Explain your answer.   (1)

2 60 British students each visited one foreign country last week.
The two-way table shows some information about these students.

	France	Germany	Spain	Total
Female			9	34
Male	15			
Total		25	18	60

a Copy and complete the two-way table.   (3)
One of these students is picked at random.

b Write down the probability that the student visited Germany last week.   (1)

(Edexcel Ltd., 2004)

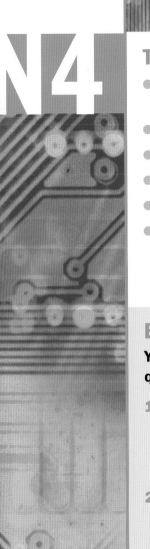

This unit will show you how to

- Describe and calculate proportions, using fractions, decimals or percentages
- Understand direct proportion and ratio
- Solve problems involving proportion and proportional change
- Solve problems involving exchange rates
- Round answers to an appropriate degree of accuracy
- Solve problems involving compound measures, including speed and density

Before you start ...

You should be able to answer these questions.

Review

1 Calculate.

   a $\frac{2}{3}$ of 60   **b** $\frac{3}{4}$ of 840

   c $\frac{3}{8}$ of 240   **d** $\frac{4}{5}$ of 250

Key stage 3

2 Three identical pens cost £5.10.
How much would 5 of the same pens cost?

Unit N2

3 Calculate the average speed of cars that made these journeys.

   a 480 miles in 12 hours

   b 85 miles in two hours

   c 27 miles in 30 minutes

   d 48 miles in 90 minutes

Key stage 3

4 **a** Increase these amounts by 10%.

     i £45   **ii** 60 mm

     iii 48 km   **iv** 4 hours

  b Decrease these amounts by a quarter.

     i 6 miles   **ii** 58 minutes

     iii 14 kg   **iv** £62

Key stage 3

This spread will show you how to:

- Describe and calculate proportions, using fractions, decimals or percentages
- Understand direct proportion and ratio

Proportions can be described and calculated using fractions, decimals or percentages.

There are 12 students in Class 3A.
$\frac{2}{3}$ of them are girls.
Number of girls $= \frac{2}{3} \times 12 = 8$

1000 ml of Quango contains 100 ml of fruit juice.
Proportion of fruit juice $= \frac{100}{1000}$
$= \frac{1}{10}$

Example

Find   **a** $\frac{3}{5}$ of 85   **b** 28% of 360   **c** 120% of 45   **d** $\frac{7}{8}$ of 86

a $\frac{3}{\cancel{5}_1} \times \cancel{85}^{17} = \frac{3}{1} \times 17 = 51$

Cancel the 5s before multiplying.

b Estimate: 30% of 400 = 120
By long multiplication, $28 \times 36 = 1008$
So 28% of 360 = 100.8

c 120% of 45 = (100% of 45) + (20% of 45)
20% of 45 = 45 ÷ 5 = 9 ▶ 120% of 45 = 45 + 9 = 54

Work out 20% of 45 and add it.

d $\frac{1}{8}$ of 86 = 86 ÷ 8 = $10 + \frac{6}{8} = 10\frac{3}{4}$ ▶ $\frac{7}{8}$ of 86 = $86 - 10\frac{3}{4} = 76 - \frac{3}{4} = 75\frac{1}{4}$

Work out $\frac{1}{8}$ of 86 and then subtract.

Example

What proportion is   **a** 4 of 32   **b** 5 of 35   **c** 12 of 100?
Write your answers as percentages.

a $\frac{4}{32} = \frac{1}{8}$
so 4 is $\frac{1}{8}$ of 32
$\frac{1}{8} = \frac{1}{8} \times 100\%$
$= 12\frac{1}{2}\%$

b $\frac{5}{35} = \frac{1}{7}$
so 5 is $\frac{1}{7}$ of 35
$\frac{1}{7} = \frac{1}{7} \times 100\%$
$= 14.3\%$

c $\frac{12}{100} = \frac{3}{25}$
so 12 is $\frac{3}{25}$ of 100
$\frac{3}{25} = \frac{3}{25} \times 100\%$
$= 12\%$

Example

A 1 kg bag of 'Grow Up' fertiliser contains 45 grams of phosphate. A 500 gram packet of 'Top Crop' fertiliser contains 20 grams of phosphate.

What is the proportion of phosphate in each fertiliser?

Proportion of phosphate in 'Grow Up' $= \frac{45}{1000} \times 100\%$
$= 4.5\%$

Proportion of phosphate in 'Top Crop' $= \frac{20}{500} \times 100\%$
$= 4\%$

Percentages are easy to compare.

Exercise N4.1

1. One tenth ($\frac{1}{10}$) of the weight of a soft drink is sugar. Find the amount of sugar in these weights of drink.

   a 750 g   **b** 45 g   **c** 1 kg   **d** 1250 g

You should be able to do all of these mentally.

2. Three fifths ($\frac{3}{5}$) of the volume of a fruit cocktail is orange juice. Find the amount of orange juice in these volumes of fruit cocktail. You should show all of your working.

   a 150 ml   **b** 380 ml   **c** 2 litres   **d** 280 cm^3

See Example 1 part **a**.

3. Calculate

   a 120% of 50 g   **b** 90% of 40 mm

   c 95% of 400 g   **d** 80% of 39 km

Try these mentally. See Example 1 part **c**.

4. What proportion is

   a 5 of 50   **b** 6 of 80   **c** 9 of 45   **d** 15 of 80?

5. Write your answers from question **4** as percentages.

6. Find these proportions, giving your answers as
      i fractions in their lowest terms   **ii** percentages.

   a 3 out of every 20   **b** 4 parts in a hundred

   c 8 out of 20   **d** 64 in every thousand

7. Find these.

   a 38.3% of 192 mm   **b** $\frac{3}{8}$ of £840

   c 19% of £52.00   **d** $\frac{7}{8}$ of 960 kg

8. A 250 ml glass of fruit drink contains 30 ml of pure orange juice. What proportion of the drink is pure orange juice? Give your answer as

   a a fraction in its lowest terms   **b** a percentage.

9. Antifreeze contains 10% concentrated antifreeze; the rest is water.

   a How much concentrate is contained in 2 litres of antifreeze?

   b How much antifreeze can you make with 300 ml of concentrate?

   c How much water would you add to 200 ml of concentrate?

10. Samantha wins £4500 in a competition. She gives $\frac{1}{3}$ to her mother and $\frac{1}{5}$ to her sister.

    a How much does she keep? Show your working.

    b What proportion of the prize money does she give away?
    Give your answer as a fraction and as a percentage.

Direct proportion

This spread will show you how to:

- Understand direct proportion and ratio
- Solve problems involving proportion and proportional change

Keywords

Direct proportion
Ratio
Variable

Quantities which can change are called **variables**.

The amount of screenwash used is a variable.

● Two variables are in **direct proportion** if the **ratio** between them stays the same as the actual values vary.

When you multiply (or divide) one of the variables by a certain number, you have to multiply (or divide) the other variable by the same number.

Water (litres)	Capfuls
1 (× 4)	4
5 (× 4)	20

5 litres of water needs 20 capfuls of screenwash

Example

A pipe 2.5 metres long weighs 35 kilograms. How much would 5.5 metres of the same pipe weigh?

Here are three different ways to solve this problem:

```
2.5 m   weighs      35
  ⇓ ÷ 5            ⇓ ÷ 5
0.5 m   weighs       7
  ⇓ × 11           ⇓ × 11
5.5 m   weighs      77 kg
```

```
2.5 m   weighs      35
  ⇓ ÷ 2.5          ⇓ ÷ 2.5
1 m     weighs     14 kg
  ⇓ × 5.5          ⇓ × 5.5
5.5 m   weighs      77 kg
```

The formula $w = kx$ tells you the weight, w kg, of a pipe x metres long. The scale factor k is the weight of 1 metre of pipe, which is $35 ÷ 2.5 = 14$ kg, so the formula is: $w = 14x$.
Substituting $x = 5.5$ gives $w = 14 × 5.5 = 77$ kg

This is an informal scaling method.   This is called the unitary method.   This is an algebraic method.

5.5 metres of pipe weighs 77 kg.

Example

The cost of 12 pencils is £2.16. Work out the cost of 9 pencils.

12 pencils cost £2.16, so 3 pencils cost 54p (dividing by 4), and 9 pencils cost £1.62 (multiplying by 3).

1   Ribbon costs £2.75 per metre. Find the cost of these lengths of ribbon.

   a 3 m   **b** 4.5 m   **c** 6.85 m   **d** 27.55 m

2   A shop sells shelving at £3.45 per metre. Find the cost of these lengths of shelving.

   a 5 m   **b** 3.45 m   **c** 2.25 m   **d** 4.85 m

3   A 2 m length of pipe weighs 8 kg. How much does a 3 m length of the same pipe weigh?

4   Four buckets of water weigh 60 kg. How much would 5 buckets of water weigh?

5   400 g of powder paint costs £2.40.

   a Find the cost of 100 g of the paint.

   b Use your answer to part **a** to find the cost of 300 g of the paint.

6   300 g of sherbet drops cost £1.20.

   a How much do 100 g of sherbet drops cost?

   b How much do 700 g of sherbet drops cost?

7   A pack of 250 tea bags contains 130 g of tea, and costs £7.50. Calculate

   a the cost of one tea bag   **b** the weight of tea in one bag.

DID YOU KNOW?

Each day Britons drink 165 million cups of tea.

8   A shop sells five different types of luxury tea. Calculate the cost of 100 g of each brand, given that

   a 200 g of brand A costs £3.75   **b** 500 g of brand B costs £7.40

   c 300 g of brand C costs £5.20   **d** 250 g of brand D costs £5.10

   e 350 g of brand E costs £6.50.

9   Alan and Barry buy sand from a builders' merchant. Alan buys 35 kg of sand for £4.55. Barry buys 28 kg of the same sand. How much does he pay? Show your working.

10 A shop sells drawing pins in two different packs.

   Pack A contains 120 drawing pins and costs £1.45.

   Pack B contains 200 of the same drawing pins, and costs £2.30.

   Calculate the cost of one drawing pin from each pack, and explain which pack is better value.

11 A store sells packs of paper in two sizes.

Regular	**Super**
150 sheets	500 sheets
Cost £1.05	Cost £3.85

   Which of these two packs gives better value for money? You must show all of your working.

This spread will show you how to:

● Solve problems involving exchange rates

You can **convert** from one currency to another using an **exchange rate**.

Example

If €1 (1 euro) is worth 63p, find the value of

a €185 in pounds   **b** £260 in euros.

a The exchange rate is €1 = £0.63, so €185 = £0.63 × 185 = £116.55

b As an estimate, £250 ÷ 0.5 = 250 × 2 = €500,
£260 = 260 ÷ 0.63 = €412.70 to the nearest cent.

To convert from
one currency to
another, simply
multiply by the
appropriate
exchange rate.

If you need to do
the 'reverse'
conversion, divide
by the given
exchange rate.

Example

Samantha travels from Ottawa to Buenos Aires, where
she changes 375 Canadian dollars into Argentine pesos.

a How many pesos does she receive, if the exchange
rate is CAN$1 = 2.4027 pesos?

b After her trip, Samantha changes 58 pesos back into
Canadian dollars, at a rate of CAN$1 = 2.5458 pesos.
How many dollars does she receive?

a Estimate 400 × 2 = 800   **b** Estimate 60 ÷ 3 = 20
$375 = 375 × 2.4027 = 901.01 pesos   58 ÷ 2.5458 = CAN$22.78

Example

a Mary changed £900 into US dollars ($), when the rate of exchange
was £1 = $1.75.

How many dollars did she receive?

b After her holiday Mary had $110 left. She changed them into pounds
at an exchange rate of £1 = $1.79. How many pounds did she get?

a £900 = $1.75 × 900 = $1575.

b Divide by the conversion rate to convert dollars back to pounds.
So, $110 ÷ 1.79 = £61.45.

Example

Use the fact that 1 inch = 2.54 centimetres to convert

a 7.25 inches to centimetres   **b** 15 cm to inches.

a 7.25 inches = 7.25 × 2.54 cm = 18.4 cm

b 15 cm = 15 ÷ 2.54 inches = 5.91 inches

Do these
conversions in the
same way as
currency
conversions.

1 The exchange rate between pounds and dollars is £1 = $1.55.
Convert these amounts from pounds to dollars.

   a £20   **b** £35   **c** £10.50   **d** £38.55

2 The exchange rate between pounds and euros is £1 = €1.49.
Convert these amounts from pounds to euros.

   a £5   **b** £30   **c** £59   **d** £264

3 One Australian dollar (A$1) is worth 42.7p.
Convert these amounts to pounds.

   a A$2.50   **b** A$45   **c** A$299   **d** A$715

4 £1 is worth 205 Japanese yen (£1 = ¥205).
Convert these amounts into pounds.

   a ¥410   **b** ¥2050   **c** ¥300   **d** ¥750

   e ¥6500   **f** ¥595

See example 1b

5 One US dollar is worth 0.8182 Canadian dollars (US$1 = CAN$0.8182).

   a Convert US$50 into Canadian dollars.

   b Convert CAN$50 into US dollars.

6 The table shows the exchange rates between two currencies.

£1 (pound) is worth €1.67
$1 (dollar) is worth €1.15

   a Alan changes £300 into euros.
How many euros does he receive?

   b Barbara changes €875 into dollars.
How many dollars does she receive?

7 One pint is exactly 0.568261 litres. Find the number of pints in one litre, giving your answer to 3 decimal places.

8 One pound weight (1 lb) is exactly 0.45359237 kg. Use this information to convert the following weights to kilograms, giving your answers to 3 decimal places.

   a 28 lb   **b** 2240 lb   **c** 375.5 lb   **d** 38.125 lb

9 One mile is exactly 1609.344 metres. Use this information to convert

   a 25 miles into kilometres   **b** 10 km into miles

   c 40 000 km into miles   **d** 4.5 miles into kilometres

10 Use the approximate conversion rate of 1 gallon = 4.5 litres to convert

   a 35 gallons into litres   **b** 50 litres into gallons

   c 12.5 gallons into litres   **d** 38.8 litres into gallons

Give your answers to a suitable degree of accuracy.

This spread will show you how to:

- Round answers to an appropriate degree of accuracy
- Solve problems involving compound measures, including speed and density

Keywords

Compound
Density
Speed

Compound measures involve a combination of measurements and units.

- The **density** of a material is its mass divided by its volume.
- **Speed** is the distance travelled divided by the time taken.

- A formula triangle can be a useful way of remembering the relationships between the different parts of a compound measure.
  For example:
  speed = distance ÷ time
  distance = speed × time
  time = distance ÷ speed

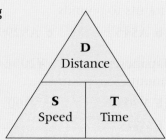

The units for density can be grams per cubic centimetre (g/cm³), or kilograms per cubic metre (kg/m³).

Speed can be measured in miles per hour (mph), kilometres per hour (kph) or metres per second (m/s).

Example

a A cube of side 3.5 cm has a mass of 600 g. Find the density of the cube in g/cm³, correct to 3 significant figures.

b A car travels 240 miles in 3 hours 45 minutes. Find the average speed of the car in miles per hour.

c Lubricating oil has a density of 0.58 g/cm³. Find
  i the mass of 2.5 litres of this oil
  ii the volume of 10 grams of the oil.

Volume of a cube = length³.

a Volume = 3.5^3 cm³ = 42.875 cm³
Density = 600 g ÷ 42.875 cm³ = 13.994 ... g/cm³ = 14.0 g/cm³ to 3 sf

Density = $\dfrac{\text{mass}}{\text{volume}}$

b Average speed = $\dfrac{\text{total distance}}{\text{total time}}$

$= \dfrac{240}{3.75}$

$= 64$ mph

Put the time in hours.

c i Mass = density × volume = 0.58 g/cm³ × 2500 cm³ = 1450 g
ii Volume = mass ÷ density = 10g ÷ 0.58g/cm³ = 17.2 cm³

1 litre ≡ 1000 cm³

Example

A metal cuboid is 95 cm long and has a length of 2 cm, width 4 cm and height 6 cm.
a Find the density of the cuboid.
b If the mass of the cuboid is 7.8 kg, find its density in g/cm³.

a Volume of cuboid = 2 × 4 × 6 = 48 cm³

b Density of cuboid = $\dfrac{\text{mass}}{\text{volume}} = \dfrac{7800}{48} = 163$ g/cm³ to 3 sf

Volume of a cuboid = length × width × height.

Change the mass into grams.

1 Rod cycles 18 miles in 2 hours. Find his average speed, in miles per hour (mph).

2 If 4 metres of fabric costs £8.40, find the price of the fabric in pounds per metre.

3 A car travels 24 miles in 45 minutes. Find the average speed of the car in miles per hour (mph).

4 A train leaves Euston at 8:57 a.m. and arrives at Preston at 11:37 a.m. If the distance is 238 miles find the average speed of the train.

5 Copy and complete the table to show speeds, distances and times for five different journeys.

Speed (kph)	Distance (km)	Time
105		5 hours
48	106	
	84	2 hours 15 minutes
86		2 hours 30 minutes
	65	1 hour 45 minutes

6 A cube of side 2 cm weighs 40 grams.

  a Find the density of the material from which the cube is made, giving your answer in g/cm^3.

  b A cube of side length 2.6 cm is made from the same material. Find the mass of this cube, in grams.

> Volume of cube = length^3.

7 A box has a length and width of 22.50 mm, and a height of 3.15 mm. It has a mass of 9.50 g.

  a Find the density of the metal from which the box is made, giving your answer in g/cm^3.

  b How many boxes can be made from 1 kg of the material?

> Volume of cuboid = length × width × height.

8 Emulsion paint has a density of 1.95 kg/litre. Find

  a the mass of 4.85 litres of the paint.

  b the number of litres of the paint that would have a mass of 12 kg.

9 A steel cable weighs 2450 kg.
The cable has a uniform circular cross-section of radius 0.85 cm.
The steel from which the cable is made has a density of 7950 kg/m^3.
Find the length of the cable.

> Volume of a cylinder = πr^2 × length.

0.85 cm ←——— l ———→

This spread will show you how to:

● Solve problems involving proportion and proportional change

Proportional change means increasing or decreasing a quantity by a certain percentage or fraction.

Daily Tabloid ⁴⁰ᵖ

Town population down a quarter

Chairman's Report

'Revenue increased by 5%'

CEO Mr James Jameson's announcement in his Review of the company's Annual Accounts for 2005-2006.

Example

Find the new amount when £350 is increased by a quarter.

£350 ÷ 4 = £87.50

£350 + £87.50 = £437.50

Work out the increase.

Add it to original amount.

When a quantity is increased by 20% its new value is 120% of the original value, for example

£50 increased by 20% = 120% of £50

When a quantity is decreased by 20% its new value is 80% of the original value, for example

£50 decreased by 20% = 80% of £50

You can work out the new amounts in one step as in this example.

Example

When a new motorway was built between Utopia and Edenlandia the 6-hour journey time was decreased by 30%.

a Find the new journey time.

The motorway from Utopia to Edenlandia is being resurfaced resulting in a 15% increase in journey time.

b Find the new journey time due to resurfacing.

a New time is 70% of old time

New time = 70% × 6 hours
= 0.7 × 6
= 4.2 hours or 4 h 12 min

100% − 30% = 70%

70% = 0.7

b New time due to resurfacing is 115% of old time.

New time = 115% × 4.2 hours
= 1.15 × 4.2
= 4.83 hours or 4 h 50 min

100% + 15% = 115%

115% = 1.15

1 Find the result when these amounts are increased by one quarter.

a 20 g **b** 160 cm **c** 6 hours **d** 500 kg

You should be able to do these mentally.

2 Decrease these amounts by one third.

a 600g **b** 30 sec **c** 45° **d** 72 hours

Again, you should be able to do these mentally.

3 Copy and complete the table to show the result of some proportional increases and decreases.

Original number	Proportional change	Result
42	Decrease by $\frac{1}{4}$	
110	Increase by $\frac{1}{5}$	
250	Increase by $\frac{1}{10}$	
450	Decrease by $\frac{2}{5}$	
965	Increase by $\frac{1}{10}$	

4 A recipe for making 16 scones includes these ingredients

60 g butter   3 teaspoons caster sugar   200 ml milk

Find the quantity of each ingredient needed to make 24 scones.

5 Increase these amounts by $\frac{1}{6}$.

a 240 g **b** 300 g **c** 200 g **d** 750 g

6 Increase these amounts by 5%.

a £120 **b** £240 **c** £500 **d** £72

7 Calculate these percentage changes.

a £450 increased by 10% **b** £600 decreased by 15%

c £900 increased by 6% **d** £740 decreased by 11%

8 Andrew earns £240 per week. He is awarded a pay rise of 4.5%.
Bella earns £260 per week. She is awarded a pay rise of 4%.
Whose weekly pay increases by the larger amount?
Show all your working.

9 Mr and Mrs Jones receive their electricity bill. The details are

Present meter reading   23 087
Previous meter reading   20 893
Charge per unit   8.2 pence
Service charge   £13.75
VAT   5%

Find the total cost of the electricity including VAT.
Show all your working.

Key objectives

- Use knowledge of operations and inverse operations, and of methods of simplification, in order to select and use suitable strategies and techniques to solve problems and word problems, including those involving ratio and proportion, fractions, percentages and measures and conversion between measures, and compound measures defined within a particular situation

1 Martha uses 150 grams of flour to bake a cake.

  a How many grams of flour will she need to bake 5 cakes? (2)

  b To bake a bigger cake she needs 25% more flour.

  How much flour does she need to bake this cake? (2)

2 Ben bought a car for £12 000.

  Each year the value of the car depreciated by 10%.

  Work out the value of the car two years after he bought it. (3)

(Edexcel Ltd., 2003)

This unit will show you how to

- Identify properties preserved under reflection, rotation and translation
- Understand congruence
- Describe reflections, using mirror lines
- Understand that rotations are specified by a centre and an (anticlockwise) angle
- Understand and use vector notation
- Describe translations by giving a distance and direction (or vector)
- Transform 2-D shapes by translation, rotation and reflection and combinations of these transformations

Before you start ...

Review

You should be able to answer these questions.

1 Write the coordinates of these points.

Key stage 3

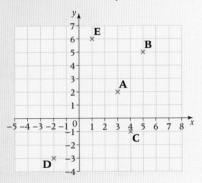

2 Write the equations of these lines.

Unit A4

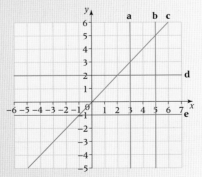

3 Draw these lines on a grid.

Unit A4

   a $x = 4$ **b** $x = -2$ **c** $y = 3$
   d $y = 0$ **e** $x = 0$

This spread will show you how to:

- Identify properties preserved under reflection
- Understand congruence
- Describe reflections, using mirror lines

Keywords
Congruent
Mirror line
Perpendicular
Reflection

Just as you can see
your reflection in a mirror,
you can reflect a shape
in a mirror line.

Corresponding points on
the object and image are
the same distance from
the **mirror line**.

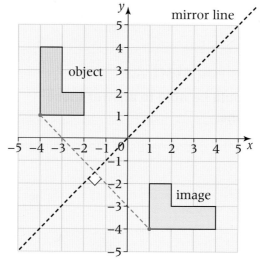

Corresponding
angles and lengths
are the same in
the image and the
object.

The object and
the image are
congruent.

Reflection flips the
shape over.

The line joining a point
and its image is
perpendicular to the
mirror line.

Example

Reflect the triangle with vertices
at (1, 2), (2, 7) and (4, 7) in
the line $x = 1$.

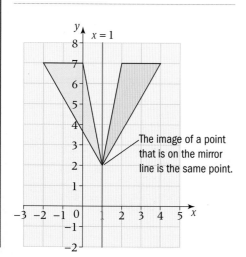

The image of a point
that is on the mirror
line is the same point.

Example

Reflect the pink triangle in the
line $y = 3$.

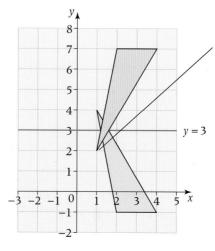

When an object
crosses the mirror
line, so does its
image.

158

1 Copy this diagram.

   a Reflect the triangle T in the line $x = 2$.
      Label the image U.

   b Reflect the triangle T in the line $y = 1$.
      Label the image V.

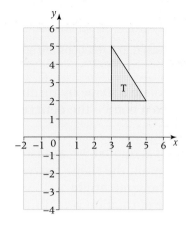

2 Copy this diagram.

   a Reflect the kite K in the line $x = -1$.
      Label the image L.

   b Reflect the kite K in the line $y = 1$.
      Label the image M.

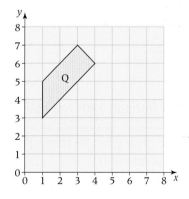

3 Copy this diagram and extend the y-axis to -8.

   a Reflect the quadrilateral Q in the x-axis.
      Label the image R.

   b Reflect the quadrilateral Q in the line $y = x$.
      Label the image S.

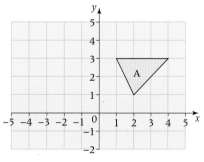

4 Copy this diagram.

   a Reflect triangle A in the y-axis.
      Label the image B.

   b Reflect triangle B in the y-axis.
      Label the image C.

   c What do you notice? Does this always happen?
      Check with some other reflections.

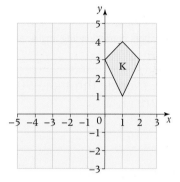

S3.2 Rotation

This spread will show you how to:

● Identify properties preserved under rotation
● Understand that rotations are specified by a centre and an (anticlockwise) angle

Keywords
Angle
Centre of
   rotation
Congruent
Rotation

You can rotate a shape by turning it about a fixed point – the **centre of rotation**.

This object is rotated 90° anticlockwise. Every point on the shape moves through the same **angle**.
The centre of rotation is (1, 0).

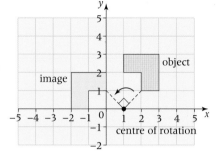

The object and the image are **congruent**.

● In a rotation, corresponding angles and lengths are the same in the image and the object.

Corresponding points on the object and image are the same distance from the centre of rotation.

Example

Rotate the triangle through −90° about centre (1, 2).

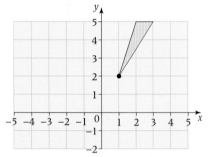

Rotate each vertex through −90° Use tracing paper to help.

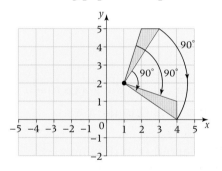

You measure angles anti-clockwise, so −90° means 90° clockwise.

A vertex at the centre of rotation does not move.

Rotation through 180° is a half turn, the same clockwise or anticlockwise.

Example

Rotate the pink triangle through 180° about centre *O*.

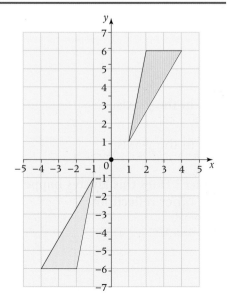

O is the origin, (0, 0).

160

1 Copy this diagram.

   a Rotate the triangle T through 180° about (0, 0). Label the image U.

   b Rotate the triangle T through 180° about (2, 2). Label the image V.

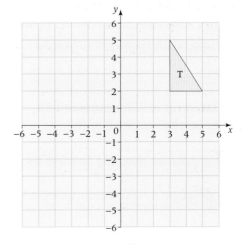

2 Copy this diagram.

   a Rotate the kite K through 180° about (1, 1). Label the image L.

   b Rotate the kite K through −90° about (2, 0). Label the image M.

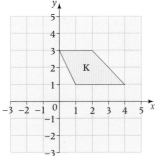

3 Copy this diagram.

   a Rotate the triangle A through 90° clockwise about (1, 0). Label the image B.

   b Rotate the triangle A through 90° anticlockwise about (0, 1). Label the image C.

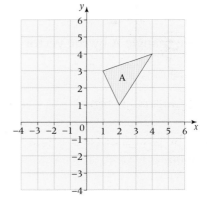

4 Copy this diagram.

   a Rotate shape P through −90° about (2, 1). Label the image Q.

   b Rotate image Q through −90° about (2, 1). Label the image R.

   c Rotate shape P through 180° about (2, 1). What do you notice? Explain why this happens.

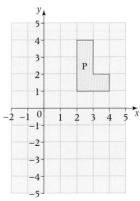

Translation

This spread will show you how to:

Keywords

Translation
Vector

- Identify properties preserved under translation
- Understand and use vector notation
- Describe translations by giving a distance and direction (or vector)

A **translation** is a sliding movement.
All points on the shape slide the same distance in the same direction.

- In a translation the object and the image are congruent.

You can describe a translation using a **vector**.

- Vector $\begin{pmatrix} a \\ b \end{pmatrix}$ means moving a units in the x-direction and b units in the y-direction.

The L-shape is translated by the vector $\begin{pmatrix} 4 \\ 3 \end{pmatrix}$.

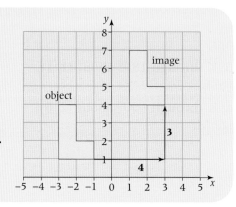

Example

Translate triangle A by the vector $\begin{pmatrix} -3 \\ 4 \end{pmatrix}$.

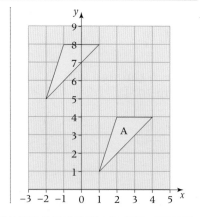

$\begin{pmatrix} -3 \\ 4 \end{pmatrix}$ means move

−3 in the
x-direction, so 3
squares to the left.

Move 4 in the
y-direction, so 4
squares up.

Example

Translate triangle T by the vector $\begin{pmatrix} 0 \\ -3 \end{pmatrix}$.

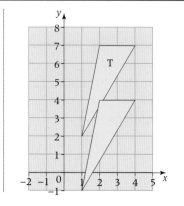

$\begin{pmatrix} 0 \\ -3 \end{pmatrix}$ means move 0

in the x-direction, so
the shape only moves
in the y-direction.

Move −3 in the
y-direction, so 3
squares down.

1 Copy this diagram.

a Translate the triangle T by the vector $\begin{pmatrix} 3 \\ 4 \end{pmatrix}$.

Label the image U.

b Translate the triangle T by the vector $\begin{pmatrix} 5 \\ -2 \end{pmatrix}$.

Label the image V.

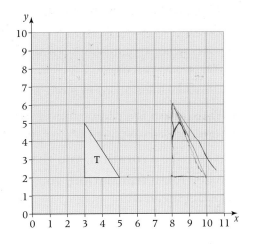

2 Copy this diagram.

a Translate the kite K by the vector $\begin{pmatrix} -4 \\ 3 \end{pmatrix}$.

Label the image L.

b Translate the kite K by the vector $\begin{pmatrix} 2 \\ -4 \end{pmatrix}$.

Label the image M.

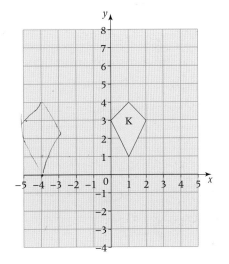

3 Copy this diagram.

a Translate shape A by the vector $\begin{pmatrix} 6 \\ -4 \end{pmatrix}$.

Label the image B.

b Translate the image B by the vector $\begin{pmatrix} -6 \\ 4 \end{pmatrix}$.

What do you notice? Does this always happen? If so, why?

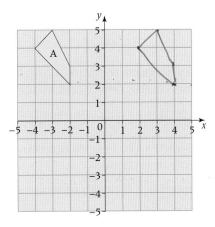

Describing transformations

This spread will show you how to:

- Describe reflections, using mirror lines
- Understand that rotations are specified by a centre and an (anticlockwise) angle
- Describe translations by giving a distance and direction (or vector)

Keywords

Maps
Reflection
Rotation
Transformation
Translation

- To describe a **reflection**, you give the equation of the line.
- To describe a **rotation**, you give the centre and the angle of rotation.
- To describe a **translation**, you give the distance and direction or you specify the vector.

Reflections, rotations and translations are all **transformations**.

Example

Describe the transformation that maps shape A on to

a shape B   **b** shape C   **c** shape D.

Maps means changes.

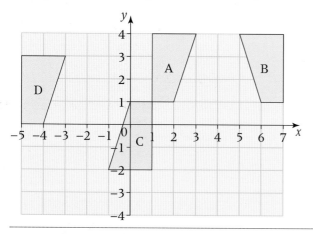

a In a reflection the mirror line bisects the line joining corresponding points on the object and image.

Shape B is a reflection of shape A in the line $x = 4$.

b The vertex (1, 1) does not move during the rotation, so it must be the centre of rotation.

Shape C is a rotation of shape A through 180°.

c Shape D is a translation of shape A by the vector $\begin{pmatrix} -5 \\ -1 \end{pmatrix}$.

1 Describe fully the transformation that maps

  a shape A onto shape B

  b shape A onto shape C

  c shape A onto shape D

  d shape B onto shape D

  e shape C onto shape D

  f shape D onto shape C.

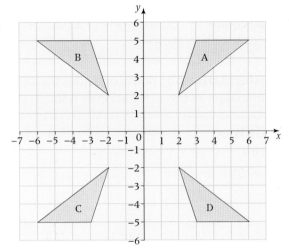

2 Describe fully the transformation that maps

  a shape J onto shape K

  b shape L onto shape K

  c shape M onto shape K

  d shape L onto shape M

  e shape J onto shape M

  f shape M onto shape J.

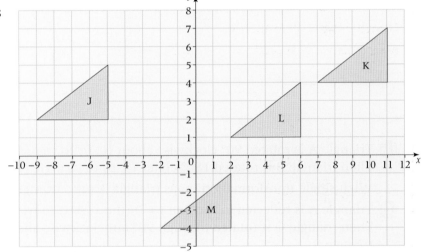

3 Describe fully the transformation that maps

  a shape W onto shape X

  b shape W onto shape Y

  c shape W onto shape Z

  d shape Z onto shape W

  e shape X onto shape Y

  f shape Z onto shape X.

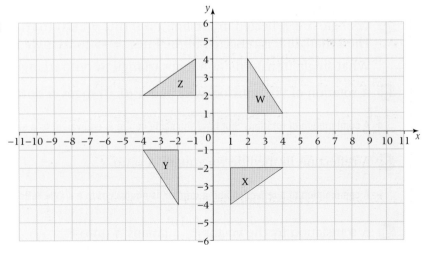

Combining transformations

This spread will show you how to:

- Transform 2-D shapes by translation, rotation and reflection and combinations of these transformations

Keywords

Reflection
Rotation
Transformations
Translation

You can combine **transformations** by doing one after the other.
You can describe a combination of transformation as a single transformation.

Example

In this diagram, triangle A undergoes three pairs of transformations.

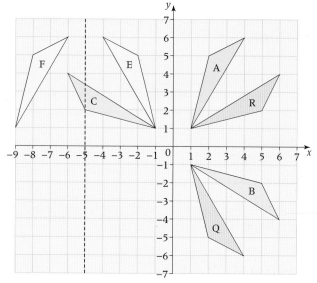

1. Triangle A is rotated 90° clockwise about (0, 0) to triangle B.
   Then triangle B is rotated 180° about (0, 0) to triangle C.

   What single transformation maps triangle A onto triangle C?

2. Triangle A is reflected in the line $y = 0$ (the x-axis) to triangle Q.
   Then triangle Q is rotated through 90° anticlockwise about (0, 0) to triangle R.

   What single transformation maps triangle A onto triangle R.

3. Triangle A is reflected in the line $x = 0$ (the y-axis) to triangle E.
   Then triangle E is reflected in the line $x = -5$ to triangle F.

   What single transformation maps triangle A onto triangle F?

1. A rotation of 90° anticlockwise about (0, 0) maps A onto C.

- A combination of rotations that have the same centre is equivalent to a single **rotation**.

2. A reflection in the line $y = x$ maps A onto R.

- A combination of a reflection and a rotation is equivalent to a single **reflection**.

3. A translation by the vector $\begin{pmatrix} -10 \\ 0 \end{pmatrix}$ maps A onto F.

- A combination of reflections is a **translation** when the mirror lines are parallel.

1 Copy this diagram.

a Rotate triangle A 90° anticlockwise about centre (0, 0). Label the image B.

b Rotate triangle B 90° anticlockwise about centre (0, 0). Label the image C.

c Describe fully the single transformation that takes triangle A to triangle C.

d Rotate triangle B 90° clockwise about centre (2, 1). Label the image D.

e Describe fully the single transformation that takes triangle A to triangle D.

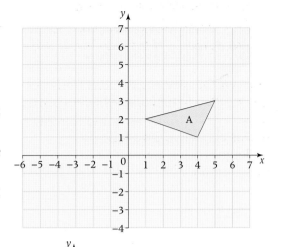

2 Copy this diagram.

a Reflect triangle E in the *y*-axis. Label the image F.

b Reflect triangle F in the *x*-axis. Label the image G.

c Describe fully the single transformation that takes triangle G to triangle E.

d Reflect triangle F in the line *x* = 2. Label the image H.

e Describe fully the single transformation that takes triangle E to triangle H.

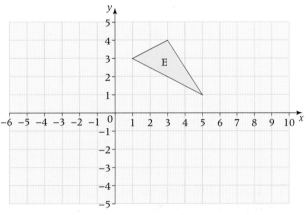

3 Copy this diagram.

a Reflect triangle J in the *x*-axis. Label it K.

b Rotate triangle K 180° about centre (0, 0). Label it L.

c Describe fully the single transformation that takes triangle L to triangle J.

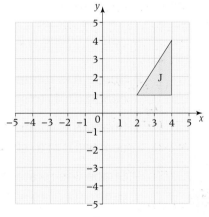

4 Copy this diagram.

a Translate trapezium R by the vector $\begin{pmatrix} 5 \\ -3 \end{pmatrix}$. Label the image S.

b Translate trapezium S by the vector $\begin{pmatrix} -1 \\ 2 \end{pmatrix}$. Label the image T.

c Describe fully the single transformation that takes trapezium T to trapezium R.

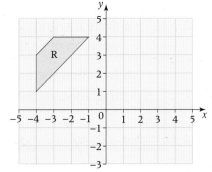

Key objectives

- Recognise and visualise rotations, reflections and translations including reflection symmetry of 3-D shapes; transform triangles and other 2-D shapes by combinations of these transformations
- Distinguish properties that are preserved under particular transformations

1 Copy the grid.

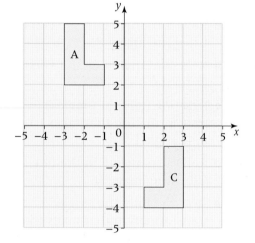

a Reflect shape A in the line $x = 0$. Label this shape B. (1)

b Describe the single transformation that takes shape B to shape C. (2)

c Draw a shape, D, so that the diagram is symmetrical in the line $x = 0$.

Describe the transformation from shape C to shape D. (2)

2 Shape A is rotated 90° anticlockwise, centre (0, 1) to shape B.

Shape B is rotated 90° anticlockwise, centre (0, 1) to shape C.

Shape C is rotated 90° anticlockwise, centre (0, 1) to shape D.

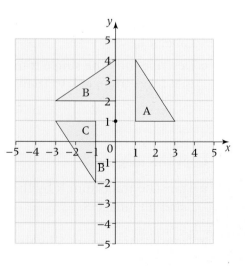

a Copy the grid and mark the position of shape D. (2)

b Describe the single transformation that takes shape C to shape A. (2)

(Edexcel Ltd., 2003)

This unit will show you how to

- Understand and use powers and index notation
- Find the highest common factor and lowest common multiple of two numbers
- Use index laws, including fractional and negative indices
- Understand and use standard form in calculations with large and small numbers
- Use calculators to calculate in standard form

Before you start ...

You should be able to answer these questions.

Review

1 Write these multiplications in power form.

For example, $4 \times 4 \times 4 = 4^3$

Key stage 3

a 3×3 **b** $4 \times 4 \times 4 \times 4 \times 4$

c $6 \times 6 \times 6$ **d** $5 \times 5 \times 5 \times 5$

2 Write these powers of numbers as multiplications in expanded form.

For example, $5^3 = 5 \times 5 \times 5$

Key stage 3

a 3^3 **b** 6^2 **c** 4^5

d 8^3 **e** 7^4 **f** 5^5

3 Evaluate.

Key stage 3

a 2^4 **b** 3^3 **c** 4^2

d 5^3 **e** 2^7 **f** 7^1

This spread will show you how to:

- Understand and use powers and index notation
- Find the highest common factor and lowest common multiple of two numbers

Keywords

HCF
Index
LCM
Power

- Repeated multiplications such as $2 \times 2 \times 2 \times 2$ can be written in **index** notation as 2^4.

You read 2^4 as 'two to the **power** 4'.

- You use powers when factorising, for example,
$$24 = 2 \times 2 \times 2 \times 3 = 2^3 \times 3$$

But note that 2^2 is 'two squared', and 2^3 is 'two cubed'.

- When powers of the same number are multiplied together, you can find the answer by adding the indices.

For example,
$$2^2 \times 2^3 = (2 \times 2) \times (2 \times 2 \times 2) = 2^5$$
$$\text{Similarly, } 3^5 \times 3^6 = 3^{(5+6)} = 3^{11}.$$

In 3^{11}, 11 is the index. The plural of index is indices.

Example

Find the value of   **a** 2^3 **b** 3^2 **c** 5^3 **d** 10^4 **e** 2^8

a $2^3 = 2 \times 2 \times 2 = 8$ **b** $3^2 = 3 \times 3 = 9$ **c** $5^3 = 5 \times 5 \times 5 = 125$
d $10^4 = 10 \times 10 \times 10 \times 10 = 100 \times 100 = 10\ 000$
e $2^8 = 2 \times 2 \times 2 \times 2 \times 2 \times 2 \times 2 \times 2 = 256$

Example

Write   **a** 625 as a power of 5   **b** 100 000 as a power of 10
c 48 as a product of prime factors in index form

a $625 = 5 \times 125 = 5 \times 5 \times 25 = 5 \times 5 \times 5 \times 5 = 5^4$
b $100\ 000 = 10 \times 10 \times 10 \times 10 \times 10 = 10^5$
c $48 = 2 \times 24 = 2 \times 2 \times 12 = 2 \times 2 \times 2 \times 6 = 2 \times 2 \times 2 \times 2 \times 3 = 2^4 \times 3$

Example

Find, in index form, the values of
a $4^3 \times 4^7$ **b** $5^2 \times 5$ **c** $3^2 \times 3^4 \times 3^3$ **d** $2^2 \times 3^4 \times 2^5 \times 3^3$

a $4^3 \times 4^7 = 4^{(3+7)} = 4^{10}$ **b** $5^2 \times 5 = (5 \times 5) \times 5 = 5^3$
c $3^2 \times 3^4 \times 3^3 = 3^{(2+4+3)} = 3^9$
d $2^2 \times 3^4 \times 2^5 \times 3^3 = 2^{(2+5)} \times 3^{(4+3)} = 2^7 \times 3^7$

Example

Find the **HCF** and **LCM** of 84 and 280.

$84 = 2 \times 2 \times 3 \times 7$
$\quad = 2^2 \times 3 \times 7$

$280 = 2 \times 2 \times 2 \times 5 \times 7$
$\quad\quad = 2^3 \times 5 \times 7$

$\text{HCF} = 2^2 \times 7 = 28$
$\text{LCM} = 2^3 \times 3 \times 5 \times 7 = 840$

HCF is the highest number that is a factor of both 84 and 280.

LCM is the lowest number that both 84 and 280 will divide into exactly.

1 Write these in index form.

a 3×3 **b** $2 \times 2 \times 2$ **c** $3 \times 3 \times 3$ **d** $5 \times 5 \times 5 \times 5$

e $7 \times 7 \times 7$ **f** $10 \times 10 \times 10$ **g** $6 \times 6 \times 6 \times 6$ **h** $5 \times 5 \times 5$

2 Write these numbers in product form. For example, $4^3 = 4 \times 4 \times 4$.

a 3^4 **b** 5^2 **c** 7^4 **d** 10^5 **e** 4^9 **f** 6^3 **g** 2^5 **h** 9^3

3 Find the value of each of these. For example, $5^3 = 5 \times 5 \times 5 = 125$.

a 4^2 **b** 4^3 **c** 2^5 **d** 10^2 **e** 10^3 **f** 3^3 **g** 2^3 **h** 3^2

4 Copy and complete the table to show the values of powers of 10.

Index form	Product	Value
10^6	$10 \times 10 \times 10 \times 10 \times 10 \times 10$	1 000 000
10^5		
10^4		
10^3		
10^2		
10^1		

5 Make a table, like the one in question **4**, to show the values of powers of 2 from 2^1 to 2^8.

6 Write

a 81 as a power of 9   **b** 125 as a power of 5

c 128 as a power of 2   **d** 100 000 as a power of 10

e 81 as a power of 3   **f** 343 as a power of 7

7 Write the answers to these multiplications in index form.

a $3^4 \times 3^2$ **b** $2^8 \times 2^1$ **c** $4^4 \times 4^4$

d $5^2 \times 5^3$ **e** $8^3 \times 8^5$ **f** $3^2 \times 3^5 \times 3^1$

g $2^3 \times 3^2 \times 2^4 \times 3^2$ **h** $5^2 \times 7^1 \times 5^2 \times 7^6$

8 Write each of these numbers as a prime number raised to a power. For example, $49 = 7^2$.

a 16   **b** 121   **c** 27   **d** 125   **e** 169   **f** 625   **g** 243   **h** 256

9 Write these numbers as products of their prime factors, using index notation. For example,
$72 = 2 \times 36 = 2 \times 2 \times 18 = 2 \times 2 \times 2 \times 9 = 2 \times 2 \times 2 \times 3 \times 3 = 2^3 \times 3^2$.

a 52   **b** 36   **c** 50   **d** 24   **e** 18   **f** 48   **g** 60   **h** 144

10 Find the HCF and LCM of

a 64 and 112   **b** 38 and 133

This spread will show you how to:

● Use index laws, including fractional and negative indices

Keywords
Index
Indices
Power

There are rules that you can use when calculating with indices.

In these **index laws**, letters are used to represent numbers.

● Add **indices** when multiplying **powers** of the same number.

$$x^a \times x^b = x^{a+b}$$

$5^4 \times 5^3 = 5^{4+3} = 5^7$

● Subtract indices when dividing powers of the same number.

$$x^a \div x^b = x^{a-b}$$

$6^5 \div 6^2 = 6^3$

Using the rule for multiplication, $7^3 \times 7^0 = 7^3$. Since multiplying by 7^0 leaves the 7^3 unchanged, 7^0 must be equal to 1.

● Any number (except 0) to the power 0 = 1: $x^0 = 1$ for any value of x, if $x \neq 0$.

0^0 is *not defined* – it doesn't actually mean *anything*!

You know that $3^2 \times 3 = (3 \times 3) \times 3 = 3^3$.
You can write this as $3^2 \times 3^1 = 3^{(2+1)} = 3^3$, so $3 = 3^1$.

● Any number to the power 1 is just the number itself: $x^1 = x$ for any value of x.

Example

Solve these, giving your answers in index form.

a $2^3 \times 2^2$ **b** $5^7 \div 5^3$ **c** $6^4 \times 6^2 \div 6^3$
d $7^2 \times 5^3 \times 7^3 \times 5^4$ **e** $(2^5 \times 3^4) \div (2^3 \times 3^2)$

a $2^3 \times 2^2 = 2^{(3+2)}$ **b** $5^7 \div 5^3 = 5^{(7-3)}$
　　　$= 2^5$ 　　　　　　　$= 5^4$

c $6^4 \times 6^2 \div 6^3 = 6^{(4+2-3)}$ **d** $7^2 \times 5^3 \times 7^3 \times 5^4 = 7^{(2+3)} \times 5^{(3+4)}$
　　　　　$= 6^3$ 　　　　　　　　　　　$= 7^5 \times 5^7$

e $(2^5 \times 3^4) \div (2^3 \times 3^2) = 2^{(5-3)} \times 3^{(4-2)}$
　　　　　　　　　　$= 2^2 \times 3^2$

Example

Write the value of **a** 19^0 **b** 8^1 **c** $(16^3 - 81 \times 17)^0$ **d** $(4.8)^1$

a $x^0 = 1$ for any value of x, so $19^0 = 1$.
b $x^1 = x$ for any value of x, so $8^1 = 8$.
c There is no need to evaluate the expression in the bracket (except to note that it is bigger than zero.)
　The power of 0 means that $(16^3 - 81 \times 17)^0 = 1$.
d $x^1 = x$ for any value of x (including decimal numbers), so $(4.8)^1 = 4.8$.

1 Write the answers to these multiplications in index form.

    a $7 \times 7 \times 7$ **b** 3×3^2 **c** 5×5^2 **d** $6 \times 6 \times 6^2$

    e $5^3 \div 5$ **f** $8^4 \times 8$ **g** $9^3 \times 9^2 \times 9$ **h** $8^7 \times 8$

2 Work out these, giving your answers in index form.

    a $6^2 \times 6^3$ **b** $4^5 \times 4^4$ **c** $2^6 \times 2^7$ **d** $11^5 \times 11^2$

    e $1^{17} \times 1^{13}$ **f** $7^8 \times 7^4$ **g** $3^6 \times 3^6$ **h** $9^9 \times 9^1$

3 Work out these, giving your answers in index form where appropriate.

    a $7^8 \div 7^6$ **b** $8^6 \div 8^2$ **c** $3^3 \div 3^2$ **d** $9^{11} \div 9^8$

    e $4^7 \div 4^1$ **f** $2^9 \div 2^9$ **g** $12^8 \div 12^6$ **h** $6^{13} \div 6^{13}$

4 Work out these, giving your answers in index form.

    a $8^6 \times 8^2 \div 8^3$ **b** $5^7 \times 5^2 \div 5^4$ **c** $2^8 \times 2^3 \div 2^5$ **d** $9^6 \times 9^3 \div 9^7$

    e $8^5 \times 8^5 \div 8^2$ **f** $7^6 \times 7^5 \div 7^4$ **g** $4^6 \times 4^8 \div 4^4$ **h** $11^2 \times 11^2 \div 11^3$

5 Write the answers to these in index form.

    a $3^4 \times 3^2 \div (3^3 \times 3^2)$ **b** $(5^6 \div 5^2) \times 5^4 \times 5^2$

    c $(4^5 \div 4^2) \div (4^6 \div 4^5)$ **d** $(7^9 \div 7^2) \div (7^2 \times 7^3)$

    e $(8^7 \div 8^4) \times 8^5 \times 8^3$ **f** $9^3 \times (9^5 \div 9^2) \times 9^4$

6 Simplify these expressions, giving your answers in index form.

    a $\dfrac{4^2 \times 4^2}{4^2}$ **b** $\dfrac{6^3 \times 6^4}{6^5}$ **c** $\dfrac{9^8}{9^2 \times 9^4}$ **d** $\dfrac{8^6 \div 8^3}{8^2}$

    e $\dfrac{5^9 \times 5^4}{5^3 \times 5^7}$ **f** $\dfrac{6^3 \times 6^4}{6^5 \div 6^3}$ **g** $\dfrac{8^9 \div 8^2}{8^7 \div 8^2}$ **h** $\dfrac{10^6 \div 10^2}{10^2 \times 10^2}$

7 Simplify these expressions as far as possible, giving your answers in index form.

    a $4^2 \times 3^3 \times 4^2$ **b** $8^5 \times 7^2 \div 8^2$ **c** $6^2 \times 5^3 \times 6^2 \times 5^3$

    d $4^5 \times 3^3 \times 3^3 \times 4^4$ **e** $5^4 \times 2^3 \div 5^2$ **f** $9^5 \times 7^2 \times 7^2 \times 9^2$

    g $8^2 \times 5^6 \times 8^3 \div 5^3$ **h** $3^4 \times 8^5 \times 3^4 \times 8^2$ **i** $9^3 \times 2^5 \div 2^3 \times 9^2$

8 Simplify these expressions, giving your answers in index form.

    a $\dfrac{5^2 \times 8^5}{8^2}$ **b** $\dfrac{6^5 \times 7^2}{6^3}$ **c** $\dfrac{6^4 \times 5^4}{6^2 \times 5^2}$ **d** $\dfrac{7^8 \times 5^6}{5^3 \times 7^2}$

    e $\dfrac{8^7 \times 3^5}{3^2 \times 8^5}$ **f** $4^3 \times \dfrac{4^5 \times 5^9}{4^3 \times 5^7}$ **g** $\dfrac{6^9 \times 7^5}{6^7 \times 7^3} \times 6^2$ **h** $4^3 \times \dfrac{7^6 \times 4^5}{4^4 \times 7^3} \times 7^2$

This spread will show you how to:

● Use index laws, including fractional and negative indices

Keywords
Fractional
Index laws
Negative
Power
Reciprocal

Indices can be fractions as well as whole numbers.

$$5^{\frac{1}{2}} \times 5^{\frac{1}{2}} = 5^{(\frac{1}{2}+\frac{1}{2})} = 5^1 = 5$$

Since $\sqrt{5} \times \sqrt{5} = 5$, $5^{\frac{1}{2}}$ must represent the square root of 5.

● In general, $x^{\frac{1}{2}} = \sqrt{x}$ for any value of x. Similarly, $x^{\frac{1}{3}} = \sqrt[3]{x}$ (the cube root of x), and so on.

A **fractional** index means a root.

$\frac{1}{5} = 1 \div 5 = 5^0 \div 5^1 = 5^{0-1} = 5^{-1}$, so, 5^{-1} means $\frac{1}{5}$

● In general, x^{-1} is equal to $\frac{1}{x}$, for any value of x. This is the **reciprocal** of x.

$\frac{1}{5^2} = 1 \div 5^2 = 5^0 \div 5^2 = 5^{0-2} = 5^{-2}$, so, 5^{-2} means $\frac{1}{5^2}$

A **negative** index means a reciprocal.

● In general, x^{-n} means $\frac{1}{x^n}$. This is the reciprocal of x^n.

$(5^2)^3 = 5^2 \times 5^2 \times 5^2 = 5^{(2+2+2)} = 5^6$
The indices are multiplied: $(5^2)^3 = 5^{2\times 3} = 5^6$

Note that zero has no reciprocal as 0^{-1} is not defined.

● In general, when finding a 'power of a power', multiply the indices: $(x^m)^n = x^{mn}$.

You can use these index laws to calculate quantities including powers.

Notice that $8 \times 8^{-1} = 1$. What about 6×6^{-1}? Or 50×50^{-1}?

Example

Evaluate these expressions.

a $16^{\frac{1}{2}}$ **b** 8^{-1} **c** $(3^3)^2$

a $16^{\frac{1}{2}} = \sqrt{16} = 4$ **b** $8^{-1} = \frac{1}{8}$ **c** $(3^3)^2 = 27^2 = 729$

Example

Write these expressions as powers of the numbers indicated.

a 2 as a power of 4   **b** 0.125 as a power of 8   **c** $\frac{1}{16}$ as a power of 2

a $2 = \sqrt{4} = 4^{\frac{1}{2}}$ **b** $0.125 = \frac{125}{1000} = \frac{1}{8} = 8^{-1}$ **c** $\frac{1}{16} = \frac{1}{2^4} = 2^{-4}$

Example

Evaluate these expressions.

a 10^{-4} **b** $2^{-\frac{1}{2}}$ **c** $100^{-\frac{1}{2}}$

a $10^{-4} = \frac{1}{10\,000} = 0.0001$ **b** $2^{-\frac{1}{2}} = \frac{1}{2^{\frac{1}{2}}} = \frac{1}{\sqrt{2}} = 0.707$ to 3 dp

(by calculator)

c $100^{-\frac{1}{2}} = \frac{1}{\sqrt{100}} = \frac{1}{10} = 0.1$

1 Find the value of each expression.

   a 5^1 **b** 6^1 **c** 6^0 **d** 7^0

   e $(4 + 88^2)^0$ **f** $(4^2 + 5^2)^1$ **g** $(92.5)^0$ **h** 0^1

2 Find the value of each expression.

   a $100^{\frac{1}{2}}$ **b** $16^{0.5}$ **c** $49^{\frac{1}{2}}$ **d** $4^{0.5}$

   e $64^{\frac{1}{2}}$ **f** $9^{0.5}$ **g** $121^{\frac{1}{2}}$ **h** $64^{0.5}$

   i $144^{\frac{1}{2}}$ **j** $8^{\frac{1}{3}}$ **k** $27^{\frac{1}{3}}$ **l** $100^{0.5}$

3 Evaluate.

   a $36^{\frac{1}{2}}$ **b** 36^1 **c** $81^{0.5}$ **d** 81^1

4 Write these in index form. For example, $\frac{1}{2} = 2^{-1}$

   a $\frac{1}{3}$ **b** $\frac{1}{5}$ **c** $\frac{1}{7}$ **d** $\frac{1}{11}$

   e 0.5 **f** 0.2 **g** 0.1 **h** $0.\dot{3}$

> Remember that x^{-1} is the reciprocal of x, and vice versa.

5 Evaluate.

   a 9^{-1} **b** 9^0 **c** $9^{0.5}$ **d** 9^1

   e 9^2 **f** 9^3 **g** 9^5 **h** 9^{-3}

6 Write these in index form. For example, $\frac{1}{5^2} = 5^{-2}$

   a $\frac{1}{7^2}$ **b** $\frac{1}{9^2}$ **c** $\frac{1}{2^2}$ **d** $\frac{1}{2^3}$

   e $\frac{1}{2^5}$ **f** $\frac{1}{3^4}$ **g** $\frac{1}{5^3}$ **h** $\frac{1}{6^4}$

7 Write these in fraction form. For example, $7^{-3} = \frac{1}{7^3}$

   a 8^{-2} **b** 7^{-3} **c** 5^{-2} **d** 9^{-4}

   e 3^{-2} **f** 9^{-3} **g** 4^{-5} **h** 6^{-6}

8 Evaluate.

   a 4^{-2} **b** 4^{-1} **c** 4^0 **d** $4^{0.5}$

   e 4^1 **f** 4^2 **g** 4^3 **h** $4^{-0.5}$

9 Write these in index form. For example, $\dfrac{1}{\sqrt{2}} = \dfrac{1}{2^{\frac{1}{2}}} = 2^{-\frac{1}{2}}$

   a $\dfrac{1}{\sqrt{3}}$ **b** $\dfrac{1}{\sqrt{5}}$ **c** $\dfrac{1}{\sqrt{7}}$ **d** $\dfrac{1}{\sqrt{11}}$

10 Simplify these expressions. For example, $(5^2)^3 = 5^{2 \times 3} = 5^6$

   a $(2^2)^2$ **b** $(2^3)^2$ **c** $(3^2)^3$ **d** $(4^{0.5})^2$

   e $(5^2)^4$ **f** $(4^{-2})^3$ **g** $(7^2)^6$ **h** $(5^2)^{-2}$

11 Calculate these, giving your answers in index form.

   a $2^2 \div 2^4$ **b** $3^5 \div 3^6$ **c** $4^3 \div 4^9$

   d $(3^4 \times 5^5) \div (3^5 \times 5^6)$ **e** $[(5^4 \times 7^3) \div (5^2 \times 7^5)]^2$

This spread will show you how to:

- Understand and use standard form in calculations with large and small numbers
- Use calculators to calculate in standard form

You can use **standard form** to represent large numbers.

- In **standard form**, a number is written as $A \times 10^n$.
  - A is a number between 1 and 10 (but not including 10). Using algebra, $1 \leqslant A < 10$.
  - The value of n is an integer.
  For example, $856 = 8.56 \times 10^2$ and $43\,994 = 4.3994 \times 10^4$.

13×10^5 is *not* in standard form, because 13 is larger than 10.

0.75×10^4 is *not* in standard form, because 0.75 is less than 1.

You can calculate with numbers in standard form.

- Multiplication works like this:
  $(3 \times 10^5) \times (4 \times 10^3) = (3 \times 4) \times 10^{(5+3)} = 12 \times 10^8 = 1.2 \times 10^9$

- Division works like this:
  $(1.4 \times 10^8) \div (7 \times 10^5) = (1.4 \div 7) \times 10^{(8-5)} = 0.2 \times 10^3 = 2 \times 10^2$

The correct version is 1.3×10^6

The correct version is 7.5×10^3

Multiplication – add the indices

Division – subtract the indices

Example

Write these numbers in standard form.

a 235 **b** 12 492 **c** 15×10^4 **d** 0.23×10^6

a $235 = 2.35 \times 10^2$ **b** $12\,492 = 1.2492 \times 10^4$
c $15 \times 10^4 = 1.5 \times 10^5$ **d** $0.23 \times 10^6 = 2.3 \times 10^5$

Example

Calculate

a $(4.2 \times 10^3) \times (2 \times 10^2)$ **b** $(3.6 \times 10^5) \div (1.2 \times 10^3)$ **c** $(5.4 \times 10^4) \times (2 \times 10^3)$

a $(4.2 \times 10^3) \times (2 \times 10^2) = (4.2 \times 2) \times (10^3 \times 10^2)$
$\qquad = 8.4 \times 10^{(3+2)} = 8.4 \times 10^5$
b $(3.6 \times 10^5) \div (1.2 \times 10^3) = (3.6 \div 1.2) \times (10^5 \div 10^3)$
$\qquad = 3 \times 10^{(5-3)} = 3 \times 10^2$
c $(5.4 \times 10^4) \times (2 \times 10^3) = (5.4 \times 2) \times (10^4 \times 10^3)$
$\qquad = 10.8 \times 10^7 = 1.08 \times 10^8$

Example

The Andromeda Galaxy has a radius of about 1 040 700 000 000 000 000 km. Write this in standard form.

$1\,040\,700\,000\,000\,000\,000 = 1.0407 \times 10^{18}$

1 Write these numbers as powers of 10.

   a 100   **b** 10   **c** 100 000   **d** 1

2 Write these numbers in standard form.

   a 200   **b** 800   **c** 9000   **d** 650

   e 6500   **f** 952   **g** 23.58   **h** 255.85

3 These numbers are in standard form. Write each of them as an 'ordinary' number.

   a 5×10^2 **b** 3×10^3 **c** 1×10^5 **d** 2.5×10^2

   e 4.9×10^3 **f** 3.8×10^6 **g** 7.5×10^{11} **h** 8.1×10^{18}

4 Although they are written as multiples of powers of 10, these numbers are not in standard form. Rewrite each of them correctly in standard form.

   a 60×10^1 **b** 45×10^3 **c** 0.65×10^1 **d** 0.05×10^8

5 Work out these calculations, giving your answers in standard form. Do not use a calculator.

   a $(2 \times 10^2) \times (2 \times 10^3)$ **b** $(3 \times 10^4) \times (3 \times 10^3)$

   c $(5 \times 10^3) \times (5 \times 10^4)$ **d** $(8 \times 10^7) \times (3 \times 10^5)$

6 Evaluate these, showing your working. Do not use a calculator; give your answers in standard form.

   a $(4 \times 10^4) \div (2 \times 10^2)$ **b** $(8.4 \times 10^9) \div (4.2 \times 10^5)$

   c $(2 \times 10^6) \div (4 \times 10^4)$ **d** $(3 \times 10^5) \div (4 \times 10^2)$

7 Use a calculator to find these. Give your answers in standard form, to 3 significant figures.

   a $(2.5 \times 10^5) \times (3.9 \times 10^4)$ **b** $(4.1 \times 10^6) \div (3 \times 10^2)$

   c $(4.95 \times 10^3) \times (8.11 \times 10^7)$ **d** $(3.7 \times 10^{11}) \div (1.8 \times 10^3)$

8 The speed of light is approximately 3×10^8 metres per second. Copy and complete the table to show the time taken for light from the Sun to reach the various planets.

Planet	Mean distance from Sun (m)	Light travel time
Mercury	5.79×10^{10}	
Earth	1.50×10^{11}	
Mars	2.28×10^{11}	
Jupiter	7.78×10^{11}	
Pluto	5.90×10^{12}	

This spread will show you how to:

- Understand and use standard form in calculations with large and small numbers
- Use calculators to calculate in standard form

Keywords
Standard form

It is often useful to write small numbers, such as 0.000415, in standard form.

- Negative powers of 10, such as 10^{-4}, represent small numbers.

    For example, $10^{-4} = \frac{1}{10^4} = \frac{1}{10\,000} = 0.0001$

- You can write any small number in standard form.

    For example, $0.00312 = 3.12 \times 0.001 = 3.12 \times 10^{-3}$

- You can calculate with small numbers expressed in standard form.

    For example, $(4.25 \times 10^{-3}) \div (3.75 \times 10^4) = (4.25 \div 3.75) \times 10^{(-3-4)} = 1.13 \times 10^{-7}$

- You can obtain a small number as a result of a calculation involving large numbers.

    For example, $(3 \times 10^5) \div (4 \times 10^8) = 0.75 \times 10^{-3} = 7.5 \times 10^{-4}$

You can also work with numbers in standard form on a scientific calculator. The button for entering the power of 10 is often marked EXP or EE; so for 4.5×10^{-3}, you enter

Example

Write these numbers in standard form.

a 0.003   **b** 0.000 000 416   **c** 0.45

a $0.003 = 3 \times 10^{-3}$
b $0.000\,000\,416 = 4.16 \times 10^{-7}$
c $0.45 = 4.5 \times 10^{-1}$

Example

Calculate.   **a** $(4.8 \times 10^4) \times (3.6 \times 10^{-5})$ **b** $(4.5 \times 10^3) \div (9.7 \times 10^8)$

These can be done directly, using a scientific calculator.
The answers, to 3 significant figures, are   **a** 1.73   **b** 4.64×10^{-6}

Example

What is wrong with this calculation? $10^4 \div 10^{-5} = 10^{-1}$

Be careful with the signs of the indices in examples like this. Because you are dividing, you need to subtract the indices. The correct power of 10 for the answer is $4 - (^-5) = 4 + 5 = 9$, so $10^4 \div 10^{-5} = 10^9$

1. Write these numbers in standard form.

   a 0.3   **b** 0.0047   **c** 0.000 078   **d** 0.4485

2. Although these numbers are written as multiples of powers of 10, they are not in standard form. Write each of the numbers correctly in standard form.

   a 28×10^{-2} **b** 0.4×10^{-1} **c** 13.5×10^{-4} **d** 12×10^{-8}

3. Write these measurements using standard form.

   a One hundredth of a kilometre   **b** Two thousandths of a gram

   c Five millionths of a metre   **d** 11 thousandths of a litre

4. Work out these calculations without a calculator.
   Give your answers in standard form.

   a $(2.5 \times 10^{-3}) \times (2 \times 10^{2})$ **b** $(4.6 \times 10^{-6}) \times (2 \times 10^{-2})$

   c $(4 \times 10^{4}) \div (2 \times 10^{6})$ **d** $(8.4 \times 10^{-2}) \div (2 \times 10^{6})$

5. Use a scientific calculator to work out these. Give your answers in standard form, and to three significant figures.

   a $(3.7 \times 10^{-4}) \times (3.1 \times 10^{-4})$ **b** $(5.3 \times 10^{5}) \div (2.9 \times 10^{8})$

   c $(3.18 \times 10^{2}) \div (6.55 \times 10^{7})$ **d** $(1.79 \times 10^{5}) \times (2.8 \times 10^{-6})$

6. Work out these calculations without using a calculator, giving your answers in standard form.

   a $(5 \times 10^{-1}) + (2 \times 10^{-2})$ **b** $(4 \times 10^{-2}) + (6 \times 10^{-3})$

   c $(2 \times 10^{-2}) + (9 \times 10^{-4})$ **d** $(1.5 \times 10^{-2}) - (2 \times 10^{-3})$

   > You may find it easier to convert the numbers to 'ordinary' numbers first, and then convert the answers back to standard form.

7. Use a scientific calculator to check your answers to question 6.

8. Use a scientific calculator to find the volume of a cube of side length 4.5×10^{-3} metres. Give your answer in m³, to 3 sf, in standard form.

9. A pack of 500 sheets of A4 paper weighs 2.65 kg. Find the mass in kg of a single sheet of paper, giving your answer in standard form.

10. A sheet of gold leaf is one ten thousandth of a millimetre thick.
    The diameter of an atom of gold is about 0.26 nanometres.
    (One nanometre is 10^{-9} metres.)
    Approximately how many atoms thick is the sheet of gold leaf?
    Show your working.

Key objectives

- Use index laws to simplify and calculate the value of numerical expressions involving multiplication and division of fractional and negative powers
- Use standard index form expressed in conventional notation and on a calculator display

1 Evaluate these expressions.

a 3^{-1} (2)

b $(\sqrt{3})^2$ (2)

c Hence evaluate $3^{-1} \times (\sqrt{3})^2$ (1)

2 A nanosecond is 0.000 000 001 second.

a Write the number 0.000 000 001 in standard form. (1)

b A computer does a calculation in 5 nanoseconds.

How many of these calculations can the computer do in 1 second?

Give your answer in standard form. (2)

(Edexcel Ltd., 2004)

This unit will show you how to

- Understand the difference between identities, formulae and equations
- Use formulae from mathematics and other subjects
- Substitute numbers into a formula after simplification
- Write a formula from given information
- Rearrange a formula in order to change its subject
- Understand the difference between a practical demonstration and a proof
- Recognise the importance of assumptions when deducing results

Before you start ...

You should be able to answer these questions.

1 You will have met some formulae before. Match each formula with the information that it is designed to find.

Key stage 3

Formula:
$A = lw$
$A = \pi r^2$
$a^2 + b^2 = c^2$
$V = lwh$
$A = \frac{1}{2}bh$

To find:
Area of triangle
Volume of a cuboid
Area of a rectangle
Area of a circle
Length of sides in a right angle triangle

2 Solve to find the value of x.

Unit A2, Unit N5

  a $3x - 2 = 15$ **b** $4(4x - 5) = 20$
  c $3x^2 = 75$ **d** $5\sqrt{x} = 20$
  e $x^3 + 1 = 9$ **f** $10 - x = 8$

3 Give an example to show that each statement is true.

Key stage 3

  a Adding an even number and an even number gives an even number.
  b Square numbers are always positive.
  c Adding a positive integer to another positive integer gives a larger positive integer.
  d Multiplying any number by two gives an even number.

Review

This spread will show you how to:

- Understand the difference between identities, formulae and equations
- Use formulae from mathematics and other subjects

Keywords
Equation
Formula(e)
Identity
Substitute

- An **identity** is true for all values of x. For example

    $x(x + 1) \equiv x^2 + x$ Whatever value of x you try, this statement is always true.

$\equiv$ means 'is identical to'

- An **equation** is only true for a limited number of values of x. For example

    $2x + 1 = 5$ is only true when $x = 2$.

- A **formula** describes the relationship between two or more variables. For example the formula for the area of a triangle is

    $A = \frac{1}{2}bh$

- You can **substitute** numbers into a formula to work out the value of a variable. For example

$A = \frac{1}{2}bh$

$ = \frac{1}{2} \times 8 \times 6$

$ = 24 \text{ cm}^2$

Example

Decide if each of these statements is an identity, an equation or a formula.

a $x^3 - 2x = x(x^2 - 2)$ **b** $5x - 1 = 2x + 3$ **c** $A = \pi r^2$

a Expand the right-hand side: $x(x^2 - 2) = x^3 - 2x$
   $x^3 - 2x$ = left-hand side so the statement is an identity.
b $5x - 3 = 2x + 3$
   $3x - 3 = 3$ This statement is only true for one value of x.
   $3x = 6$
   $x = 2$
   $5x - 3 = 2x + 3$ is an equation.
c $A = \pi r^2$ is a formula showing the relationship between the radius and the area of a circle.

If you know the value of r, you can find A from the formula and vice versa.

Example

Use the formula $V = \pi r^2 h$ to find V, the volume of a can of beans, when $h = 11$ cm and $r = 7.5$ cm.

$V = \pi r^2 h$

$ = \pi \times 7.5^2 \times 11$

$ = 1943.860 \text{ cm}^3$

$ = 1944 \text{ cm}^3$ to 2 sf

Write the formula.

Substitute the values of r and h.

Give the answer to 2 sf to match the accuracy of the data in the question.

1 Copy these statements and say whether they are identities, equations or formulae.

a $c = 2\pi r$	**b** $3x(x+1) = 3x^2 + 3x$	**c** $3x + 1 = 10$
d $y \times y = y^2$	**e** $2x + 5 = 3 - 7x$	**f** $A = \frac{1}{2}(a + b)h$
g $a^2 + b^2 = c^2$	**h** $20 - x = -(x - 20)$	**i** $2x^2 = 50$

2 A campsite charges for the pitching of a tent and the number of people, p, that stay in it. If C is the total cost in pounds, the formula used by the campsite is $C = 2p + 5$

a Work out the cost of pitching a tent for 6 people.

b If the cost is £23, how many people are sleeping in the tent?

c Explain why the cost could never be £26.

3 Use these formulae to work out the required information.

a $F = \frac{9}{5}C + 32$.   Find 12° Celsius in degrees Fahrenheit

b $P = \frac{1}{4}t - 8$.   Find P when $t = 32$

c $A = \dfrac{bh}{2}$.   Find h when $A = 10$ and $b = 5$

d $W = 3d^2$.   Find W when d is 6 and d when W is 75

e $C = 2a - b$.   Find C when a is −8 and b is −2

f $P + 2r = K$.   Find K when P is $\frac{3}{4}$ and r is $\frac{1}{8}$

4 a The formula for the area of a trapezium is $A = \frac{1}{2}(a + b)h$, where a and b represent the lengths of the parallel sides and h is the perpendicular height.
Find the area of this trapezium.

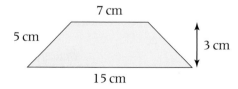

7 cm
5 cm
3 cm
15 cm

b If another trapezium with the same perpendicular height has area 15 cm², suggest as many possibilities for the lengths of the parallel sides as you can.

This spread will show you how to:
- Write a formula from given information

Keywords
Derive
Formula

- You can **derive** a **formula** from information you are given.

Example

Derive a formula for the perimeter of this pentagon.

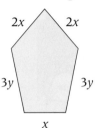

Let P represent the perimeter.
$$P = 2x + 2x + 3y + x + 3y$$
$$P = 2x + 2x + 3y + 3y + x$$
$$P = 5x + 6y$$

Perimeter is the distance all of the way around a shape.

Write formulae as simply as possible, using the rules of algebra.
For example

Write $S = \dfrac{D}{T}$, not $S = D \div T$ and $A = lw$, not $A = l \times w$

Example

a Write a formula to show your total amount of pocket money P (£s), if you receive £3 per month with an extra £2 for every job (j) you do at home.

b Write a formula for the total area of this shape. Let the total area be A.

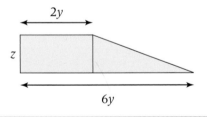

a If I don't do any jobs, I get £3.
If I do 4 jobs I get £3 plus $4 \times £2$, which is £11.
If I do 10 jobs I get £3 plus $10 \times £2$, which is £23.
So, if I do j jobs I get £3 plus $j \times £2$.
Hence
$$P = 3 + 2j$$

Try the situation with various numbers, before putting it into algebra.

Write $2j$ not $j \times 2$

b $A =$ area of a rectangle plus the area of a triangle
$$= (z \times 2y) + \tfrac{1}{2}((6y - 2y) \times z)$$
$$= 2yz + \tfrac{1}{2}(4y \times z)$$
$$= 2yz + \tfrac{1}{2}(4yz)$$
$$= 2yz + 2yz$$
$$= 4yz$$

The area of a rectangle is $A = lw$ and the area of a triangle is $A = \tfrac{1}{2}bh$

1 Write your own formula to represent these quantities.

a The perimeter, P, of a square.

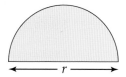

b The area, A, of a semicircle.

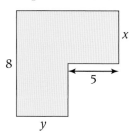

c The perimeter, P, of this hexagon.

d The time in minutes, T, to complete my homework if I take 10 minutes to get my books organised and then 35 minutes to do each piece, p.

2 Megan is writing a formula to work out the volume of this prism.
Her formula is

$$V = \frac{abc}{2}$$

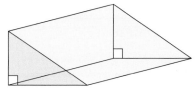

Explain what a, b and c stand for.
Explain why Megan's formula is correct.

3 This cylinder has base radius r and height h.

a Write a formula for V, the volume of the cylinder.

b Nicholas has written this formula for the surface area, A, of the cylinder.

$$A = 2\pi r^2 + 2\pi rh$$

Explain what the $2\pi r^2$ part represents. Repeat for the $2\pi rh$ term.

> Remember a cylinder is a prism.
> Volume = area of cross section × height.

185

This spread will show you how to:

● Rearrange a formula in order to change its subject

Keywords

Rearrange
Subject

● The **subject** of a formula is the variable before the equals sign.

For example, in $A = \pi r^2$, A is the subject of the formula.

● You can **rearrange** a formula in order to change its subject.

Example

Rearrange the formula $A = \pi r^2$ to make r the subject.

$A = \pi r^2$ Divide both sides by π.

$\dfrac{A}{\pi} = r^2$ Square root both sides.

$\sqrt{\dfrac{A}{\pi}} = r$ r is now the subject.

$r = \sqrt{\dfrac{A}{\pi}}$ Put r on the left-hand side.

Start by writing the formula.
Use inverse operations to get r on its own.

Example

Make h the subject of the formula $A = \frac{1}{2}bh$.

$A = \frac{1}{2}bh$ Clear this fraction by multiplying both sides by 2.

$2A = bh$ Divide both sides by b.

$\dfrac{2A}{b} = h \rightarrow h = \dfrac{2A}{b}$

Example

Make x the subject of these formulae

a $V = u + bx$ **b** $M = axy - c^2$

a $V = u + bx$ Subtract u from both sides.

$V - u = bx$ Divide both sides by b.

$\dfrac{V - u}{b} = x$

$x = \dfrac{V - u}{b}$

b $M = axy - c^2$ Add c^2 to both sides.

$M + c^2 = axy$

$\dfrac{M + c^2}{ay} = x$ Divide both sides by ay.

$x = \dfrac{M + c^2}{ay}$

1 Make x the subject of each formula.

a $C = ax + b$ **b** $M = x - b - c$ **c** $K = \dfrac{x}{t} - q$ **d** $W = t + xy$

e $H = \dfrac{x + z}{p}$ **f** $D = p(x - q)$ **g** $AB = x - ct$ **h** $Y = mx + c$

2 James and Sebastian are rearranging the formula $C = a(x - b)$ in order to make x the subject. They both come up with solutions that look different but are, in fact, correct. Can you explain why?

James' solution	Sebastian's solution
$\dfrac{C}{a} + b = x$	$\dfrac{C + ab}{a} = x$

3 Make y the subject of each formula.

a $c = y^2$ **b** $k = \frac{1}{4}y - 2$ **c** $M = xyz + t$ **d** $2x = \sqrt{y}$

e $p = y^3 + 2$ **f** $T = ky^2$ **g** $R = \frac{1}{3}ayz$ **h** $\sqrt[3]{y} = p$

4 Richard has made a mistake with his rearranging whilst trying to make p the subject of this formula. Copy his working and explain where he has gone wrong.

$$K = mp^3$$
$$\sqrt[3]{k} = mp$$
$$\dfrac{\sqrt[3]{k}}{m} = p$$

5 These are the stages in changing the subject of the formula $c = \dfrac{8(D + k)}{ab}$.

Put them in order.

$D + k = \frac{1}{8}abc$	$\dfrac{8(D + k)}{ab} = c$	$D = \frac{1}{8}abc - k$	$8(D + k) = abc$

6 A formula to change from degrees Celsius to degrees Fahrenheit is

$$F = \dfrac{9(C + 40)}{5} - 40$$

a Use this to change 30°C into °F.

b Rearrange to make C the subject of the formula.

c Use your new formula to find the Celsius equivalent of −32°F.

Rearranging harder formulae

This spread will show you how to:

- Rearrange a formula in order to change its subject

Keywords

Rearrange
Subject

Some formulae can be difficult to **rearrange**.
This is the case when

1 The new **subject** is subtracted in the original formula.
For example to make x the subject of the formula

$$p - x = k \qquad \text{Start by adding } x \text{ to both sides.}$$
$$p = k + x \qquad \text{Now subtract } k \text{ from both sides.}$$
$$p - k = x$$

2 The new subject is in the denominator in the original formula.
For example to make x the subject of the formula

$$\frac{p}{x} = k \qquad \text{Start by multiplying both sides by } x.$$
$$p = kx \qquad \text{Now divide both sides by } k.$$
$$\frac{p}{k} = x$$

Example

a Make x the subject of the formula $t(p - ax) = y$.

b Make y the subject of the formula $\dfrac{p}{y} + k = w$.

---

a $t(p - ax) = y$

$$p - ax = \frac{y}{t} \qquad \text{Divide both sides by } t.$$

$$p = \frac{y}{t} + ax \qquad \text{Add } ax \text{ to both sides.}$$

$$p - \frac{y}{t} = ax \qquad \text{Subtract } \frac{y}{t} \text{ from both sides.}$$

$$\frac{p - \dfrac{y}{t}}{a} = x \qquad \text{Divide both sides by } a.$$

You can go one step further and tidy up the numerator.

$$\frac{p - \dfrac{y}{t}}{a} = x \quad \Longrightarrow \quad \frac{\dfrac{pt - y}{t}}{a} = x \quad \Longrightarrow \quad \frac{pt - y}{at} = x$$

b $\dfrac{p}{y} + k = w$

$$\frac{p}{y} = w - k \qquad \text{Subtract } k \text{ from both sides.}$$

$$p = y(w - k) \qquad \text{Multiply both sides by } y.$$

$$\frac{p}{w - k} = y \qquad \text{Divide both sides by } (w - k).$$

1 Find a counter-example to show that each of these statements is untrue.

a When you subtract 7 from a number, the answer is always odd.

b When you square a number, the answer is always even.

c When you treble a prime number, the answer is always odd.

d When you find the product of two consecutive numbers, the answer is always odd.

2 The sum of any five consecutive integers is always a multiple of 5.

a Demonstrate, with a few examples of your own, that this statement is true.

b By letting the numbers be n, $n + 1$, $n + 2$, $n + 3$ and $n + 4$, prove the statement is true.

3 Repeat question **3** for these statements.

a The sum of two even numbers is even.

b Squaring an even number gives a number in the four times table.

c The sum of an odd number and an even number is always odd.

d If two consecutive numbers are multiplied, and the smaller number is subtracted from the result, you always get a square number.

4 a Can you find a counter-example to this statement?

> Squaring a number will always give you a value greater than the number you started with.

b What range of values does not support this statement?

5 Prove that these statements are true.

a When taking three consecutive integers, the square of the middle integer is always one more than the product of the other two.

b The answer to each equation in this pattern will always be 4.

$$5 \times 8 - 4 \times 9 = 4$$
$$6 \times 9 - 5 \times 10 = 4$$
$$7 \times 10 - 6 \times 11 = 4 \dots$$

c The difference between two square numbers will always be equal to the product of their sum and of their difference.

6 Prove that the angle in a semicircle is always a right angle.

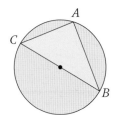

Join A to O.
Let angle ACB be x and angle ABC be y.

Exam review

Key objectives

- Use formulae from mathematics and other subjects
- Substitute numbers into a formula
- Change the subject of a formula
- Generate a formula
- Understand the difference between a practical demonstration and a proof

1 The diagram shows a triangle with lengths given in terms of x.

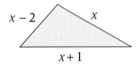

a Write an expression, in terms of x, for the perimeter, P, of the triangle. (2)

b Rearrange the expression from part **a** to make x the subject. (1)

c Hence find the value of x given that the perimeter, P, is 8 cm. (1)

2 John says 'For all prime numbers, n, the value of $n^2 + 3$ is always an even number.'

Give an example to show that John is not correct. (2)

(Edexcel Ltd., 2004)

This unit will show you how to

- Understand and explain congruence and symmetry
- Recall the definitions of special types of quadrilaterals
- Classify quadrilaterals by their geometric properties
- Use angle properties of equilateral, isosceles and right-angled triangles
- Understand, recall and use Pythagoras' theorem in 2-D problems
- Find the coordinates of the midpoint of the line segment AB, and its length, given the points A and B
- Understand and use coordinates to represent points in three dimensions

Before you start ...

You should be able to answer these questions.

1 Copy these shapes and draw in their lines of symmetry.

   a   **b**   **c**

2 Calculate

  a 7^2 **b** $4^2 + 6^2$

  c $3^2 + 5^2$ **d** $8^2 - 4^2$

  e $7^2 - 2^2$ **f** $9^2 - 1^2$

Review

Key stage 3

Unit N5

Congruence and symmetry

This spread will show you how to:

● Understand and explain congruence and symmetry

Keywords

Congruent
Corresponding
Symmetry

● In **congruent** shapes, **corresponding** lengths are equal and corresponding angles are equal.

Example

Explain why these triangles are congruent.

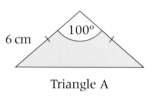

Triangle A

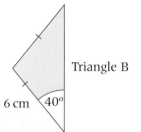

Triangle B

Both triangles are isosceles, with two equal sides of 6 cm.

Triangle A: angle 100° between equal sides, so other two angles must each be 40°.

Triangle B: base angles both 40°, so angle between two equal sides is 100°.

Side opposite 100° angle must be same length as side opposite 100° angle in Triangle A.

Three angles and three sides same in both triangles, so congruent.

Congruent shapes may be reflections, rotations or translations of each other.

Base angles of isosceles triangle are equal.
180° − 100° = 80°
80° ÷ 2 = 40°

180 − (2 × 40) = 100°

Example

Explain why these triangles are not congruent.

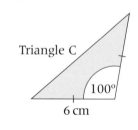

Triangle C

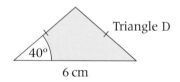

Triangle D

Both triangles are isosceles.

Both triangles have angles 40°, 40°, 100°.

Triangle D: lengths of the two equal sides not given. If they were 6 cm, as in Triangle C, then Triangle D would be equilateral.

Triangle D is not equilateral, so its equal sides are not 6 cm.

So the two triangles are not congruent.

Using base angles of isosceles triangle, as in the first example.

In an equilateral triangle, angles are 60°, 60°, 60°.

They have equal angles, but the side lengths are *not* equal.

A line of **symmetry** divides a shape into two congruent shapes.

A plane of symmetry divides a 3-D shape into two congruent 3-D shapes.

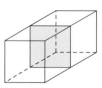 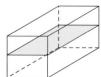 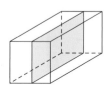

1 Explain whether or not these pairs of triangles are congruent.

a b  c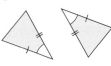

2 The dotted lines show two different ways of splitting a rectangle into two congruent shapes.

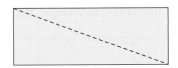

a Draw two copies of the rectangle.
Draw dotted lines to show two more ways of splitting the rectangle into congruent shapes.

b For each, state whether or not the dotted line is also a line of symmetry.

3 Copy these shapes.
Draw dotted lines to split them into congruent shapes.

You may need more than one copy of each shape.

4 Copy these 3-D shapes.
Show how each 3-D shape can be split into two congruent 3-D shapes.

You may need more than one copy of each shape.

5 How many planes of symmetry divide a cylinder into congruent shapes?

195

Quadrilaterals

This spread will show you how to:

- Recall the definitions of special types of quadrilaterals
- Classify quadrilaterals by their geometric properties

Keywords
Kite
Parallelogram
Quadrilateral
Rectangle
Rhombus
Square
Trapezium

- A **quadrilateral** is a shape with four straight sides.
- Angles in a quadrilateral add up to 360°.

Properties of quadrilaterals

	Square	Rhombus	Rectangle	Parallelogram	Trapezium	Kite
1 pair opposite sides parallel	✓	✓	✓	✓	✓	
2 pairs opposite sides parallel	✓	✓	✓	✓		
Opposite sides equal	✓	✓	✓	✓		
All sides equal	✓	✓				
All angles equal	✓		✓			
Opposite angles equal	✓	✓	✓	✓		
Diagonals equal	✓		✓			
Diagonals perpendicular	✓	✓				✓
Diagonals bisect each other	✓	✓	✓	✓		
Diagonals bisect the angles	✓	✓		✓		
2 pairs of adjacent sides equal	✓	✓				✓

Example

Work out the area of this rhombus.

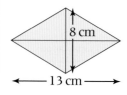

The diagonals bisect each other and are perpendicular.
Opposite angles are equal.
Diagonals bisect the angles.

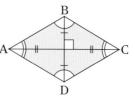

So the diagonal AC is a line of symmetry and the triangles ABC and ACD are congruent.

Each triangle has base 13 cm and height 4 cm.
Area of each triangle = $\frac{1}{2} \times 13 \times 4 = 26$ cm^2
Area of rhombus = $2 \times 26 = 52$ cm^2

Recap properties of a **rhombus**.

A line of symmetry divides a shape into two congruent shapes.

Area of triangle $= \frac{1}{2}bh$

Example

Are these statements true or false? Give reasons for your answer.

a All squares are rhombuses

b Parallelograms are rectangles

a True, all properties of a rhombus are also properties of a square.

b False, a **parallelogram** does not have all its angles equal, nor are its diagonals perpendicular.

A **square** is a special type of rhombus with all angles equal.

A **rectangle** is a special type of parallelogram with all angles equal.

1 The lengths of the diagonals of a rhombus are 5 cm and 9 cm.
Find the area of the rhombus.

2 A square has diagonals 10 cm long.

  a Sketch the square.

  b Find the area of the square.

3 Jenny draws these three kites each with diagonals 6 cm and 14 cm.

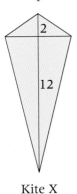
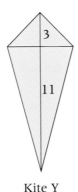
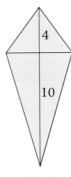

Kite X   Kite Y   Kite Z

DID YOU KNOW?
The kite is a rare bird of prey that hovers in the wind, which influenced the naming of the toy kite. This in turn influenced the naming of the kite shape.

  a Find the areas of kites X, Y and Z.

  b Describe how to work out the area of any kite.

4 The lengths of the diagonals of a kite are 8 cm and 15 cm.
Find the area of the kite.

Use your method from question **3**.

8 cm

15 cm

5 Here is a diagram of a parallelogram, P, and a rhombus, R.

P   R

One diagonal been drawn inside each shape.
Has either shape has been split into two congruent triangles?
Give a reason for your answer.

6 Are these statements true or false? Give reasons for your answers.

  a All squares are rectangles.

  b All kites are rhombuses.

  c All rhombuses are rectangles.

Triangles and Pythagoras' theorem

This spread will show you how to:

- Use angle properties of equilateral, isosceles and right-angled triangles
- Understand, recall and use Pythagoras' theorem in 2-D problems

Keywords
Equilateral
Hypotenuse
Isosceles
Pythagoras'
   theorem
Right-angled
   triangle
Scalene

You need to know the properties of these triangles.

Equilateral	Isosceles	Scalene	Right-angled
All sides equal All angles equal	Two sides equal Two angles equal	No sides equal No angles equal	One angle 90°

In a right-angled triangle the **hypotenuse** is the longest side. It is opposite the right-angle.

In a right-angled triangle, the area of the square of the hypotenuse equals the sum of the areas of the squares of the other two sides.

- This is **Pythagoras' theorem.**

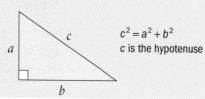

$c^2 = a^2 + b^2$
c is the hypotenuse

Example

Work out the missing sides in these triangles.

a 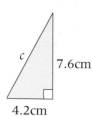   **b** 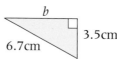   **c**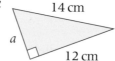

To find the hypotenuse: square, add and square root.

a $c^2 = a^2 + b^2$
$\quad = 4.2^2 + 7.6^2$
$\quad = 17.64 + 57.76$
$\quad = 75.4$
$c = \sqrt{75.4}$
$\quad = 8.7 \text{ cm}$

b $c^2 = a^2 + b^2$
$b^2 = c^2 - a^2$
$\quad = 6.7^2 - 3.5^2$
$\quad = 44.89 - 12.25$
$\quad = 32.64$
$b = \sqrt{32.64}$
$\quad = 5.7 \text{ cm}$

c $c^2 = a^2 + b^2$
$a^2 = c^2 - b^2$
$\quad = 14^2 - 12^2$
$\quad = 196 - 144$
$\quad = 52$
$a = \sqrt{52}$
$\quad = 7.2 \text{ cm}$

To find a shorter side: square, subtract and square root.

1 Find the hypotenuse in each of these right-angled triangles.

Give answers in this exercise to 1 dp where appropriate.

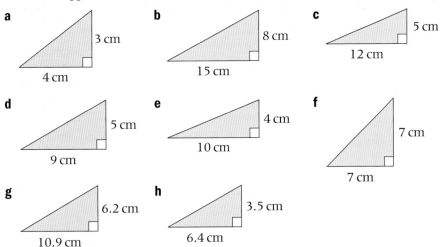

a 3 cm, 4 cm

b 8 cm, 15 cm

c 5 cm, 12 cm

d 5 cm, 9 cm

e 4 cm, 10 cm

f 7 cm, 7 cm

g 6.2 cm, 10.9 cm

h 3.5 cm, 6.4 cm

2 In some of the triangles in question **1**, all three sides have integer (whole number) values.
Such sets of three numbers are called Pythagorean triples.
Write the Pythagorean triples from question **1**.

3 Find the missing side in each of these right-angled triangles.

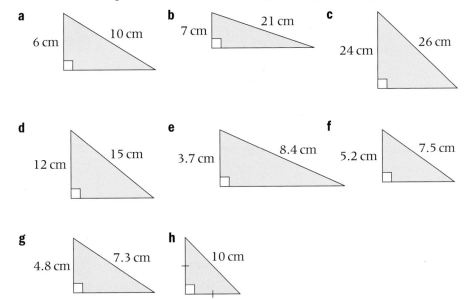

a 6 cm, 10 cm

b 7 cm, 21 cm

c 24 cm, 26 cm

d 12 cm, 15 cm

e 3.7 cm, 8.4 cm

f 5.2 cm, 7.5 cm

g 4.8 cm, 7.3 cm

h 10 cm

DID YOU KNOW?

Pythagoras was a Greek mathematician most famous for his theorem, who taught his students that absolutely everything was related to mathematics.

4 Some of the triangles in question **3** are Pythagorean triples.

a Write down the Pythagorean triples in question **3**.

b Compare these with your answers to question **2**.

c Comment on anything you notice.

Problem solving using Pythagoras' theorem

This spread will show you how to:

● Understand, recall and use Pythagoras' theorem in 2-D problems

Keywords

Hypotenuse
Pythagoras'
    theorem
Right-angled
    triangle

● In a **right-angled triangle**, the square on the **hypotenuse** equals the sum of the squares of the other two sides.

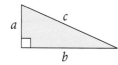

$c^2 = a^2 + b^2$
c is the hypotenuse

This is **Pythagoras' theorem.**
The hypotenuse is always opposite the right-angle.

To solve problems using Pythagoras' theorem

– sketch a diagram and label the right-angle

– label the unknown side

– round your answer to a suitable degree of accuracy.

Unless the question tells you otherwise, round to 2 dp.

Example

A rectangle measures 5 cm by 8 cm.
Find the length of its diagonal.

Diagonal, d, is the hypotenuse of
a right-angled triangle.
$d^2 = 5^2 + 8^2$
$d^2 = 89$
$d = 9.43$ cm

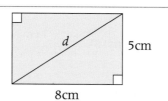

Mark all the facts you know on your diagram.

Example

Find the length AD.

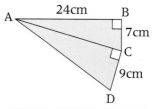

First find AC $AC^2 = 7^2 + 24^2$
            $AC^2 = 625$
            $AC = 25$ cm

AC is the hypotenuse of the right-angled triangle *ABC*.

In triangle ACD $AD^2 = 9^2 + 25^2$
            $AD^2 = 706$
            $AD = 26.57$ cm

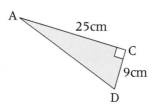

AD is the hypotenuse of the right-angled triangle *ACD*.

Exercise S4.4

1 Find the length of the diagonal of each rectangle.

a
8.3 cm
2.4 cm

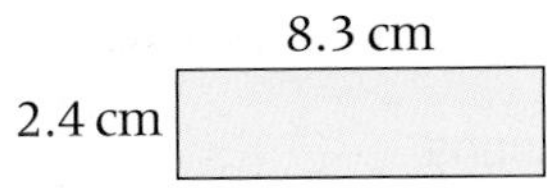

b
6 cm
5.2 cm

Give answers in this exercise to 2 dp where appropriate.

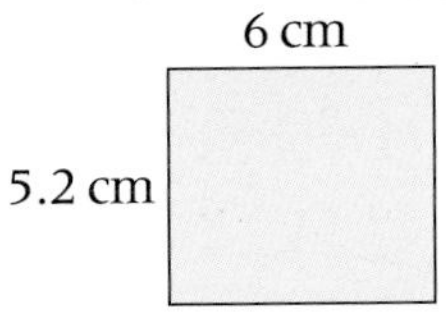

2 A rectangle has one side 4 cm and diagonal 10.4 cm.
Find the length of the other side.

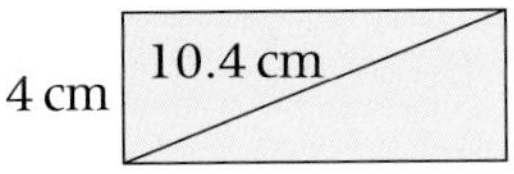

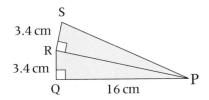

4 cm   10.4 cm

3 Find the length of the diagonal of a square with side length 8 cm.

4 Find the length of the side of a square with diagonal length 8 cm.

5 PQR and PRS are right-angled triangles.
Find the length PS.

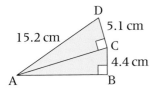

S
3.4 cm
R
3.4 cm
Q   16 cm   P

6 ABC and ACD are right-angled triangles.
Find the length AB.

D
5.1 cm
15.2 cm
C
4.4 cm
A   B

7 A ladder of length 5.5 m leans against a wall.
The foot of the ladder is 1 m from the wall.

How far up the wall does the ladder reach?

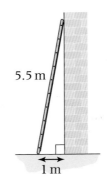

5.5 m

1 m

This spread will show you how to:

- Find the coordinates of the midpoint of the line segment *AB*, and its length, given the points *A* and *B*
- Understand and use coordinates to represent points in three dimensions

Coordinates identify a point on a grid.

A (2, 1)   B (4, 4)

- The **midpoint** of two points is the mean of their coordinates.

- Midpoint of (*a*, *b*) and (*s*, *t*) is $\left(\frac{a+s}{2}, \frac{b+t}{2}\right)$

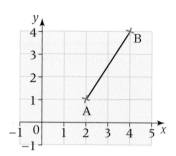

You can use Pythagoras' theorem to find the length of a line joining two points on a grid.

Example

Find the midpoint and length of the line joining A and B.

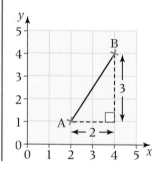

Midpoint

$\left(\frac{2+4}{2}, \frac{1+4}{2}\right) = (3, 2.5)$

Length

Draw a right-angled triangle and label the lengths of the two shorter sides.

$AB^2 = 2^2 + 3^2$
$AB^2 = 13$
$AB = 3.61 \ (2 \ dp)$

Using Pythagoras' theorem

- You can use coordinates to identify a point in a 3-D grid.

The point (2, 5, 3) is located
2 units along the *x*-axis,
5 units along the *y*-axis, and
3 units up the *z*-axis.

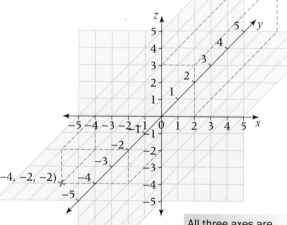

The point (−4, −2, −2) is located
4 units along the negative *x*-axis,
2 units along the negative *y*-axis,
and 2 units down the negative *z*-axis.

All three axes are perpendicular to each other.

1 Find the midpoints of these line
   segments.

   a A (6, 4) and B (2, 7)

   b C (−2, 5) and D (3, −3)

   c E (0, 6) and F (4, 0)

   d G (7, −1) and H (−4, 5)

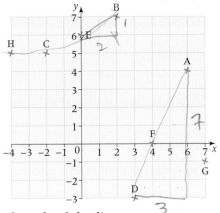

Use the diagram to
check your
answers.

Make a sketch to
show the position
of the points.

2 Use Pythagoras' theorem to find the length of the line segment
   joining each pair of points.

   a (2, 5) and (6, 8)   **b** (7, 1) and (2, 8)

   c (0, 7) and (1, −3)   **d** (−4, 5) and (4, −2)

3 Write the coordinates
   of these points.

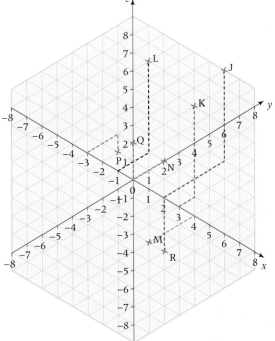

4 Here is a grid in three dimensions.
   The points on the grid have coordinates of

   A(1, 2, 5)   B(1, 4, 2)   C(2, 3, 1)   D(5, 3, 4)

   a Find the midpoints of these line segments.

   i AB   **ii** CD   **iii** BC   **iv** AD

   b Use Pythagoras' theorem to find the length of each
   line segment in part **ai** and **aii**.

   c Can you extend the use of Pythagoras' theorem to
   find the length of BC? You will need to use it twice.

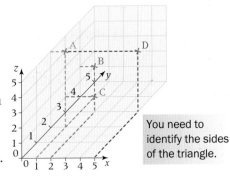

You need to
identify the sides
of the triangle.

203

Key objectives

- Understand congruence
- Use angle properties of equilateral, isosceles and right-angled triangles
- Recall the essential properties of special types of quadrilateral
- Classify quadrilaterals by their geometric properties
- Understand, recall and use Pythagoras' theorem in 2-D problems
- Find the coordinates of the midpoint of the line segment *AB*, given points *A* and *B*
- Understand that three coordinates identify a point in space, using the term '3-D'

1 a Find the length of the line joining points *A* and *B*. (2)

b Find the midpoint of line *AB*. (2)

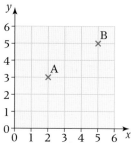

2 Here are four quadrilaterals labelled **A**, **B**, **C** and **D**.

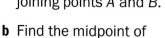

A rectangle **B** parallelogram **C** rhombus **D** kite

a Write down the letter of the quadrilateral which has

i exactly **one** line of symmetry,

ii **no** lines of symmetry,

iii **both** diagonals as lines of symmetry. (3)

b Write down the letter of the quadrilateral which **does not** have rotational symmetry of order 2. (1)

(Edexcel Ltd., 2005)

This unit will show you how to

- Perform the operations within a calculation in the correct order
- Estimate answers to calculations, using these to check the solution
- Round numbers to sensible degrees of accuracy
- Use surds and π in exact calculations
- Use mental and written methods to multiply and divide with decimals
- Understand place value and where to place the decimal point
- Use multiplicative inverses
- Use calculators effectively and efficiently, knowing when and when not to round the display
- Use calculators to calculate the upper and lower bounds of calculations

Before you start ...

You should be able to answer these questions.

Review

1 Evaluate these expressions.

   a $4 \times 3 + 2$ **b** $8 - 2 \times 3$

   c $(6 - 3) \times 5$ **d** $5^2 - 3^2$

   Key stage 3

2 Calculate.

   a $\sqrt{4}$ **b** $\sqrt{256}$

   c $\sqrt{13^2 - 12^2}$ **d** $\sqrt{4} \times \sqrt{25}$

   Unit N5

3 Round these numbers to one decimal place.

   a 45.56 **b** 9.906

   c 104.008 **d** 16.09

   Unit N2

4 Write notes to show how you would find a mental estimate for each of these calculations.

   a 19.2×28.9 **b** $355.72 \div 58.91$

   c $1206 - 816$ **d** $6987 + 6039$

   Unit N2

This spread will show you how to:

- Perform the operations within a calculation in the correct order
- Estimate answers to calculations, using these to check the solution
- Round numbers to sensible degrees of accuracy

Keywords
BIDMAS
Operation
Order

When a calculation involves a number of steps, or **operations**, you need to do them in the right order.

The order in which operations are carried out is:

- Brackets – start by working out the contents of any brackets
- Powers and indices – for example, squares, cubes or square roots – come next
- Multiplication and division are done next
- Addition and subtraction are done last.

$$(3 + 2) \times 4^2 - 6$$
$$= 5 \times 4^2 - 6$$
$$= 5 \times 16 - 6$$
$$= 80 - 6$$
$$= 74$$

BIDMAS (**B**rackets, **I**ndices or powers, **D**ivision, **M**ultiplication, **A**ddition, **S**ubtraction) will help you to remember this.

Example

Evaluate   **a** $4 + 3 \times 2$ **b** $5 + 3^2$ **c** $\sqrt{(5 + 4 \times 11)}$ **d** $\sqrt{(5^2 - 4^2)}$

a $4 + 3 \times 2 = 4 + 6 = 10$
b $5 + 3^2 = 5 + 9 = 14$
c $\sqrt{(5 + 4 \times 11)} = \sqrt{(5 + 44)} = \sqrt{49} = 7$
d $\sqrt{(5^2 - 4^2)} = \sqrt{(25 - 16)} = \sqrt{9} = 3$

Examiner's tip
Using brackets in parts **c** and **d** shows that **all** the values are contained in the square root.

Example

Estimate the answer to $\dfrac{6.3 + \sqrt{9.7^2 - 17}}{149}$

Round all the numbers to a sensible amount.

$$\frac{6.3 + \sqrt{9.7^2 - 17}}{149} \approx \frac{6 + \sqrt{10^2 - 20}}{150} = \frac{6 + \sqrt{80}}{150} \approx \frac{6 + 9}{150} = \frac{15}{150} = 0.1$$

Deal with the numerator and denominator separately.

Example

Adam explained how he would calculate $\dfrac{5 + \sqrt{9}}{11}$

What is the problem here?

> There is a root, an addition and a division. I'll do the root first, then divide, and then add.

Even though there are no brackets in this expression, the whole of the 'top line' is divided by 11, so you need to find the square root, then add, then divide. The expression could be written as $(5 + \sqrt{9}) \div 11$.

1 Find the value each of these.

a 5 + 6 × 9   **b** 4 × 9 + 1   **c** 8 − 3 − 3   **d** 4 × 3 + 7 × 2

2 Find the value of each of these.

a 9 − 4 × 2   **b** (9 − 4) × 2   **c** 5 × 7 + 4 × 2   **d** 5 × (7 + 4) × 2

3 Copy and complete these equations, replacing the □ with the correct operation.

a 5 □ 3 × 7 = 26   **b** 4 × 6 □ 2 = 22

c 4 □ 7 + 1 = 29   **d** 17 □ 2 + 2 = 6^2

4 Copy these calculations, inserting brackets to make the answers correct.

a 11 − 1 × 5 = 50   **b** 12 + 3 ÷ 3 = 5

c 12 − 4 − 1 = 9   **d** 8 ÷ 4 + 4 + 1 = 2

5 Copy and complete these equations, replacing the ● with the correct number.

a (● + 2) × 9 = 36   **b** 64 ÷ (● + 3) = 8

c $\sqrt{(● − 10)} = 5 × 2$ **d** $\sqrt{● − (5 × 2)} = 2$

6 Find the values of these expressions.

a $(5^2 + 3) × 7$ **b** $(9 − 7)^2$

c $(5 − 3) × (4^2 − 7)$ **d** $(5^2 − 8)^2$

7 Calculate the values of these expressions.

a $(4 + 7)^2$ **b** $(6 + 7) × 9 ÷ 3$

c $\dfrac{6 × (5^2 − 13)}{4}$ **d** $\sqrt{100 − 2 × 6^2}$

8 Find the values of these expressions.

a $\dfrac{28}{4} + \sqrt{100 − (9^2 + 5 × 2)}$ **b** $\sqrt{28 + 4^2} − (10 − 2) + 4 × 3$

9 Estimate the value of each of these expressions without using a calculator. Show all of your working.

a $\dfrac{5.2 × (4.8^2 − 12)}{3.9}$ **b** $73 × \dfrac{202 − 11}{38 × 5} + 29$

c $18.4^2 − \dfrac{592}{11.4}$ **d** $\sqrt{26.4 × \dfrac{12.5 + 7.4}{5.36}}$

This spread will show you how to:

- Use surds and π in exact calculations

Keywords

Approximation
Surds
Surd form

- Numbers like $\sqrt{2}$ and $\sqrt{5}$ are called **surds**.

When calculating, first do the written calculation using surds.
Then use a calculator to find approximate values at the end, if required.

- You can separate the square roots of a factorised number.

For example:

- $36 = 4 \times 9$, so $\sqrt{36} = \sqrt{(4 \times 9)} = \sqrt{4} \times \sqrt{9} = 2 \times 3 = 6$
- $80 = 5 \times 16$, so $\sqrt{80} = \sqrt{(5 \times 16)} = \sqrt{5} \times \sqrt{16} = \sqrt{5} \times 4 = 4\sqrt{5}$

Decimal **approximations** are useful in practical contexts.

You can carry out written calculations with surds without converting them to decimals.

If you wanted to mark out a square of area 5 m^2, you would give the side length as 2.24 m, not $\sqrt{5}$ m.

Example

Evaluate   **a** $\sqrt{2}(1 + \sqrt{2})$ **b** $2\pi(4^2 + 4 \times 6)$

a Multiply each term in the bracket by $\sqrt{2}$.

$$\sqrt{2}(1 + \sqrt{2}) = \sqrt{2} \times 1 + \sqrt{2} \times \sqrt{2} = \sqrt{2} + 2$$

You can then find a decimal approximation:

$$\sqrt{2} = 1.414 \text{ to 3 dp so } 2 + \sqrt{2} \approx 3.414$$

b First simplify, leaving π in the expression.

$$2\pi(4^2 + 4 \times 6) = 2\pi(16 + 24) = 2\pi \times 40 = 80\pi$$

You can now find a decimal approximation:

$$\pi = 3.14 \text{ to 2 dp so } 80\pi \approx 251 \text{ to 3 sf.}$$

This is normally written as $2 + \sqrt{2}$.

Leaving an answer in **surd form** is more accurate than a decimal approximation.

Example

Given that $\sqrt{3} = 1.73$ and that $\sqrt{7} = 2.65$ (both to 3 sf), find approximate values for   **a** $\sqrt{27}$ **b** $\sqrt{243}$ **c** $\sqrt{28}$ **d** $\sqrt{21}$,   without using the square root function on your calculator. Show your working, and give your answers to a suitable degree of accuracy.

a $\sqrt{27} = \sqrt{(9 \times 3)} = \sqrt{9} \times \sqrt{3} = 3\sqrt{3} = 3 \times 1.73 = \underline{5.19}$ to 3 sf

b $243 = 3 \times 81 = 3 \times 9 \times 9$

$\sqrt{243} = \sqrt{(3 \times 9 \times 9)} = \sqrt{3} \times \sqrt{9} \times \sqrt{9} = \sqrt{3} \times 3 \times 3 = 9\sqrt{3} = \underline{15.6}$ to 3 sf

c $\sqrt{28} = \sqrt{(4 \times 7)} = \sqrt{4} \times \sqrt{7} = 2\sqrt{7} = \underline{5.30}$ to 3 sf

d $\sqrt{21} = \sqrt{(3 \times 7)} = \sqrt{3} \times \sqrt{7} = \underline{4.58}$ to 3 sf

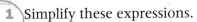

1. Simplify these expressions.

   a $\sqrt{3} + \sqrt{3}$ b $\sqrt{5} + \sqrt{5}$ c $\sqrt{9} + \sqrt{4}$ d $\sqrt{7} + \sqrt{7} + \sqrt{7}$

2. Simplify these expressions.

   a $\pi + \pi$ b $2\pi + \pi$ c $4 + \pi - 2 + \pi$

   d $\pi(4^2 - 6)$ e $2\pi(4^2 - 7)$ f $4\pi(2 + \sqrt{4})$

3. Simplify these expressions.

   a $\sqrt{16} + \sqrt{3}$ b $\sqrt{49} + \sqrt{2} - \sqrt{16}$

   c $3\sqrt{7} + \sqrt{49} - \sqrt{7}$ d $17 + \sqrt{17} - \sqrt{9}$

4. Simplify these expressions.

   a $\pi(6^2 - 4)$ b $\sqrt{49} + \pi(7 - 5)$ c $4(7 + \sqrt{2})$ d $\pi(8^2 - \sqrt{4})$

5. Work out these, giving your answers in surd form where necessary.

   a $\sqrt{2} \times \sqrt{2}$ b $\sqrt{5} \times \sqrt{5}$ c $\sqrt{3}(\sqrt{3} + 3)$ d $\sqrt{4}(\sqrt{3} + 4)$

6. Use a calculator to find an approximate decimal value for each of these expressions. Give your answers to 2 decimal places.

   a $4\sqrt{2}$ b $\sqrt{5} + 1$ c $2 + \sqrt{5}$ d $36\pi - 7$

7. Simplify these expressions, then use a calculator to find approximate decimal values for each one.
   Give your answers correct to 3 significant figures.

   a $4(3 + \sqrt{5})$ b $\sqrt{5}(5^2 - 5)$ c $\sqrt{7}(2 + \sqrt{7})$ d $\pi(2^3 + \sqrt{5})$

8. Simplify these expressions.

   a $\sqrt{20}$ b $\sqrt{125}$ c $\sqrt{18}$ d $\sqrt{98}$

9. You are told that $\sqrt{2} \approx 1.414$, $\sqrt{3} \approx 1.732$ and $\sqrt{5} \approx 2.236$. Use this information to estimate the value of each of these, without using the square root button on your calculator. Show your working, and give your answers to 4 significant figures.

   a $\sqrt{2} + \sqrt{3}$ b $\sqrt{10}$ c $\sqrt{125}$ d $\sqrt{24}$

10. Use the square root key on your calculator to work out the answers to question **9**. Compare your answers, and explain which answers are more accurate.

11. Simplify these expressions by multiplying out the brackets.

    a $(3 + \sqrt{2})(4 + \sqrt{2})$ b $(4 + \sqrt{5})(3 + \sqrt{5})$

    c $(6 + \sqrt{3})(3 - \sqrt{3})$ d $(5 + \sqrt{5})(5 - \sqrt{5})$

This spread will show you how to:

- Use mental and written methods to multiply and divide with decimals
- Understand place value and where to place the decimal point

Keywords

Inverse
Place value

- Multiplying a positive number by a number between 0 and 1 makes it smaller.

  $6 \times 0.5 = 3$

- Dividing a positive number by a number between 0 and 1 makes it bigger.

  $6 \div 0.5 = 12$

Always start calculations with an estimate.

- Work out a mental calculation using the significant digits from the question; for example, for $12.5 \div 0.05$, work out $125 \div 5$.

- Finally, use your initial estimate to check your answer and adjust the **place value**.

- Use **inverses** where possible.

  $10 \div 0.2 = 10 \times 5$ and $10 \times 0.25 = 10 \div 4$

$12.5 \div 0.05$
$\approx 10 \div 0.05 = 200$

$125 \div 5 = 25$

$12.5 \div 0.05 = 250$

Example

Use a mental method to work out

a 320×0.4 **b** $320 \div 0.4$ **c** $3.2 \div 0.4$

a $320 \times 0.4 \approx 300 \times 0.4$
$\qquad = 120$
$\quad 32 \times 4 = 64 \times 2$
$\qquad = 128$
$\quad 320 \times 0.4 = 128$

b $320 \div 0.4 \approx 320 \div \frac{1}{2} = 640$
$\qquad 32 \div 4 = 8 \quad$ so
$\quad 320 \div 0.4 = 800$
c $3.2 \div 0.4 \approx 3 \div \frac{1}{2} = 6$
$\qquad 32 \div 4 = 8 \quad$ so
$\quad 3.2 \div 0.4 = 8$

Estimate first.

Compare with estimate to get final answer.

Example

Write a multiplication that is equivalent to $34.5 \div 0.25$.

Dividing by a quarter is the same as multiplying by 4.
$34.5 \div 0.25 = 34.5 \times 4$

$34.5 \div 0.25 = 138$
$34.5 \times 4 = 138$

Example

Calculate mentally.   **a** $72.5 \div 0.05$ **b** 340×0.3 **c** $8.46 \div 0.2$

a $72.5 \div 0.05 \approx 70 \times 200 = 1400$
$\qquad 725 \div 5 = 145 \quad$ so
$\quad 72.5 \div 0.05 = 1450$
c $8.46 \div 0.2 \approx 9 \times 5 = 45$
$\qquad 8.46 \div 2 = 4.23 \quad$ so
$\quad 8.46 \div 0.2 = 42.3$

b $340 \times 0.3 \approx 300 \div 3 = 100$
$\qquad 340 \times 3 = 1020 \quad$ so
$\quad 340 \times 0.3 = 102$

$0.05 = \frac{1}{200}$

$0.3 \approx \frac{1}{3}$

$0.2 = \frac{1}{5}$

1 Use a mental method to find these.

   a 5×0.2 **b** 4×0.3 **c** 0.5×3 **d** 16×0.5

   e 7×0.2 **f** 30×0.25 **g** 40×0.4 **h** 0.6×25

2 Use a mental method to find these.

   a $8 \div 0.2$ **b** $4 \div 0.4$ **c** $6 \div 0.3$ **d** $32 \div 0.4$

   e $0.8 \div 4$ **f** $0.3 \div 0.03$ **g** $0.4 \div 0.04$ **h** $50 \div 0.01$

3 Use a mental method to find these. Start with an estimate, and show your working.

   a 2×0.4 **b** 20×0.04 **c** 3×7 **d** 0.3×0.7

   e 12×0.4 **f** $12 \div 0.3$ **g** $3.6 \div 4$ **h** $3.6 \div 0.9$

4 Use a calculator to check your answers to question **3**.

5 Write a division that is equivalent to each of these multiplications.

   a 4×0.5 **b** 6×0.2 **c** 12×0.2 **d** 2×0.001

6 Write a multiplication that is equivalent to each of these divisions.

   a $4 \div 0.5$ **b** $6 \div 0.25$ **c** $16 \div 0.01$ **d** $15 \div 0.05$

7 Use a mental method to find these. Show your method.

   a $8 \div 0.25$ **b** $15 \div 0.5$ **c** $7 \div 0.2$ **d** 4×0.25

   e 16×0.02 **f** $24 \div 0.12$ **g** 20×0.05 **h** $18 \div 0.025$

   i 3×0.125 **j** $7 \div 0.25$ **k** 40×0.025 **l** 8×0.875

8 Use a mental method to find an estimate for each of these.

   a $37 \div 0.47$ **b** $319 \div 0.3$ **c** 3.8×134 **d** $17 \div 0.031$

9 Use a calculator to find the exact answers to question **8**.

10 Use a mental method to work out these. Start each one with an estimate, and show your method.

   a 31×0.3 **b** $49 \div 0.07$ **c** $3.66 \div 0.3$ **d** $4.24 \div 0.4$

   e $13.9 \div 0.03$ **f** $3.9 \div 0.03$ **g** $171 \div 0.3$ **h** 5.2×0.125

11 Use a calculator to check your answers to question **10**.

Written calculations

This spread will show you how to:

- Use mental and written methods to multiply and divide with decimals
- Estimate answers to calculations, using these to check the solution

Keywords

Check
Estimate
Place value

For column methods of addition and subtraction, use the full decimal numbers as given in the question.

Use the decimal point to ensure that the columns are aligned correctly.

Example

Calculate

a $135.23 + 27.8$

b $34.56 - 18.729$

a Estimate $140 + 30 = 170$

```
    1 3 5 . 2 3
  +   2 7 . 8
    1 6 3 . 0 3
        1   1
```

b Estimate $30 - 20 = 10$

```
  ²3̸ ¹³4̸ . ¹5 ⁵6̸ 10
 - 1   8 . 7   2 9
   1   5 . 8   3 1
```

Use your estimates to **check** your answers.

For standard methods of multiplication and division, work with the significant digits from the numbers in the question.

Use an estimate to adjust the place value correctly.

Example

Calculate

a 18.5×7.9

b $47.592 \div 1.8$

a Estimate $20 \times 8 = 160$

```
      1 8 5
    ×   7 9
    1 6₇ 6₄ 5
  1 2₅ 9₃ 5 0
  1 4₁ 6₁ 1 5
```

Use your **estimate** to check your answer.

$18.5 \times 7.9 = 146.15$

146.15 is close to 160, so this is about right.

$20 \times 8 = 160$

$50 \div 2 = 25$

b Estimate $50 \div 2 = 25$

This could be done by long division (see unit N2.4) or by short division. Using short division you get:

```
           2 6 4 4
  18 )4 7 ¹¹5 ⁷9 ⁷2
```

Use your estimate to adjust the **place values**.

So $47.592 \div 1.8 = 26.44$

You should show your working for the questions in this exercise.

1 Use a written method to calculate these.

  a $24.72 + 14.04$ **b** $1.52 + 1.09$ **c** $6.149 + 2.052$ **d** $6.64 + 15.88$

2 Use a written method to calculate these.

  a $5.23 - 3.11$ **b** $17.45 - 13.26$ **c** $6.41 - 4.37$ **d** $23.6 - 17.9$

3 Use a written method to calculate these.

  a $1.09 + 154$ **b** $0.09 + 0.36$ **c** $14.52 + 9.8$ **d** $13.92 + 0.8$

4 Use a written method to calculate these.

  a $4.5 - 0.53$ **b** $3.085 - 2.99$ **c** $16.3 - 3.86$ **d** $112.14 - 53.8$

5 Use a written method to evaluate these.

  a $11.1 - 8.29$ **b** $2.09 - 1.333$ **c** $102.8 - 14.79$ **d** $978 + 148.72$

6 Use a calculator to check your answers to questions **1** to **5**.

7 Use a written method to work out these multiplications.

  a 15.9×4 **b** 17.9×0.3 **c** 16.9×0.8 **d** 0.048×0.07

8 Calculate these divisions, using a written method.

  a $1.36 \div 0.8$ **b** $3.01 \div 7$ **c** $19.2 \div 0.4$ **d** $13.45 \div 0.05$

9 Use long multiplication (or an equivalent written method)
  to evaluate these.

  a 8.8×1.9 **b** 190×0.054 **c** 189×4.2 **d** 214×0.037

10 Use long division (or an equivalent written method)
   to evaluate these.

  a $211.68 \div 24$ **b** $133.98 \div 0.66$ **c** $292.38 \div 0.33$ **d** $5.913 \div 0.27$

11 Use a calculator to check your answers to questions **7** to **10**.

This spread will show you how to:

- Use calculators effectively and efficiently, knowing when and when not to round the display
- Use calculators to calculate the upper and lower bounds of calculations

Keywords

Display
Effective
Efficient
Round

You need to know how to enter calculations into a calculator, and how to interpret the calculator **display**.

Calculator answers often need to be **rounded** to a suitable degree of accuracy.

Measurements (like lengths and weights) are correct to a certain degree of accuracy. You can use a calculator to find the upper and lower bounds of calculations involving measurements.

The length of a pipe given as 3.95 m is rounded to the nearest cm. The actual length is between 3.945 m and 3.955 m.

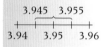

Example

Use a calculator to find the value of $(3.59 - 1.68) \div (2.4 \times 6.9)$. Write all of the digits on the calculator display, and round your answer to a suitable degree of accuracy.

A possible key sequence is:

(3 . 5 9 − 1 . 6 8) ÷ (2 . 4 × 6 . 9) =

The answer on a 10-digit display is

```
(3.59-1.68)÷■
0.115338164
```

Use 2 sf as a suitable degree of accuracy:

the answer is 0.12 (2 sf)

Do not round before the end of a calculation. You should not give an answer with more significant figures than there were in the question.

Example

A model boat travels 3.9 metres in 7.3 seconds. Both measurements are correct to 1 dp. Find the upper and lower bounds of the speed of the boat in metres per second.

Speed = distance ÷ time.

The distance 3.9 m is a rounded value in the range 3.85 m to 3.95 m, and 7.3 s is in the range 7.25 s to 7.35 s.

To get the maximum value of the speed, divide the largest possible distance by the shortest possible time:

The upper bound of the calculation is
$3.95 \div 7.25 = 0.544827 \ldots$ metres per second

To get the minimum value of the speed, divide the smallest possible distance by the greatest possible time.

The lower bound of the calculation is
$3.85 \div 7.35 = 0.523809 \ldots$ metres per second

$\text{Speed}_{\text{Max}} = \text{Distance}_{\text{Max}} \div \text{Time}_{\text{Min}}$

Similarly,

$\text{Speed}_{\text{Min}} = \text{Distance}_{\text{Min}} \div \text{Time}_{\text{Max}}$

1 Use a calculator to evaluate these, giving each answer to 2 sf.

    a $3.2 \times (2.8 - 1.05)$ **b** $2.8^2 \times (9.4 - 0.083)$

    c $16 \div (5.1^2 - 7.2)$ **d** $(3.8 + 8.9) \times (2.2^2 - 7.6)$

    e $1.8^3 + 4.7^3$ **f** $52 \div (4.6 - 1.8^2)$

2 Work out these, giving your answers to a suitable degree of accuracy.

    a The total weight of two people weighing 68 kg and 73 kg.

    b The area of a rectangle with sides 2.2 m and 3.8 m.

    c The cost of 2.37 m of material at £5.75 per metre.

    d The time taken to travel 1 mile at a speed of 80 mph.

3 Write the upper and lower bounds of these measurements, which are all given to 3 significant figures.

> You can draw a number line to help you.

    a 4.75 m   **b** 12.6 s   **c** 150 cm

    d 24.5 kg   **e** 8.07 g   **f** 4.33 s

4 A model car was rolled down a track. The length of the track was measured as 2.55 m, to the nearest centimetre. The time for the journey was 1.7 seconds, measured to the nearest tenth of a second.

    a Write the upper and lower bounds for the length of the track.

    b Write the upper and lower bounds for the time of the journey.

    c Use the relationship

$$\text{Speed}_{Max} = \text{Distance}_{Max} \div \text{Time}_{Min}$$

    to find the maximum possible average speed of the car.

    d Use the relationship

$$\text{Speed}_{Min} = \text{Distance}_{Min} \div \text{Time}_{Max}$$

    to find the minimum possible average speed of the car.

5 Find the maximum and minimum area of squares with side lengths given as

> Give your answers to 3 sf.

    a 8.00 m   **b** 6.40 cm   **c** 1.05 m

    d 3.00 mm   **e** 3.75 m   **f** 9.99 cm

6 The numbers in these calculations all relate to measurements. Find the upper and lower bound of each calculation.

    a $6.5 \times (1.2 + 4.6)$ **b** $1.2 - 0.8$

    c $(3.77 - 3.22) \times (2.43 - 1.75)$ **d** $\dfrac{(3.2 - 1.9)^2}{4.5 + 8.8}$

Exam review

Key objectives

- Use the hierarchy of operations
- Use calculators effectively and efficiently
- Develop a range of strategies for mental calculation
- Use standard column procedures for multiplication of integers and decimals
- Use calculators, or written methods, to calculate the upper and lower bounds of calculations
- Check and estimate answers to problems

1 Work out

  a i 250×5

  ii 250×0.5 (2)

  b Write a multiplication equivalent to $250 \div 0.5$.   (1)
  Explain your answer.

2 a Use your calculator to work out $(2.3 + 1.8)^2 \times 1.07$.   (2)

  Write down all the figures on your calculator display.

  b Put brackets in the expression below so that its
  value is 45.024.   (1)

  $1.6 + 3.8 \times 2.4 \times 4.2$

(Edexcel Ltd., 2003)

This unit will show you how to

- Solve linear equations after simplifying them
- Solve quadratic equations by factorisation
- Use trial and improvement to find approximate solutions to equations
- Substitute numbers into expressions
- Round answers to appropriate degrees
- Solve simultaneous equations by eliminating a variable
- Use notation and symbols correctly and consistently within a given problem
- Write equations in order to solve word problems

Before you start ...

Review

You should be able to answer these questions.

1 Solve these equations.

Unit A2

a $3x - 11 = 16$ b $4x - 2 = 2x + 5$

c $3(2 - 4x) = 5(7 - 2x)$ d $\frac{3}{x} - 8 = 4$

e $10 - 2x = 4$ f $\frac{3x + 2}{4} = \frac{4x - 1}{5}$

2 Expand and simplify the sets of double brackets.

Unit A1

a $(x + 5)(x + 2)$ b $(y + 7)(y - 3)$

c $(w + 4)^2$ d $(y - 6)(y - 9)$

e $(2w + 4)(3w + 1)$ f $(4y - 5)^2$

3 Factorise.

Unit A1

a $4x + 2$ b $xy + y^2$

c $5a + 10ab$ d $4x - 8y$

e $9x^3 + 3x^5$ f $2xy + 4xz - 6x^2$

4 Factorise into double brackets.

Unit A1

a $x^2 + 5x + 6$ b $x^2 - 2x - 24$

c $x^2 - 7x + 12$ d $x^2 - 6x + 9$

e $x^2 - 100$ f $2x^2 + 5x + 3$

This spread will show you how to:

● Solve linear equations after simplifying them

Keywords
Expand
Negative term

When solving an equation with brackets, **expand** the brackets first.

$$4(x + 2) = 7(x - 1)$$ Expand the brackets.

$$4x + 8 = 7x - 7$$ Subtract 4x from both sides.

$$8 = 3x - 7$$ Add 7 to both sides.

$$15 = 3x$$ Divide both sides by 3.

$$x = 5$$

An equation with a negative x-term is easier to solve if you get rid of the **negative term** first.

$$5 - 8x = 45$$ Add 8x to both sides to remove the −8x

$$5 = 45 + 8x$$ Subtract 45 from both sides.

$$-40 = 8x$$ Divide both sides by 8.

$$x = -5$$

Example

Solve this equation. $(x + 3)(x + 5) = (x + 2)^2$

$$(x + 3)(x + 5) = (x + 2)^2$$

$$x^2 + 3x + 5x + 15 = (x + 2)(x + 2)$$ Remove brackets first (FOIL): *multiply* the terms.

$$x^2 + 8x + 15 = x^2 + 2x + 2x + 4$$ Collect like terms.

$$x^2 + 8x + 15 = x^2 + 4x + 4$$ Subtract x^2 from each side.

$$8x + 15 = 4x + 4$$

$$4x + 15 = 4$$

$$4x = -11$$

$$x = \tfrac{-11}{4} \text{ or } -2\tfrac{3}{4}$$

Don't use a calculator – fractions are exact. Decimals that need rounding, make the **solution** less accurate.

Example

The square and the rectangle have the same area.
Find the value of x.

x $x + 5$

$x - 2$

Area of square = Area of rectangle

$$x^2 = (x + 5)(x - 2)$$ Remember brackets first (FOIL)

$$x^2 = x^2 + 5x - 2x - 10$$ Collect like terms.

$$x^2 = x^2 + 3x - 10$$ Subtract x^2 from each side.

$$0 = 3x - 10$$

$$-3x = -10$$ Divide by −1.

$$3x = 10$$

$$x = 3\tfrac{1}{3}$$

1 Solve these equations.

   a $4x - 9 = 15$ **b** $\frac{y}{7} + 2 = 4$

   c $3x + 9 = 2x + 15$ **d** $5p - 9 = 3p + 7$

   e $10 - 3t = 8t - 12$ **f** $3 - 7b = 9 - 10b$

2 Solve these equations by first expanding brackets and/or collecting like terms.

   a $3x + 7x - 3 + 9 = 17$ **b** $3(2y - 1) = 4(7y + 6)$

   c $2(3 - 8z) = 4(5 - 6z)$ **d** $2a + 3(2a - 7) = 20$

   e $3a - (9 - 7a) = 34$ **f** $6 - (3x - 9) = -2(5 - x)$

   g $(x + 3)(x + 4) = (x + 7)(x - 2)$ **h** $(y - 7)^2 = (y + 5)^2$

3 Solve these equations containing negative terms.

   a $10 - 4x = 2$ **b** $20 - 3y = 11$ **c** $15 - 8z = 12$

   d $14 - 7a = 19$ **e** $18 = 17 - b$ **f** $36 = 2(3 - 6c)$

4 In each case, write an equation to represent the information given. Solve the equation to find the starting number.

   a I think of a number, treble it and subtract this *from* 40. My answer is 22.

   b I think of a number, add 6 and double the result. This gives me the same answer as when I subtract 5 from the number and multiply the result by 6.

   c I think of a number, add 7 and square it. This gives me the same answer as when I take 5 from the number and square it.

5 In each case, form an equation to represent the information given. Solve it to find x.

   a

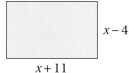

$x - 4$

$x + 11$

The area of the square and rectangle are equal.

$x + 3$

   b

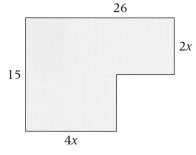

26

$2x$

15

The lengths of the unmarked sides are equal.

$4x$

This spread will show you how to:

- Solve quadratic equations by factorisation

Keywords
Factorise
Product
Quadratic
Solution
Sum

- A **quadratic** equation contains an x^2 term as the highest power, for example
$$x^2 + 5x + 6 = 0$$

Many quadratic equations can be solved by **factorising**.

For example, solve the equation $x^2 + 5x + 6 = 0$.

$x^2 + 5x + 6 = 0$ The two factors will each start with x.

$(x + \ )(x + \ ) = 0$ Now find two numbers with a **sum** of 5 (for $5x$) and a **product** of 6.

$(x + 3)(x + 2) = 0$ Check; $(x + 3)(x + 2) = x^2 + 3x + 2x + 6 = x^2 + 5x + 6$

x^2 term, so equation is a quadratic.

The two factors have a **product** of zero. This means that at least one of them is equal to zero.

Either $x + 3 = 0$ or $x + 2 = 0$
If $x + 3 = 0$, $x = -3$ and if $x + 2 = 0$, $x = -2$ Check $(-3)^2 + 5(-3) + 6 = 9 - 15 + 6 = 0$
The **solutions** are $x = -3$ and $x = -2$.   and $(-2)^2 + 5(-2) + 6 = 4 - 10 + 6 = 0$

- The key to solving quadratic equations is that the product of the two factors is zero, so you can say that one or other (or both) of the factors is zero. This leads to the two solutions.

Example

Solve these equations.

a $x^2 - 7x + 12 = 0$ **b** $x^2 + 3x = 0$

a $x^2 - 7x + 12 = 0$ To factorise, you need two negatives to multiply to make +12
    $(x - 3)(x - 4) = 0$ and add to make -7.
    Either $x - 3 = 0$ or $x - 4 = 0$
             $x = 3$ $x = 4$

b $x^2 + 3x = 0$
    $x(x + 3) = 0$
    Either $x = 0$ or $x + 3 = 0$
                        $x = -3$

Don't forget to look out for common factors as well as double bracket style factorisation.

Example

Solve $x^2 = 2x + 15$

$x^2 - 2x - 15 = 0$ Find two numbers with sum -2 and product -15.
$(x + 3)(x - 5) = 0$
Either $x + 3 = 0$ or $x - 5 = 0$
        $x = -3$ $x = 5$

First make the left hand side of the equation equal to zero

1 Here are six equations.

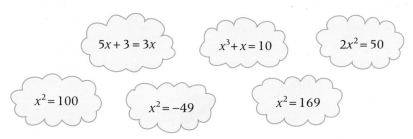

$$5x + 3 = 3x$$

$$x^3 + x = 10$$

$$2x^2 = 50$$

$$x^2 = 100$$

$$x^2 = -49$$

$$x^2 = 169$$

a Write the four quadratic equations.

b From the quadratic equations, find three that have two solutions. Write these three equations and their solutions.

2 Factorise these quadratic expressions.

a $x^2 + 8x + 12$ **b** $x^2 + 11x + 24$ **c** $x^2 + 13x + 36$

d $x^2 + 16x + 55$ **e** $x^2 - 7x + 12$ **f** $x^2 - 10x + 24$

g $x^2 + 3x - 28$ **h** $x^2 - 2x - 15$

3 Solve these quadratic equations by factorising them into double brackets.

a $x^2 + 7x + 12 = 0$ **b** $x^2 + 8x + 12 = 0$ **c** $x^2 + 10x + 25 = 0$

d $x^2 + 5x - 14 = 0$ **e** $x^2 - 4x - 5 = 0$ **f** $x^2 - 5x + 6 = 0$

g $x^2 - 10x + 21 = 0$ **h** $x^2 = 3x + 40$

4 Solve these quadratic equations by factorising them into a single bracket.

a $x^2 - 8x = 0$ **b** $x^2 + 4x = 0$ **c** $x^2 - 6x = 0$

d $y^2 + 5y = 0$ **e** $x^2 = 9x$ **f** $x^2 - 12x = 0$

g $2x^2 + 8x = 0$ **h** $6x - x^2 = 0$

5 Explain why $x^2 + 7x + 11 = 0$ cannot be solved by factorisation.

6 **a** Vicky is trying to solve $(x + 4)(x - 2) = 7$. Here is her attempt.

What is wrong with her method?

> $(x + 4)(x - 2) = 7$
> Either $x + 4 = 7$ or $x - 2 = 7$
> $x = 3$ or $x = 9$

b Can you solve this equation correctly to show that the two solutions should really be $x = 3$ and $x = -5$?

7 Simplify and then solve this equation.

$$x^2 + 4x - 5 = 2x(x - 1)$$

This spread will show you how to
- Use trial and improvement to find approximate solutions of equations
- Substitute positive and negative numbers into expressions

Keywords
Approximation
Cubic
Systematic
Trial and
improvement

Some equations don't have exact solutions and can't be solved by an algebraic method. This includes most **cubic** equations.

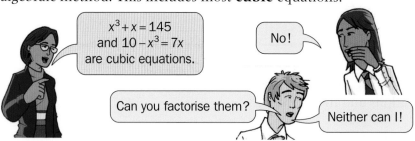

$x^3 + x = 145$ and $10 - x^3 = 7x$ are cubic equations.

No!

Can you factorise them?

Neither can I!

- You can find an approximate solution, correct to several decimal places if necessary, using **trial and improvement**.

Example

Solve the equation $x^3 + x = 145$.
Give your answer correct to 1 dp.

Put your trials in a table.
Be **systematic**.

Start with whole numbers then find a better approximation.

x	x^3	$x^3 + x$	
5	125	130	Too small
6	216	222	Too big
5.5	166.375	171.875	Too big
5.1	132.651	137.751	Too small
5.2	140.608	145.808	Too big
5.15	136.59088	141.74088	Too small

The solution is between $x = 5.1$ and $x = 5.2$.

Test the x-value half way between 5.1 and 5.2.

The solution is between $x = 5.15$ and $x = 5.2$, so $x = 5.2$ correct to 1 dp.

You need to work to one more dp than is required in the answer.

Example

The volume of this cuboid is 20 cm³.
Find its dimensions, correct to 1 dp.

x

x

$x + 1$

Volume $= x \times x \times (x + 1) = x^2(x + 1) = x^3 + x$
So $x^3 + x = 20$

The solution is between $x = 2.55$ and $x = 2.6$, so $x = 2.6$ to 1 dp.
The dimensions of the cuboid are 3.6 cm, 2.6 cm and 2.6 cm.

x	x^3	$x^3 + x$	
2	8	10	Too small
3	27	30	Too big
2.5	15.625	18.125	Too small
2.6	17.576	20.176	Too big
2.55	16.581375	19.131375	Too small

1 Solve the equation, $x^3 - x^2 = 200$, by trial and improvement to find the solution correct to 1 dp.

2 Write an equation to solve each problem. Use trial and improvement to find the exact answer.

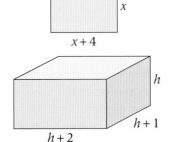

   a The area of this rectangle is 77 cm². What are the length and width?

   b The volume of this cuboid is 990 cm³. What are the dimensions?

3 **a** Solve $x^2 + 2x = 15$ using trial and improvement in order to find the *exact* solution of this equation.

   b Repeat **a** using an algebraic method and comment on which method you feel is most efficient.

4 Use trial and improvement to solve these equations, giving your answer to the stated degree of accuracy.

   a $x^2 - x = 69$ (to 1 decimal place)

   b $2x^3 + x = 197$ (to 1 decimal place)

   c $p(p + 1) = 100$ (to 2 decimal places)

   d $w^3 - 2w = 70$ (to 2 decimal places)

5 Write an equation to represent the given information. Use trial and improvement to find the value of the unknown in each case.

   a The product of three consecutive numbers is 85 140.

   b The area of a rectangle is 57 cm². Its length is 2 cm more than its width. Find the width to one decimal place.

   c The surface area of this cuboid is 500 mm². Find x to 1 decimal place.

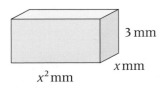

6 Find the number, to 2 dp, that satisfies this statement.

> The square of this number is 20 times its cube root.

This spread will show you how to:
- Solve simultaneous equations by eliminating a variable

Keywords

Eliminate
Simultaneous
Solution

- Two equations that have the same solution are called **simultaneous** equations.
  For example,
  $$x + y = 8$$
  $$\text{and } x - y = 2$$
- You can solve simultaneous equations using algebra.

The solution is two numbers that add to 8 and with a difference of 2. They must be $x = 5$ and $y = 3$

Example

Solve the simultaneous equations: $3x + 2y = 12$ and $3x + 8y = 30$.

$3x + 2y = 12$ (1)	Label the equations (1) and (2).
$3x + 8y = 30$ (2)	The x terms are identical so you can **eliminate** them.

$$(3x + 8y) - (3x + 2y) = 30 - 12 \quad \text{Subtract equation (1) from equation (2).}$$
$$8y - 2y = 18$$
$$6y = 18$$
$$y = 3$$

$$3x + 6 = 12 \qquad \text{Substitute 3 for } y \text{ in equation (1).}$$
$$3x = 6$$
$$x = 2 \qquad \text{Check: } 3 \times 2 + 2 \times 3 = 12 \text{ and } 3 \times 2 + 8 \times 3 = 30$$

Write the equations one under the other so that you can compare them.

Subtract because the x terms have the **s**ame **s**ign. (SSS)

Example

Solve $4x - 3y = 5$ (1)
         $8x + 3y = 1$ (2)

$$(4x - 3y) + (8x + 3y) = 5 + 1 \qquad \text{In equation (1) } 3y \text{ is negative, in equation (2) } 3y \text{ is positive,}$$
$$12x = 6 \qquad\qquad\qquad \text{so add the equations to eliminate the } y\text{-terms.}$$
$$x = \tfrac{1}{2}$$
$$8\tfrac{1}{2} + 3y = 1$$
$$4 + 3y = 1 \qquad \text{Substitute } \tfrac{1}{2} \text{ for } x \text{ in equation (2).}$$
$$3y = -3 \text{ so } y = -1 \quad \text{Check in equation (1): } 4 \times \tfrac{1}{2} - 3 \times (-1) = 5$$

Example

In a sweet shop, I spend £3.20 on three cans of soft drink and four bars of chocolate. The next day, I buy a can of soft drink and four bars of chocolate for £2. How much does each item cost?

Choose letters for the variables.

Let c be the number of cans of drink I buy and b be the chocolate.

$$3c + 4b = 320 \quad (1) \qquad\qquad c + 4b = 200 \quad (2)$$

The differences between the two equations are 2 cans and 120p.
So, $2c = 120$ and $c = 60$.

$$180 + 4b = 320 \qquad \text{Substitute 60 for } c \text{ in (1).}$$
$$4b = 140, \text{ so } b = 35 \quad \text{Subtract (2) from (1).}$$

A can of soft drink costs 60p and a bar of chocolate costs 35p.

Change from £ to pence to avoid working with decimals

Exercise A6.4

1 Which pairs of equations have the solution $x = 2$ and $y = 7$?

a
$$x + y = 9$$
$$x - y = -5$$

b
$$2x + y = 11$$
$$3x - y = -1$$

c
$$2x + 2y = 15$$
$$4x - y = 1$$

d
$$x + 2y = 16$$
$$x - 2y = 8$$

e
$$5x - y = 3$$
$$y - x = 5$$

2 Solve these pairs of simultaneous equations by subtracting one equation from the other.

a $3x + y = 15$
  $x + y = 7$

b $6x + 2y = 6$
  $4x + 2y = 2$

c $x + 5y = 19$
  $x + 7y = 27$

d $5x + 2y = 16$
  $x + 2y = 4$

e $m + 3n = 11$
  $m + 2n = 9$

f $4x + 3y = -5$
  $7x + 3y = -11$

3 Solve these pairs of simultaneous equations by adding one equation to the other.

a $3x + 2y = 19$
  $8x - 2y = 58$

b $5x + 2y = 16$
  $3x - 2y = 8$

c $7a - 3b = 24$
  $2a + 3b = 3$

d $2x + 3y = 19$
  $-2x + y = 1$

e $4x - 7y = 15$
  $2x + 7y = 4\frac{1}{2}$

f $6p - 2q = -2$
  $6p + 2q = 26$

4 Solve these pairs of simultaneous equations by either adding or subtracting in order to eliminate one variable.

a $x + y = 3$
  $3x - y = 17$

b $5x - 2y = 4$
  $3x + 2y = 12$

c $5a + b = -7$
  $5a - 2b = -16$

d $3v + w = 14$
  $3v - w = 10$

e $20p - 4q = 32$
  $7p + 4q = 22$

f $3x - 2y = 11$
  $3x + 4y = 23$

5 Solve these simultaneous equations.

$$a = 2b + 7$$
$$a + b - 1 = 10$$

$$6w = 38 - 2v$$
$$5w = 6 + 2v$$

6 Solve these problems by using simultaneous equations.

a How much does a lemon cost?

| 3 lemons |
| 4 oranges |
| £1.27 |

| 4 oranges |
| 5 lemons |
| £1.61 |

b The perimeter of this triangle is 30 cm. How long is the base?

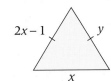

This spread will show you how to:
- Solve simultaneous equations by eliminating a variable
- Use notation and symbols correctly and consistently within a given problem

Keywords

Coefficient
Eliminate
Simultaneous
Solution

Sometimes the **coefficients** of the x-terms (or the y-terms) in simultaneous equations are not the same.

You have to adjust the equations before you can **eliminate** the x-terms (or the y-terms).

$$3x + 5y = 4 \quad (1)$$
$$2x - y = 7 \quad (2)$$

If I multiply equation (2) by 5, I'll have $-5y$.

Now you can add the equations to eliminate the y-terms.

$$3x + 5y = 4$$
$$10x - 5y = 35$$

Example

Solve the simultaneous equations $3x + 5y = 4$
$$2x - y = 7$$

$3x + 5y = 4$	(1)	Label the equations.
$2x - y = 7$	(2)	
$10x - 5y = 35$	(3)	Multiply equation (2) by 5.
$(3x + 5y) + (10x - 5y) = 4 + 35$		Add equations (1) and (3) to eliminate the y-terms.
$13x = 39$		
$x = 3$		
$9 + 5y = 4$		Substitute 3 for x in equation (1).
$y = -1$		

To get $5y$ in both equations

Check the **solution** in equation (2)
$2 \times 3 - (-1) = 6 + 1 = 7$

Example

The difference between two numbers is 4. Treble the smaller number, subtract double the larger number is 1. What are the numbers?

Let the larger number be x and the smaller be y.

$x - y = 4$	(1)	
$3y - 2x = 1$	(2)	
$-2x + 3y = 1$	(2)	Rearrange equation (2).
$3x - 3y = 12$	(3)	Multiply equation (1) by 3.
$(-2x + 3y) + (3x - 3y) = 1 + 12$		Add equations (2) and (3), to eliminate the y-terms.
$x = 13$		
$13 - y = 4$		Substitute 13 for x in equation (1).
$y = 9$		

To get $3y$ in both equations

Check in equation (2)
$3 \times 9 - 2 \times 13 = 1$

The numbers are 9 and 13.

1 Solve these simultaneous equations.

a $2x + y = 8$
  $5x + 3y = 12$

b $3x + 2y = 19$
  $4x - y = 29$

c $8a - 3b = 30$
  $3a + b = 7$

d $2v + 3w = 12$
  $5v + 4w = 23$

e $9p + 5q = 15$
  $3p - 2q = -6$

f $3x - 2y = 11$
  $2x - y = 8$

2 Make as many pairs of simultaneous equations as you can using these three cards. Solve your pairs.

| $2x + y = 12$ | $y - x = 15$ | $3x - 4y = 7$ |

3 Solve these simultaneous equations.
Remember to clear the fractions first.

a $\frac{x}{3} - \frac{y}{4} = \frac{3}{2}$
  $2x + y = 14$

b $\frac{a}{2} + 3b = 1$
  $5a - 7b = 47$

c $p - \frac{2q}{3} = \frac{26}{3}$
  $\frac{p}{4} + 3q + 1 = 0$

4 Write a pair of simultaneous equations to solve each problem.

a Two numbers have a sum of 41 and a difference of 7.
What numbers are they?

b One number is 6 more than another. Their mean average is 20.
What numbers are they?

c 230 students and 29 staff are going on a school trip. They travel by large and small coaches. The large coaches seat 55 and the small coaches seat 39. If there are no spare seats and five coaches are to make the journey, how many of each coach are used?

 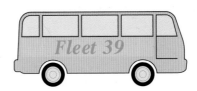

d In an isosceles triangle, the largest angle is 30° more than double the equal angles. What are the angles in the triangle?

e Uncle Jack gave me a £25 book token at Christmas. At Firestone's Bookshop I can use the token to buy exactly 3 paperbacks and 1 hardback book or 1 paperback and 2 hardbacks.

Find the cost of a paperback and a hardback.

Exam review

Key objectives

- Manipulate algebraic expressions; factorise quadratic expressions, including the difference of two squares, and cancel common factors in rational expressions
- Solve quadratic equations by factorisation
- Use systematic trial and improvement to find approximate solutions of equations
- Solve exactly, by elimination of an unknown, two simultaneous equations in two unknowns

1 Solve these simultaneous equations. (4)

$$5x - 2y = 16$$

$$3x + y = 14$$

2 The equation $x^3 - 2x = 67$

has a solution between 4 and 5.

Use a trial and improvement method to find this solution.

Give your answer correct to one decimal place.

You must show **ALL** your working. (4)

(Edexcel Ltd., 2004)

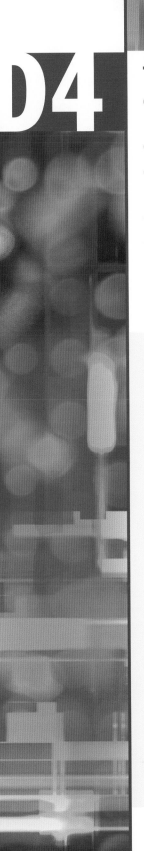

This unit will show you how to

- Use frequency tables to find the averages and range of a data set
- Use grouped frequency tables
- Use estimates of averages and range to summarise large data sets
- Draw frequency polygons
- Use frequency polygons to compare two data sets
- Draw time series graphs
- Calculate moving averages

Before you start ...

You should be able to answer these questions.

Review

1 Write the midpoint of

  a 5 and 15   **b** 20 and 30

  c 0 and 40   **d** 10 and 15

Key stage 3

2 Work out

  a $\frac{1}{2}$ of 32   **b** $\frac{1}{2}$ of 45

  c $\frac{1}{2}$ of $(27 + 1)$ **d** $\frac{1}{2}$ of $(36 + 1)$

Key stage 3

3 Work out

  a $\frac{1}{4}$ of 24   **b** $\frac{1}{4}$ of 18

  c $\frac{1}{4}$ of $(27 + 1)$ **d** $\frac{1}{4}$ of $(41 + 1)$

Key stage 3

4 Work out

  a $\frac{3}{4}$ of 36   **b** $\frac{3}{4}$ of 44

  c $\frac{3}{4}$ of $(27 + 1)$ **d** $\frac{3}{4}$ of $(13 + 1)$

Key stage 3

Large data sets – averages and range

This spread will show you how to:

- Use frequency tables to find the averages and range of a data set
- Use estimates of averages and range to summarise large data sets

Keywords

Frequency table
IQR
Mean
Median
Mode
Range

- You can put large amounts of data in a **frequency table**.
- You use the averages and the **range** to summarise the data.

Example

The table shows the length of the words in the answers to a crossword puzzle.

For these data, work out the

a mode
b median
c mean
d IQR
e range.

Word length	Frequency
4	3
5	5
6	7
7	8
8	3
9	1

The techniques covered in this unit may be useful for your statistical coursework task.

IQR means interquartile range (see **D1.4**)

Word length	Frequency	Word length × frequency
4	3	$4 \times 3 = 12$
5	5	$5 \times 5 = 25$
6	7	$6 \times 7 = 42$
7	8	$7 \times 8 = 56$
8	3	$8 \times 3 = 24$
9	1	$9 \times 1 = 9$
Total	**27**	**168**

Total number of words.
Total number of letters.

a Mode = 7   Words with 7 letters have the highest frequency.

b $\frac{1}{2}(27 + 1) = 14$ so the 14th value is the median
The 14th value is in the 'Word length 6' group.
Median = 6

c Mean = $\dfrac{\text{Total number of letters}}{\text{Total number of words}}$
    $= 168 \div 27 = 6.2$

d Lower quartile is $\frac{1}{4}(27 + 1)$th value = 7th value = 5

Upper quartile is $\frac{3}{4}(27 + 1)$th value = 21st value = 7

IQR = 7 − 5
    = 2

e Range = 9 − 4 = 5   Longest − shortest word length

1 The tables of data give information about the length of words in four different crosswords.

For each table, copy each table, add an extra working column and find the

i mode   **ii** median   **iii** mean   **iv** range   **v** interquartile range.

a

Word length	Frequency
4	2
5	5
6	4
7	2
8	2

b

Word length	Frequency
3	3
4	4
5	9
6	5
7	2

c

Word length	Frequency
4	6
5	3
6	5
7	4
8	2
9	5
10	2

d

Word length	Frequency
3	5
4	4
5	6
6	7
7	7
8	4
9	2

2 Jo had 16 boxes of matches.
She counted the number of matches in each box.
The table gives her results.

Number of matches	Frequency
41	2
42	7
43	4
44	3

Work out the mean number of matches in a box.

3 Brian played in 24 hockey matches one season.
The table gives information about the number of goals scored in these matches.

Number of goals scored	Frequency
1	6
2	9
3	4
4	2
5	3

Work out the mean number of goals scored.

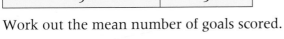

This spread will show you how to:

● Use grouped frequency tables

Keywords
Grouped
  frequency table
Mean
Median
Midpoint
Modal class

● You can put large amounts of continuous data into a **grouped frequency table**.

A grouped frequency table does not tell you the actual data values so you can only find estimates of the averages.

● You use estimates of averages to summarise the data.

The table shows the time taken, to the nearest minute, by a group of students to solve a crossword puzzle.

Time, t, minutes	Frequency
$5 < t \leqslant 10$	2
$10 < t \leqslant 15$	14
$15 < t \leqslant 20$	13
$20 < t \leqslant 25$	6
$25 < t \leqslant 30$	1

For these data, work out an estimate for the

a modal class
b class containing the median
c mean

The **modal class** is the class with the greatest frequency.

a Modal class = $10 < t \leqslant 15$

b Class containing median = $15 < t \leqslant 20$

c

Time t minutes	Frequency	Midpoint	Word length × frequency
$5 < t \leqslant 10$	2	7.5	$7.5 \times 2 = 15$
$10 < t \leqslant 15$	14	12.5	$12.5 \times 14 = 175$
$15 < t \leqslant 20$	13	17.5	$17.5 \times 13 = 227.5$
$20 < t \leqslant 25$	6	22.5	$22.5 \times 6 = 135$
$25 < t \leqslant 30$	1	27.5	$27.5 \times 1 = 27.5$
Total	**36**		**580**

Total number of students   Total time

Put two extra columns in the table.

Find the totals of the Frequency and Word length × frequency columns.

$$\text{Mean} = \frac{\text{Estimated total time}}{\text{Total number of students}}$$

$$\text{Mean} = \frac{580}{36} = 16.1$$

1 The grouped frequency tables give information about the time taken to solve four different crosswords.
For each table, copy the table, add extra working columns and find

   i   the modal class   **ii** the class containing the median

   iii an estimate of the mean

a

Time, *t*, minutes	Frequency
$5 < t \leqslant 10$	2
$10 < t \leqslant 15$	14
$15 < t \leqslant 20$	13
$20 < t \leqslant 25$	6
$25 < t \leqslant 30$	1

b

Time, *t*, minutes	Frequency
$0 < t \leqslant 10$	3
$10 < t \leqslant 20$	6
$20 < t \leqslant 30$	4
$30 < t \leqslant 40$	5
$40 < t \leqslant 50$	2

c

Time, *t*, minutes	Frequency
$5 < t \leqslant 10$	8
$10 < t \leqslant 15$	5
$15 < t \leqslant 20$	7
$20 < t \leqslant 25$	4
$25 < t \leqslant 30$	0
$30 < t \leqslant 35$	1

d

Time, *t*, minutes	Frequency
$5 < t \leqslant 15$	3
$15 < t \leqslant 25$	9
$25 < t \leqslant 35$	7
$35 < t \leqslant 45$	8
$45 < t \leqslant 55$	2
$55 < t \leqslant 65$	1

2 Alfie kept a record of his monthly mobile phone bills for one year.

Phone bill, *b*, pounds	Frequency
$10 < b \leqslant 20$	6
$20 < b \leqslant 30$	2
$30 < b \leqslant 40$	3
$40 < b \leqslant 50$	1

a Write the class interval that contains the median.

b Calculate an estimate for the mean cost of Alfie's mobile phone bill.

3 The heights of 50 Year 10 students were measured. The results are shown in the table.

Height, *h*, cm	Number of students
$150 \leqslant h < 155$	3
$155 \leqslant h < 160$	5
$160 \leqslant h < 165$	15
$165 \leqslant h < 170$	25
$170 \leqslant h < 175$	2

a What is the modal group?

b Estimate the mean height.

c Which class interval contains the median?

This spread will show you how to:

Keywords
Class interval
Frequency
  polygon
Midpoint
Modal

- Draw frequency polygons
- Use frequency polygons to compare two data sets

- You can represent grouped data in a **frequency polygon**.

To draw a frequency polygon for continuous data you plot the **midpoint** of each **class interval** against the frequency.

Example

The tables show the ages of people attending concerts to see the bands Badness and Cloudplay.

Badness

Age, a, years	Frequency
$20 < a \leqslant 30$	1600
$30 < a \leqslant 40$	4300
$40 < a \leqslant 50$	2100
$50 < a \leqslant 60$	1000

Cloudplay

Age, a, years	Frequency
$10 < a \leqslant 20$	2800
$20 < a \leqslant 30$	4600
$30 < a \leqslant 40$	3300
$40 < a \leqslant 50$	1200

a Draw frequency polygons for these data.
b Make comparisons, with reasons, between the ages of people attending these concerts.

a

Midpoint	25	35	45	55
Frequency	1600	4300	2100	1000

b

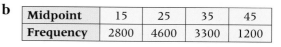

Midpoint	15	25	35	45
Frequency	2800	4600	3300	1200

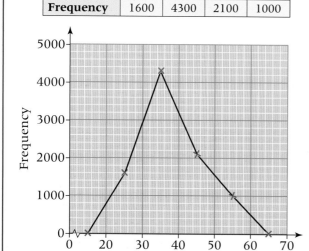

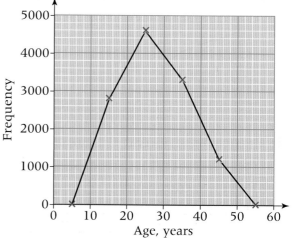

b The **modal** age is greater at Badness concerts than Cloudplay concerts. The highest frequency for Badness is in the class interval 30–40 years old, whereas the highest frequency for Cloudplay is in the interval 20–30 years old.

1 The tables show the ages of the first 100 people to visit a garden centre on a weekday and a Sunday.

Weekday

Age, *a*, years	Frequency
$0 < a \leqslant 20$	16
$20 < a \leqslant 40$	28
$40 < a \leqslant 60$	32
$60 < a \leqslant 80$	24

Sunday

Age, *a*, years	Frequency
$0 < a \leqslant 20$	22
$20 < a \leqslant 40$	45
$40 < a \leqslant 60$	18
$60 < a \leqslant 80$	15

a Draw frequency polygons for these data.

b Find, for each data set, the class which contains the modal age.

c Compare the ages of people at the garden centre on a weekday and a Sunday.

2 Jayne kept a daily record of the number of miles she travelled in her car during two months.

December

Miles travelled, *m*	Frequency
$0 < m \leqslant 20$	3
$20 < m \leqslant 40$	8
$40 < m \leqslant 60$	10
$60 < m \leqslant 80$	6
$80 < m \leqslant 100$	4

January

Miles travelled, *m*	Frequency
$0 < m \leqslant 20$	0
$20 < m \leqslant 40$	5
$40 < m \leqslant 60$	12
$60 < m \leqslant 80$	8
$80 < m \leqslant 100$	6

a Draw frequency polygons for these data.

b Find, for each month, the class which contains the modal number of miles

c Compare the number of miles Jayne travelled in December and January.

3 David carried out a survey to find the time taken by 120 teachers and 120 office workers to travel home from work.

Teachers

Time taken, *t*, minutes	Frequency
$0 < t \leqslant 10$	12
$10 < t \leqslant 20$	33
$20 < t \leqslant 30$	48
$30 < t \leqslant 40$	20
$40 < t \leqslant 50$	7

Office workers

Time taken, *t*, minutes	Frequency
$10 < t \leqslant 20$	2
$20 < t \leqslant 30$	21
$30 < t \leqslant 40$	51
$40 < t \leqslant 50$	28
$50 < t \leqslant 60$	18

a Draw frequency polygons for these data.

b Work out for each data set the class which contains the modal time taken.

c Make comparisons between the time taken by the teachers and office workers to travel home from work.

This spread will show you how to:

- Draw time series graphs

- **Time series** data are collected over a period of time at regular intervals.

For example,
Electricity and gas bills are produced
every quarter (three months).
Mobile phone bills are generated each month.
Unemployment rates are published each month.

- You can plot time series data on a graph with time on the horizontal axis.

Example

Jenny's quarterly gas bills over a period of two years are shown in the table.

	Jan–March	April–June	July–Sept	Oct–Dec
2003	£65	£38	£24	£60
2004	£68	£42	£30	£68

Draw a time series graph to represent these data.

Draw axes on graph paper with time on the horizontal axis.
Plot the coordinates as crosses on the grid.
Join them up with straight lines.

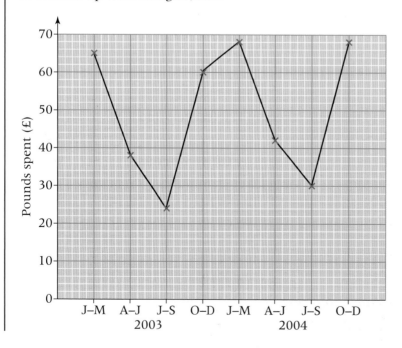

On the horizontal axis,
J–M means Jan–March.

Exercise D4.4

1 The table shows Ken's monthly mobile phone bills.

Jan	Feb	Mar	April	May	June	July	Aug	Sept	Oct	Nov	Dec
£16	£12	£15	£18	£16	£18	£12	£10	£12	£15	£16	£20

Draw a time series graph for these data.

2 The table shows Mary's quarterly electricity bills over a two-year period.

	Jan–March	April–June	July–Sept	Oct–Dec
2004	£45	£20	£15	£48
2005	£54	£24	£18	£50

Draw a time series graph to represent these data.

You will need the graphs for questions **2, 5** and **6** for Exercise D4.5.

3 The table shows monthly ice-cream sales at Angelo's shop during one year.

Jan	Feb	Mar	April	May	June	July	Aug	Sept	Oct	Nov	Dec
£16	£12	£15	£18	£38	£48	£52	£58	£18	£15	£16	£40

Draw a time series graph for these data.

4 A town council carried out a survey over a number of years to find the percentage of local teenagers who used the town's library. The table shows the results.

Year	1998	1999	2000	2001	2002	2003	2004	2005
%	14	18	24	28	25	20	18	22

Draw a time series graph for these data.

5 Christabel kept a record of how much money she had earned from babysitting during three years.

	January–April	May–August	Sept–Dec
2001	£12	£18	£30
2002	£21	£33	£60
2003	£39	£42	£72

Draw a time series graph for these data.

6 Steve kept a record of his quarterly expenses over a period of two years.

	Jan–March	April–June	July–Sept	Oct–Dec
2003	£35	£56	£27	£12
2004	£39	£68	£29	£18

Draw a time series graph to represent these data.

Time series and moving averages

This spread will show you how to:

- Calculate moving averages

- **Time series** data can vary depending on the month or season.

For example,

Electricity and gas bills are likely to be higher in the winter months.
Unemployment rates can be higher in the summer when students leave school.

- You can find a **moving average** of time series data to help you spot a trend.

- You calculate moving averages over a whole cycle.

Example

Jenny drew this table for her quarterly gas bills over a period of two years.

	Jan–March	April–June	July–Sept	Oct–Dec
2003	£65	£38	£24	£60
2004	£68	£42	£30	£68

a Calculate 4-point moving averages for these data.
b Plot the moving averages on a graph.
c Describe the trend shown by the moving averages.

A year (4 quarters) is one cycle, so 4-point moving averages are calculated.

a

Jan–Mar 2003	Apr–June 2003	July–Sept 2003	Oct–Dec 2003	Jan–Mar 2004	Apr–June 2004	July–Sept 2004	Oct–Dec 2004
£65	£38	£24	£60	£68	£42	£30	£68

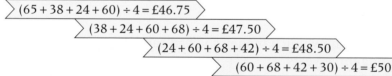

$(65 + 38 + 24 + 60) \div 4 = £46.75$
$(38 + 24 + 60 + 68) \div 4 = £47.50$
$(24 + 60 + 68 + 42) \div 4 = £48.50$
$(60 + 68 + 42 + 30) \div 4 = £50$
$(68 + 42 + 30 + 68) \div 4 = £52$

To calculate 4-point averages, calculate the average of each group of 4 consecutive values.

b

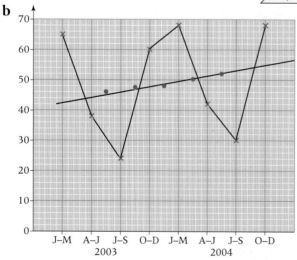

trend line

Plot each value at the midpoint of the group of 4.

c The trend shown by the moving averages is that, in general, gas bills are increasing.

1 For the data given in Exercise D4.4 question **1**, calculate the three-point moving averages.

2 **a** For the data given in Exercise D4.4 question **2**, calculate four-point moving averages.

b Plot the moving averages on the graph you have already drawn for Exercise D4.4 question **2**.

c Describe the trend shown by the moving averages.

3 For the data given in Exercise D4.4 question **3**, calculate four-point moving averages.

4 For the data given in Exercise D4.4 question **4**, calculate

a two-point moving averages

b five-point moving averages.

5 **a** For the data given in Exercise D4.4 question **5**, calculate three-point moving averages.

b Plot the moving averages on the graph you have already drawn for Exercise D4.4 question **5**.

c Describe the trend shown by the moving averages.

6 **a** For the data given in Exercise D4.4 question **6**, calculate four-point moving averages.

b Plot the moving averages on the graph you have already drawn for Exercise D4.4 question **6**.

c Describe the trend shown by the moving averages.

7 Audrey recorded the number of cups of tea sold over three weeks to office staff at afternoon break.

	Monday	Tuesday	Wednesday	Thursday	Friday
Week 1	62	55	36	47	68
Week 2	60	54	35	47	70
Week 3	61	55	37	50	73

a Draw a time series graph to represent these data.

Audrey wants to find out about the trend in the number of cups of tea sold.

b Explain why it would be appropriate for Audrey to calculate five-point moving averages.

c Calculate the five-point moving averages.

d Plot the moving averages on the graph.

e Describe the trend shown by the moving averages.

Exam review

Key objectives

- Draw and produce pie charts for categorical data, and diagrams for continuous data
- Identify the modal class for grouped data
- Find the median, quartiles and interquartile range for large data sets
- Calculate the mean for large data sets with grouped data
- Calculate an appropriate moving average

1 The table shows the number of pets owned by a class of 20 students.

Number of pets	Number of students
0	3
1	7
2	5
3	4
4	1

Calculate the mean number of pets owned. (3)

2 A garage keeps records of the costs of repairs to its customers' cars. The table gives information about the costs of all repairs which were less than £250 in one week.

Cost (£C)	Frequency
$0 < C \leqslant 50$	4
$50 < C \leqslant 100$	8
$100 < C \leqslant 150$	7
$150 < C \leqslant 200$	10
$200 < C \leqslant 250$	11

a Find the class interval in which the median lies. (2)

There was only one further repair that week, not included in the table. That repair cost £1000.
Dave says 'The class interval in which the median lies will change.'

b Is Dave correct? Explain your answer. (1)

c The garage also sells cars. It offers a discount of 20% off the normal price for cash. Dave pays £5200 cash for a car.
Calculate the normal price of the car. (3)

(Edexcel Ltd., 2003)

This unit will show you how to

- Use geometry to solve problems involving bearings
- Use and interpret maps and scale drawings
- Use a ruler and compasses to draw standard constructions
- Find loci by reasoning and using diagrams

Before you start ...

You should be able to answer these questions.

1 Use a protractor to measure these angles.

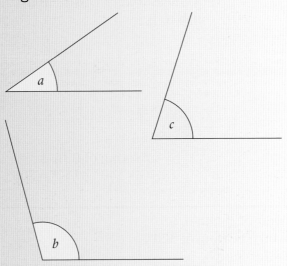

Review

Key stage 3

2 Use compasses to draw

a a circle with radius 3 cm

b an arc with radius 5 cm.

Key stage 3

3 a Draw a square *ABCD* with sides 4 cm.

b Draw eight arcs, each with radius 3 cm, centred on each of the four corners of *ABCD*.

c Join together the points where the arcs cut the sides of the square to form an octagon.

Key stage 3

This spread will show you how to:
- Use geometry to solve problems involving bearings
- Use and interpret maps and scale drawings

Keywords
Angle
Bearing

• A bearing is an **angle** measured in a clockwise direction from north.

You always write bearings with 3 digits, for example 070°, 190°, 230°

To find the bearing of A from B

- Imagine you are standing at B, facing north.
- Turn clockwise until you face A.

The angle you have turned through is the bearing of A from B.

The bearing of A from B is 256°.

Example

The bearing of G from B is 028°. Find the bearing of B from G.

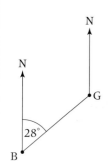

Bearing of B from G is the angle at G measured clockwise from north to B.

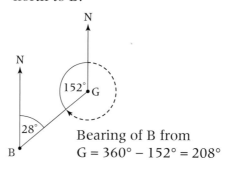

Bearing of B from G = 360° − 152° = 208°

Interior angles are supplementary.

Example

A church, C, is 10 km due west of a school, S.
Joe is 6 km from the school on a bearing of 320°.
He wants to walk directly to the church.

Draw a diagram to show the positions of Joe, the church and the school, and use it to find the bearing Joe should take.
Use a scale of 1 cm to 2 km.

Label point S.
Draw C 5 cm west of S.
Draw the north line at S.
Measure and draw the 320° bearing from S and, 3 cm from S, mark a point, J, to show Joe's position.

Draw the line JC.
Draw the north line at J.
Measure the clockwise angle between the north line and JC.
The bearing Joe needs is 233°.

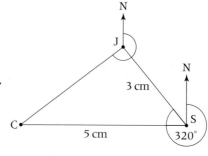

Scale 1 cm to 2 km, so represent 10 km by a line 5 cm long.

Represent 6 km by a line 3 cm long.

Draw and measure lengths and angles carefully or your answer will be inaccurate.

1 These diagrams are drawn accurately.
Measure the bearing of T from S in each.

a N ⋅T
S

b T⋅ N
S

2 P and Q are points 2 cm apart. Draw diagrams to show the position
of points P and Q where the bearing of Q from P is

a 070°   **b** 155°   **c** 340°   **d** 260°

3 These diagrams have *not* been drawn accurately.
Find the bearing of X from Y in each case.

a N
136°
X
⋅Y

b N
95°
X   ⋅Y

c N
X
248°
Y⋅

4 Find these bearings.

a The bearing of A from B is 104°. Work out the bearing of B from A.

b The bearing of E from F is 083°. Work out the bearing of F from E.

c The bearing of J from K is 297°. Work out the bearing of K from J.

> Draw a sketch to help you.

5 A youth club (Y) is 4 km due east of a school (S).
Hazel leaves school and walks 5 km on a bearing of 042° to her
house (H).

a Make a scale drawing to show the position of Y, S and H.
Use a scale of 1 cm to 1 km.

b Hazel walks directly from her house to the youth club.
What bearing does she take?

6 A lighthouse, L, is 6 km on a bearing of 160° from a point H at
the harbour.
A boat, B, is 3 km from L on a bearing of 125°.

a i Make a scale drawing to show the positions of L, H and B.

ii What bearing should B travel to go directly to H?

The boat moves 4 km due west.

b i Mark on your drawing the new position of B.

ii What bearing should B now travel to go directly to H?

DID YOU KNOW?

At Fleetwood in
Lancashire, boats
navigate from the
River Wyre to the
sea by being
positioned in relation
to three lighthouses.

Constructing triangles

This spread will show you how to:

● Use a ruler and compasses to draw standard constructions

Keywords
Arc
Compasses
Radius
Triangle

You can **construct** a unique triangle when you know

| Two sides and the angle between them (SAS) | or | Two angles and a side (ASA) | or | Right angle, the hypotenuse and a side (RHS) | or | Three sides (SSS) |

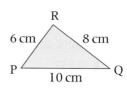

You will need a ruler and a protractor for SAS, ASA and RHS triangles.

You will need a ruler and compasses for SSS triangles.

Example

a Construct this equilateral triangle ABC with side length 4 cm.

b Construct another equilateral triangle with base AB and side length 4 cm.

c What special quadrilateral have you drawn?

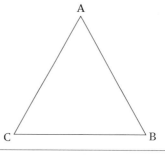

a

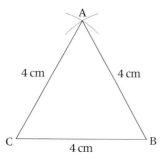

Draw BC 4 cm long.

Draw arcs of **radius** 4 cm from B and C to intersect at A.

Join AC and BC.

The construction **arcs** show your method, so do not erase them.

You can use this method to construct an angle of 60° (angles in an equilateral triangle = 60°).

b

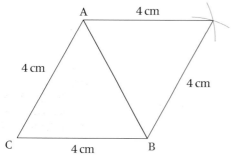

Draw arcs of radius 4 cm from A and B to intersect at D.
Join AD and DB.

Four equal sides, two pairs of parallel sides, opposite angles equal.

c A rhombus

1 Use a straight edge and compasses or a protractor to construct these triangles.

It helps to draw a rough sketch first.

  a Sides 8 cm, 4 cm, 7 cm (SSS)   **b** 3 cm, 30°, 4 cm (SAS)

  c Sides 10 cm, 7.5 cm, 6 cm (SSS)   **d** 8 cm, 2 cm, 90° (RHS)

  e Sides 6 cm, 9 cm, 5 cm (SSS)   **f** 45°, 4 cm, 45° (ASA)

2 **a** Explain why you cannot construct a triangle with sides 9 cm, 4 cm, 3 cm.

Try to construct the triangles to see what happens.

  b Explain what happens when you try and construct a triangle with sides 9 cm, 4 cm, 5 cm.

3 Without drawing the construction, write whether these sets of 3 sides will make a triangle.

  a Sides 5 cm, 5 cm, 9 cm   **b** Sides 2 cm, 2 cm, 2 cm

  c Sides 29 cm, 26 cm, 4 cm   **d** Sides 22 cm, 12 cm, 10 cm

  e Sides 20 cm, 7 cm, 9 cm   **f** Sides 14 cm, 8 cm, 6 cm

  g Sides 15 cm, 60 mm, 100 mm   **h** Sides 120 mm, 8 cm, 9 cm

4 Construct isosceles triangles with sides

  a 7 cm, 7 cm, 5 cm   **b** 5 cm, 5 cm, 7 cm

5 **a** Construct an equilateral triangle ABC with sides 5 cm.

  b Construct a second equilateral triangle with base AB and sides 5 cm to get a rhombus.

  c Follow the steps in **a** and **b** to construct a rhombus with sides 3.5 cm.

6 **a** Construct a triangle with sides 5 cm, 12 cm, 13 cm.

  b What type of triangle is this?

7 **a** Construct a triangle ABC with sides AB = 3 cm, BC = 4 cm, CA = 5 cm.

  b Construct triangle ADC with side CA from the triangle in part **a**, and side AD = 4 cm and side DC = 3 cm.

  c What special type of quadrilateral is this?

8 **a** Construct a triangle ABC such that BC = 12 cm, AC = 7 cm and ∠B = 30°.

This triangle is called SSA because you have two sides and the non-included angle.

  b Now try to draw a second, **different** triangle with the same measurements. (Move the position of A.)

  c Are SSA triangles unique?

This spread will show you how to:

- Use a ruler and compasses to draw standard constructions

Keywords
Bisect
Equidistant
Perpendicular

- **Bisect** means cut into two equal parts.

You can use a straight edge and compasses to construct an angle bisector.

- To bisect angle ABC

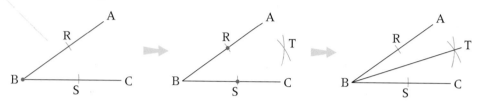

Use the same compass radius throughout the construction. Start at the red dots.

BRTS is a rhombus.

- All points on the angle bisector are equidistant from the arms of the angle.

Equidistant means equal distance from.

- The **perpendicular** bisector of a line bisects the line at right angles.

- To construct the perpendicular bisector of line AB

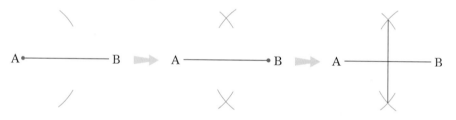

Use the same compass radius throughout the construction.

Start at the red dots.

- All points on the perpendicular bisector of AB are equidistant from A and B.

Example

Construct an angle of 45°.

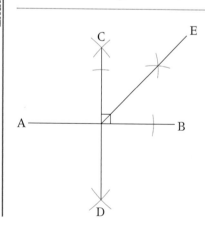

Draw a line AB
Construct the perpendicular bisector CD.
Construct the angle bisector of ∠BCD.
∠BCE = 45°

45° is $\frac{1}{2}$ of 90°. Construct a perpendicular bisector to AB (90°) and then bisect the angle.

1 Trace these angles.
Construct the angle bisector of each angle, using ruler and compasses.

a   **b**   **c**

You will need to extend the lines.

2 **a** Construct an equilateral triangle with sides 5 cm.

  b Construct the angle bisector of each angle of the triangle.

  c What do you notice about the three angle bisectors?

See the example in S5.2 for help with constructing an equilateral triangle.

3 Follow these steps to construct an angle of 30°.

  a Construct an equilateral triangle with sides 4 cm.

  b Bisect one of the base angles.

Angles in an equilateral triangle = 60°.

4 Draw lines AB for these lengths and construct their perpendicular bisectors.

  a 6 cm   **b** 9 cm   **c** 5.6 cm   **d** 10 cm   **e** 11.2 cm

  Check by measuring that each bisector intersects the line AB at its midpoint.

5 **a** Construct an equilateral triangle with sides 5 cm.

  b Construct the perpendicular bisectors of each side of the triangle.

  c Compare your diagram with the one for question **2**. Write down what you notice.

6 **a** Construct two triangles with sides 8 cm, 5 cm, 7 cm. Label them triangle A and triangle L.

  b On triangle A construct the angle bisector of each internal angle of the triangle.

  c On triangle L construct the perpendicular bisectors of each side of the triangle.

  d Compare and comment on your answers to **b** and **c**.

7 **a** Draw a line AB, 8 cm long, and construct its perpendicular bisector.

  b Construct an angle of 45° where the perpendicular bisector intersects AB.

  c What other angles have you created in this construction?

Further constructions

This spread will show you how to:

● Use a ruler and compasses to draw standard constructions

Keywords
Perpendicular

You can construct a **perpendicular** from a point to a line or from a point on a line.

● To construct a perpendicular from a point X to a line YZ

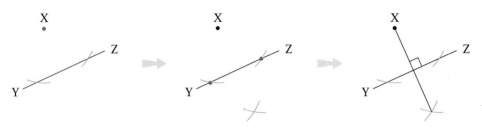

Start at the red dots.

Keep the same compass radius throughout the construction.

● To construct a perpendicular from a point E on a line DF

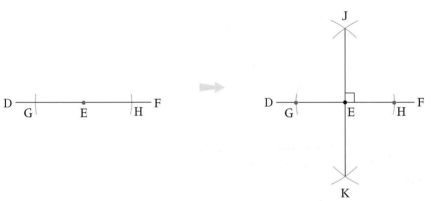

Start at the red dots.

Change your radius for the second part of the construction to a larger one.

● The shortest distance from a point to a line is the perpendicular distance.

Example

Construct a right-angled triangle with sides 3 cm, 4 cm and 5 cm.

Draw a line longer than 4 cm.
Mark the points A and B, 4 cm apart.

Construct a line perpendicular to A.
Mark point C at 3 cm above A on this line.

Draw the third side of the triangle.

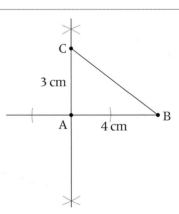

Start by drawing one of the shorter sides.

In a right-angled triangle, the hypotenuse is the longest side.
So the other two sides (3 cm and 4 cm here) meet at right angles.

1 Trace these lines and the points marked X.
For each, use ruler and compasses to construct
a perpendicular from the point X to the line.
Check your constructions using a protractor.

Leave in the
construction lines
and arcs.

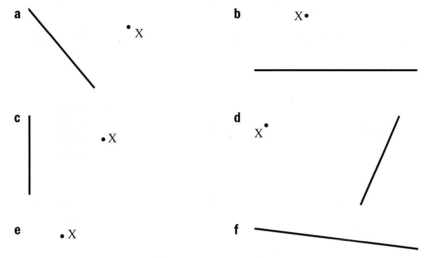

2 Trace these lines.
For each, use ruler and compasses to construct the perpendicular
from the point X on the line. Show all construction lines.
Check your constructions using a protractor.

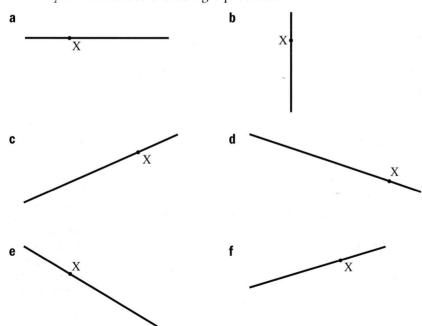

This spread will show you how to:

● Find loci by reasoning and using diagrams

Keywords

Bisector
Equidistant
Locus
Perpendicular

A **locus** is the path traced out by a moving point.

Loci is the plural of locus.

● The locus of a point which is a constant distance from another point is a circle.

● The locus of a point at a constant distance from a fixed line is a parallel line.

● The locus of a point that is **equidistant** from two other fixed points is the **perpendicular bisector** of the line joining the fixed points.

● The locus of a point equidistant from two intersecting lines is the angle bisector of the lines.

Example

P and Q are two points 2.5 cm apart.

P   Q

Shade in the region that satisfies all these conditions:

● Right of the perpendicular to the line at point P.
● Closer to P than to Q.
● More than 1 cm from P.

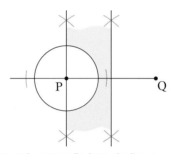

Construct the perpendicular to the line at point P.

Construct the perpendicular bisector of PQ. Points to the left are nearer to P than Q.

Draw a circle radius 1 cm, centre P. Points outside are more than 1 cm from P.

Example

ABCD is the plan of a garden.

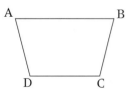

A tree is to be planted in the garden so that it is

● nearer to BC than to BA
● nearer to AD than AB.

Shade the region where the tree may be planted.

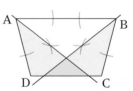

Construct the angle bisector of ∠ABC. Shade the region between this line and BC.

Construct the angle bisector of ∠BAD. Shade the region between this line and AD.

The tree can be planted in the region where the shadings overlap.

1 a Draw points A and B, 6 cm apart.

  b Shade in the region that satisfies both these conditions.

    i Closer to A than to B

    ii Less than 4 cm from B

2 a Draw points J and K, 5 cm apart.

  b Shade the region that satisfies both these conditions.

    i More than 4 cm from J

    ii More than 3 cm from K

3 a Trace the points X, Y and Z.   X•

  b Shade the region that satisfies all three of these conditions.

    i Closer to X than to Y

    ii Closer to the line XZ than to the line XY

                                    Z•   Y•

    iii More than 1 cm from X

4 a Draw a rectangle PQRS where PQ = 5 cm and QR = 3 cm.

  b Shade the region of the rectangle that is within 4 cm of P and within 2.5 cm of R.

5 a Construct a right-angled triangle ABC where angle ABC = 90°, AB = 6 cm, BC = 4.5 cm.

  b Shade the region that satisfies all three of these conditions.

    i Closer to A than to B

    ii Less than 4 cm from A

    iii Less than 4 cm from C

6 The diagram shows the rectangular garden of a house.

There are two trees, *T*, in the garden.

A radio mast is to be placed in the garden.

It must be more than 5 m from the rear of the house.

It must be more than 3 m from a tree.

Using a scale of 1 cm : 2 m, draw a scale diagram and shade the possible site for the radio mast.

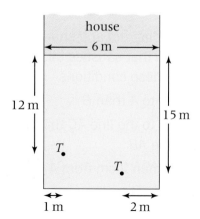

Key objectives

- Understand angle measure using the associated language
- Use and interpret maps and scale drawings
- Use a straight edge and compasses to do standard constructions
- Find loci, both by reasoning and by using ICT to produce shapes and paths

1 John lives in Smalltown. He walks 2 km to school every morning. After school every Thursday he goes to his friend Katie's house for dinner and then walks home.

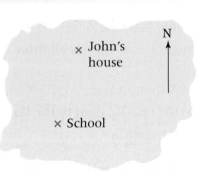

Scale: 4 cm = 1 km

Copy the diagram of Smalltown.

a Measure and write the bearing of the school from John's house. (1)

b Katie's house is 1 km from the school at a bearing of 315°.

  i Draw a cross to show the position of Katie's house. (2)

  ii What bearing, to 3 significant figures, does John take to return home from Katie's house? (1)

2 Copy the diagram and shade in the region that satisfies all three of these conditions.

  i Closer to A than B

  ii Closer to the line AC than to the line AB

  iii More than 2 cm from A (4)

× C

× A   × B

(Edexcel Ltd., (Spec.))

N7 Fraction and percentage calculations

This unit will show you how to

- Calculate with fractions effectively following simplification
- Use fractions to express an exact answer
- Calculate a fraction and a percentage of a quantity
- Solve problems involving percentage increase and decrease
- Calculate simple and compound interest
- Express a number as a percentage of another
- Calculate the original amount before a percentage increase or decrease

Before you start ...

You should be able to answer these questions.

Review

1 Change these decimal numbers to percentages. — Key stage 3

a 0.3 **b** 0.65
c 0.725 **d** 1.05
e 0.06

2 Change these percentages to decimals. — Key stage 3

a 15% **b** 6.5%
c 12.5% **d** 97.5%
e 108%

3 Calculate the percentage of each amount. — Key stage 3

a 10% of £350 **b** 5% of 400 m
c 15% of 360° **d** 20% of 600 cm
e 18% of £300 **f** 45% of 200 mm

This spread will show you how to:

- Calculate a fraction of a quantity
- Calculate with fractions effectively following simplification
- Use fractions to express an exact answer

Keywords

Cancel
Common factor
Exact

- You find fractions of a quantity by multiplying.

  For example, two thirds of $5 = \dfrac{2}{3} \times 5 = \dfrac{2 \times 5}{3} = \dfrac{10}{3} = 3\frac{1}{3}$

- When the quantity and the denominator of the fraction have a **common factor**, **cancel** this factor before multiplying.

  For example, $\dfrac{2}{{}_1 3} \times 24^8 = \dfrac{2}{1} \times 8 = 16$

Notice that you cannot give $3\frac{1}{3}$ as an **exact** answer in decimals because $\frac{1}{3}$ is a recurring decimal.

Example

Calculate   **a** $\frac{3}{4}$ of 28   **b** $\frac{5}{8}$ of 6   **c** $\frac{4}{9}$ of 12   **d** $\frac{5}{9}$ of 25

a $\dfrac{3}{{}_1 4} \times 28^7 = 3 \times 7 = 21$

b $\dfrac{5}{{}_4 8} \times 6^3 = \dfrac{5 \times 3}{4} = \dfrac{15}{4} = 3\dfrac{3}{4}$

c $\dfrac{4}{{}_3 9} \times 12^4 = \dfrac{4 \times 4}{3} = \dfrac{16}{3} = 5\dfrac{1}{3}$

d $\dfrac{5}{9} \times 25 = \dfrac{125}{9}$

      $= 13\dfrac{8}{9}$

Cancel by common factor 4 before multiplying.

You *can* also write $5\frac{1}{3}$ as $5.\dot{3}$, but it is simpler to leave it as a fraction.

In part **d**, you cannot cancel by 5. You can only cancel common factors where one is in a numerator (or a whole number) and the other is in a denominator.

You can extend this method to finding a fraction of a fraction of a quantity.

Example

Tom has £42. He spends $\frac{1}{3}$ of it on Monday. On Tuesday he spends $\frac{3}{4}$ of the remainder. How much does he spend on Tuesday?

Tom spends $\frac{1}{3} \times$ £42 on Monday so he has $\frac{2}{3} \times$ £42 on Tuesday.
On Tuesday he spends

$$\dfrac{\cancel{3}}{4\,\cancel{2}_1} \times \dfrac{{}^1\cancel{2}}{\cancel{3}_1} \times \cancel{42}^{21} = \text{£21}$$

1 Calculate these. Give your answers as fractions or mixed numbers.

   a $\frac{1}{2}$ of 7   **b** $\frac{1}{5}$ of 8   **c** $\frac{1}{3}$ of 10   **d** $\frac{1}{3}$ of 2   **e** $\frac{2}{5}$ of 6

2 Work out these. Show how common factors can be cancelled in each.

   a $\frac{1}{2}$ of 20   **b** $\frac{1}{4}$ of 84   **c** $\frac{1}{3}$ of 36   **d** $\frac{1}{5}$ of 65   **e** $\frac{2}{3}$ of 33

You should show all of your working for the questions **3**, **4** and **5**.
In particular, show how the calculations can be simplified by
cancelling common factors.

3 A reel holds 60 m of wire when new. $\frac{2}{5}$ of the wire has been used.

   a What length of wire has been used?

   b What length of wire is left on the reel?

4 Calculate the amount of liquid in these containers.

   a A 40 litre barrel that is $\frac{3}{8}$ full.   **b** A 240 cl jar that is $\frac{3}{4}$ full.

   c A 120 cl glass that is $\frac{2}{5}$ full.

   d A 750 ml litre bottle that is $\frac{2}{3}$ empty.

5 Calculate these.

   a $\frac{5}{8}$ of 48 m   **b** $\frac{2}{9}$ of 36 km   **c** $\frac{4}{7}$ of 28 mm   **d** $\frac{3}{4}$ of 120 m

The answers to questions **6** and **7** will not be whole numbers. You should
show your working as before, and give your answers as fractions or
mixed numbers.

6 Work out these.

   a $\frac{1}{6}$ of 10   **b** $\frac{1}{4}$ of 22   **c** $\frac{3}{10}$ of 15   **d** $\frac{1}{12}$ of 8   **e** $\frac{4}{9}$ of 21

7 Calculate these lengths.

   a $\frac{1}{9}$ of 24 miles   **b** $\frac{5}{6}$ of 40 miles   **c** $\frac{5}{18}$ of 45 miles   **d** $\frac{3}{20}$ of 25 miles

8 Calculate these and convert your answers to (approximate) decimal
   numbers.

   a $\frac{5}{12}$ of 16 m   **b** $\frac{4}{9}$ of 12 mm   **c** $\frac{3}{22}$ of 64 cm   **d** $\frac{3}{14}$ of 104 km

9 An empty swimming pool is to be filled with water. It takes 12 hours
   to fill the pool, and the full pool contains 98 m^3 of water. How much
   water will the pool contain after 5 hours? Show your working.

10 Calculate these times using fractions, and then convert each of your
   answers into hours and minutes.

   a $\frac{7}{12}$ of 9 hours   **b** $\frac{5}{8}$ of 22 hours   **c** $\frac{7}{10}$ of 24 hours   **d** $\frac{7}{18}$ of 63 hours

11 Calculate these.

   a $\frac{2}{3}$ of £7   **b** $\frac{5}{12}$ of £40   **c** $\frac{3}{8}$ of £34   **d** $\frac{4}{9}$ of £66

This spread will show you how to:
- Calculate a percentage of a quantity

Keywords

Decimal
equivalent

You often need to calculate a percentage of a quantity.

- Use mental methods to find simple percentages.

75% of £38: work out one quarter of £38 (which is £9.50), and multiply by 3.

- For more complicated examples, you can find 1% of the quantity, and then multiply by the required percentage.

27% of 48 m: first find 1% (which is 0.48 m), and multiply by 27.

- A quick method, especially when using a calculator, is to multiply by the appropriate decimal number.

To find 38% of a quantity, multiply by 0.38.

Example

Calculate   **a** 45% of 60 cm   **b** 34% of 85 kg   **c** 16% of £25

a 50% of 60 = 30   50% is a half.
   5% of 60 = 3
  45% of 60 cm = 30 cm − 3 cm   = 27 cm

b $34\% = \frac{34}{100} = 0.34$ 0.34 is the **decimal equivalent** of 34%.
  34% of 85 kg = 0.34 × 85
                 = 28.9 kg   By calculator
                 = 29 kg to 2 sf.

c $16\% \text{ of } £25 = \frac{{}^{4}\cancel{16}}{\cancel{100}_{4}} \times \cancel{25}^{1}$
              = £4

Example

At Fitz High School, 95% of the students have never been absent.
There are 1180 students at the school.
How many of them have a perfect attendance record?

        95% = 0.95
95% of 1180 = 0.95 × 1180
              = 1121

So 1121 students have a perfect attendance record

Example

Tom is saving 6% of his salary in a pension fund.
His current salary is £25 000.
How much will he save this year?

10% of £25 000 = £2500
 5% of £25 000 = £1250
  1% of £2500 = £250
  6% of £2500 = £1250+£250
            = £1500

This calculation is easy to do mentally.

1 Find the percentages of these numbers mentally.

  a 25% of 42   **b** 90% of 140   **c** 20% of 1200

  d 60% of 500   **e** 30% of 440   **f** 11% of 900

2 Calculate these, using a mental method wherever possible.

  a 30% of £750   **b** 55% of 1800 m

  c 90% of 2800 kg   **d** 60% of €240

3 Use an appropriate method to work out these. Show all your working, and do not use a calculator.

  a 9% of 1500   **b** 13% of 700   **c** 31% of 2400

  d 36% of 50   **e** 43% of 900   **f** 6% of 3200

4 Calculate these, using an appropriate mental or written method. Show all your working, and do not use a calculator.

  a 23% of 4800 mm   **b** 61% of 3200 kg

  c 39% of €3700   **d** 17% of £2900

5 Write a decimal number equivalent to each percentage.

  a 50%   **b** 60%   **c** 25%   **d** 51%   **e** 64%

  f 22%   **g** 15%   **h** 70%   **i** 7%   **j** 8.5%

6 Calculate these, using an appropriate method. Use a calculator where necessary.

  a 15% of 38   **b** 25% of 800   **c** 27% of 59

  d 96% of 104   **e** 41% of 41   **f** 80% of 25

7 Calculate these, rounding your answers to the nearest penny.

  a 16% of £24   **b** 63% of £85   **c** 93% of £15

  d 42% of £405   **e** 88% of £32   **f** 6% of £265

8 Mrs Jones has a conservatory built which costs £12 000. She pays an initial deposit of 15%. The remainder is to be paid in 24 equal monthly instalments. How much is each of these instalments?

9 A restaurant adds a 12% service charge to the bill. What will be the total cost for a meal that is £65.80 before the service charge is added?

10 A school has 1248 pupils, and 48% of them are girls. How many boys are there in the school? Show your working.

11 Julia earns a salary of £47 800 per year. She is awarded a 2.7% pay rise. Calculate her new salary.

This spread will show you how to:

● Solve problems, involving percentage increase and decrease

Keywords
Decimal
equivalent
Decrease
Increase

You often need to work out problems involving percentage **increase** and **decrease**.

● **To work out the new amount after a percentage increase or decrease,**
  - **first find the increase or reduction**

    Increase £25 000 by 5%
    5% of £25 000 = £1250

  then simply add or subtract

    £25 000 + £1250 = £26 250

  - **Using a calculator, you can find the new amount by multiplying by the decimal equivalent.**

    To find the result of a 12% decrease, multiply the original amount by 0.88.
    To find the result of a 23% increase, multiply the original by 1.23.

Find the result when

a 45 cm is increased by 20%   **b** 44 kg is decreased by 17%.

a 20% of 45 cm = 9 cm,
new length 45 cm + 9 cm = 54 cm

b 100% − 17% = 83%
44 kg × 0.83 = 36.52 kg

New mass is 83% of original.

This is how Jackie worked out 22% of 85 cm.

What is wrong here?

Ok – 22% off from 85 leaves 78%, so I'll work out 85 × 0.78 …

Examiner's tip
The words 'of' and 'off' can cause confusion – especially when you are in a hurry in an exam!

This question simply asks her to work out 22% **of** 85 cm – she is not being asked to work out a percentage reduction.

22% of 85 cm = 0.22 × 85
= 18.7 cm

Alan's car depreciates by 15% every year.
It is valued at £9750 now. What will be its value in one year's time?

Value after depreciation = 100% − 15% = 85% of original value
0.85 × 9750 = £8287.50 = £8300 to nearest £100

85% = 0.85

1 Calculate the results when these amounts are increased by the percentages given.

You should be able to do all of these mentally.

   a 72 by 50%   **b** 60 by 20%   **c** 45 by 10%

   d 600 by 3%   **e** 480 by 25%   **f** 500 by 40%

2 Decrease each of these amounts by the percentages given.

Again, you should be able to do these mentally.

   a 28 by 10%   **b** 45 by 20%   **c** 60 by 15%

   d 75 by 50%   **e** 380 by 40%   **f** 65 by 1%

3 Use a written method to calculate each of these percentage increases. Show your working, and do not use a calculator.

   a 300 by 14%   **b** 200 by 32%   **c** 800 by 26%

   d 250 by 30%   **e** 750 by 83%   **f** 940 by 18%

4 Use a written method to find these percentage decreases.

   a 800 by 16%   **b** 700 by 24%   **c** 400 by 32%

   d 450 by 33%   **e** 350 by 61%   **f** 260 by 52%

5 Write the decimal number you must multiply by to find these percentage increases.

   a 20%   **b** 30%   **c** 45%   **d** 85%   **e** 6.5%

6 Write the decimal number you must multiply by to find these percentage decreases.

   a 40%   **b** 60%   **c** 35%   **d** 72%   **e** 18.5%

7 Use a calculator to find these percentage increases and decreases.

   a Increase 53 by 7%   **b** Decrease 42 by 4%

   c Increase 620 by 16%   **d** Decrease 300 by 18%

8 Calculate the new salaries after these pay increases.

   a £32 000 increased by 5%   **b** £18 450 increased by 4.7%

   c £26 500 increased by 3.2%   **d** £52 850 increased by 6%

9 Calculate the new prices after these price cuts.

   a £450 decreased by 22%   **b** £860 decreased by 35%

   c £1250 decreased by 42%   **d** £740 decreased by 3.5%

10 Alan says, 'The cost of computer chips increased by 120% last year'.
Billy says, 'That's not possible. They couldn't have gone up by more than 100%'.
Explain why Billy is wrong.

11 Carina says, 'The cost of computer memory fell by 120% last year.'
Deni says, 'That's not possible. Prices couldn't have fallen more than 100%'.
Explain why Deni is correct.

This spread will show you how to:

- Calculate simple and compound interest.

Keywords

Borrow
Compound interest
Interest
Invest
Principal
Rate
Simple interest

You earn **interest** when you **invest** in a savings account at a bank. However, you pay interest if you **borrow** money for a mortgage.

Interest is either

- **simple interest** – it is not added to the principal, or
- **compound interest** – added to the principal and will itself earn interest.

> The original sum you invest is called the **principal**.

- To calculate simple interest, use the interest **rate** to work out the amount earned.
- If simple interest is paid for several years, the amount paid each time stays the same, because the interest is paid elsewhere and the **principal** stays the same.
- To calculate compound interest, work out the interest in the same way, but add the interest earned to the principal.
- If compound interest is paid for several years, the amount of interest earned each year increases, because the principal increases.

Calculate the interest when £1000 is invested for 4 years at

a 5% simple interest (SI)   **b** 5% compound interest (CI).

a SI for 1 year = 5% × £1000
$\qquad\qquad\quad$ = £50

SI for 4 years = 5% × £1000 × 4
$\qquad\qquad\quad\;$ = £200
Total SI = £200
Principal + SI = £1200

b 1st year's CI = 5% × £1000
$\qquad\qquad\qquad$ = £50
new principal = £1000 + £50
$\qquad\qquad\quad\;$ = £1050
2nd year's CI = 5% × £1050
$\qquad\qquad\qquad$ = £52.50
new principal = £1102.50
3rd year's CI = 5% × £1102.50
$\qquad\qquad\qquad$ = £55.13
new principal = £1157.63
4th year's CI = 5% × £1157.63
$\qquad\qquad\qquad$ = £57.88
new principal = £1215.51

> Money invested at compound interest earns more than the same sum invested at simple interest.

- When you have to calculate compound interest over a number of years you can do it quickly using a calculator.

£2000 is invested at 6.5% compound interest.
Find the principal after 15 years.

Principal at end of 1 year = £2000 × 1.065 = £2130

Principal at end of 2nd year = (£2000 × 1.065) × 1.065
$\qquad\qquad\qquad\qquad\qquad$ = £2000 × 1.065^2 = £2268.45

After 15 years principal = £2000 × 1.065^{15} = £5143.68

> To increase by 6.5% multiply by decimal equivalent of 6.5% = 1.065

> Each year the principal increases by a factor of 1.065

1   Find the total interest earned when these amounts of money are
    invested at these annual rates of **simple interest**.

    a £100 at 5% for 1 year   **b** £200 at 6% for 2 years

    c £1400 at 7.5% for 3 years   **d** £650 at 3.5% for 10 years

2   Find the final value of each of these amounts invested
    at **compound interest**.

    a £250 at 5% for 1 year   **b** £400 at 2% for 2 years

    c £1200 at 6% for 2 years   **d** £1000 at 2% for 3 years

3   A sum of money is invested at 5% compound interest.
    - To find the amount after one year, multiply the principal by 1.05.
    - To find the amount after two years, multiply the principal
      by $1.05^2 = 1.1025$

    a What decimal number do you need to multiply the principal by
    to work out the amount after earning interest for one year at a
    rate of 6%?

    b What decimal number do you need to multiply the principal by
    to work out the amount after earning compound interest for
    two years at 6%?

4   Use your answers to question **3** to work out the final amount when a
    principal of £500 is invested.

    a for one year at an interest rate of 6%

    b for two years at 6% compound interest.

5   Find the decimal number you should multiply the principal by
    to find the final amount after earning compound interest at
    these rates.

    a 5% per year for 3 years   **b** 6.5% per year for 5 years

6   Use your answers to question **5** to work out the final amount when

    a £5000 is invested for 3 years at 5% compound interest.

    b £800 is invested for 5 years at 6.5% compound interest.

7   Which of these options earns the most interest when £5000 is
    invested for

    a 8 years at 6% simple interest

    b 6 years at 8% compound interest?

    Explain your answer.

8   **a** What decimal number do you multiply by, to find a 15% reduction?

    b Explain how you would then find a second 15% reduction.

This spread will show you how to:

- Express a number as a percentage of another.
- Calculate the original amount before a percentage increase or decrease.

- To write one number as a percentage of another, start by writing the first number as a fraction of the other.
  - Write '23 as a percentage of 25' as $\frac{23}{25}$.
- Now convert the fraction to a percentage.
  - $\frac{23}{25} = \frac{23 \times 4}{25 \times 4} = \frac{92}{100} = 92\%$.
  - Sometimes you will need to use a calculator. For example, to find 9 as a percentage of 17: $\frac{9}{17} = (9 \div 17) \times 100\% = 52.9\%$ (to 1 decimal place).

- To find the original amount before a percentage change, use the final amount to work out 1% of the original amount, and then find the original amount by multiplying.

Example

Find

a 13 as a percentage of 20   **b** 39 as a percentage of 75
c 8 as a percentage of 23

a $\frac{13}{20} = \frac{13 \times 5}{20 \times 5} = \frac{65}{100} = 65\%$

b $\frac{39}{75} = \frac{13}{25} = \frac{52}{100} = 52\%$

c $\frac{8}{23} = (8 \div 23) \times 100\% = 34.8\%$

Example

Find the **original price** of a denim jacket reduced by 15% to £32.30.

85% of original price = £32.30
1% of original price = £32.30 ÷ 85
100% of original price = £32.30 ÷ 85 × 100
             = £38

100% − 15% = 85%

Example

Following a 5% price increase, a car radio costs £168. How much did it cost before the increase?

105% of original price = £168
1% of original price = £168 ÷ 105 = £1.60
100% of original price = £1.60 × 100 = £160

1 For these pairs of numbers, find the smaller number as a percentage of the larger one. You should be able to do all of these mentally.

   a 16 and 20   **b** 7 and 25   **c** 6 and 50

   d 20 and 40   **e** 17 and 34

2 Use a written method to find

   a 12 as a percentage of 20   **b** 36 as a percentage of 75

   c 24 as a percentage of 40

3 Use a calculator to find

   a 19 as a percentage of 37   **b** 42 as a percentage of 147

   c 8 as a percentage of 209

4 Jason scored 53 out of 60 in a science test, and 39 out of 45 in a maths test.

   a Convert each of his marks to a percentage.

   b Jason says, 'I did better in maths, because I only dropped 6 marks; I dropped 7 marks in science.' Explain why Jason is wrong.

5 A book costs £4 after a 20% price reduction. How much did it cost before the reduction? Show your working.

6 These numbers are the results when some amounts were increased by 10%. For each one, find the original number.

   a 55   **b** 44   **c** 88   **d** 121

7 Find the original cost of the following items.

   a A vase that costs £7.20 after a 20% price increase.

   b A table that costs £64 after a 20% decrease in price.

8 **a** Francesca earns £350 per week. She is awarded a pay rise of 3.75%. Frank earns £320 per week. He is awarded a pay rise of 4%. Who gets the bigger pay increase? Show all your working.

   b Bertha's pension was increased by 5.15% to £82.05. What was her pension before this increase?

9 Carys is trying to find the original price of an item that is on sale. Its sale price is £56 and it was reduced by 20%.
   Carys' working is shown to the right.
   What is wrong with her working?

   | 20% of 56 = 11.2 |
   | 56 + 11.2 = 67.2 |
   | The answer is £67.20 |

Exam review

Key objectives

- Use efficient methods to calculate with fractions, including cancelling common factors before carrying out the calculation, recognising that, in many cases, only a fraction can express the exact answer.

- Solve percentage problems, including percentage increase and decrease

1 Tom received a total of £80 for his birthday.
He put $\frac{2}{5}$ of this money into his bank account and spent the rest.

  a How much money did he put into his bank account? (1)

  b Tom spent 25% of his spending money on a t-shirt.

  How much did the t-shirt cost? (2)

  c How much of his total birthday money did Tom have left to spend after buying the t-shirt

  i in pounds (2)

  ii as a fraction of his total birthday money (1)

  iii as a percentage of his total birthday money? (1)

2 (Calculator allowed.)

Simon repairs computers. He charges £56.80 for the first hour he works on a computer and £42.50 for each extra hour's work.

Yesterday Simon repaired a computer and charged a total of £269.30.

  a Work out how many hours Simon worked yesterday on this computer. (2)

Simon reduces his charges by 5% when he is paid promptly.
He was paid promptly for yesterday's work on the computer.

  b Work out how much he was paid. (3)

(Edexcel Ltd., 2003)

This unit will show you how to

- Draw cumulative frequency polygons for grouped data
- Find the modal class of a data set
- Estimate the median and upper and lower quartiles from a cumulative frequency diagram
- Use cumulative frequency diagrams to compare two data sets
- Draw box plots
- Compare two data sets using box plots

Before you start ...

You should be able to answer these questions.

1 Calculate.

   a $\frac{1}{2}$ of 124   **b** $\frac{1}{2}$ of 140   **c** $\frac{1}{4}$ of 240

   d $\frac{1}{4}$ of 180   **e** $\frac{3}{4}$ of 136   **f** $\frac{3}{4}$ of 144

Use this graph showing the cost per day to hire a power tool for questions **2** and **3**.

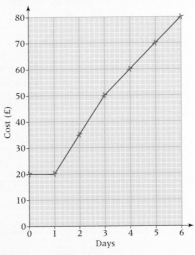

2 How much does it cost to hire the power tool for

   a 3 days   **b** 5 days?

3 Mike has £40.

   What is the maximum number of days he can hire the power tool?

Review

Unit N7

Unit D2

Unit D2

This spread will show you how to:
- Draw cumulative frequency polygons for grouped data
- Find the modal class of a data set

Keywords
Cumulative
    frequency
Grouped
    frequency
Modal class
Upper bound

If you have a large amount of data, you can group the data in a **grouped frequency** table.

- You can represent grouped data on a **cumulative frequency** diagram.

Examiner's tip
The diagrams shown in this unit may be useful for your statistical coursework task.

The heights of 120 boys are given in the table.

Height, h, cm	$145 \leqslant h < 150$	$150 \leqslant h < 155$	$155 \leqslant h < 160$	$160 \leqslant h < 165$	$165 \leqslant h < 170$
Frequency	8	27	48	31	6

a Draw a cumulative frequency table for these data.
b Use the table to draw a cumulative frequency diagram.
c Write the **modal class** interval.

a

Height, h, cm	<150	<155	<160	<165	<170
Cumulative frequency	8	35	83	114	120

Upper bound of each class.

b

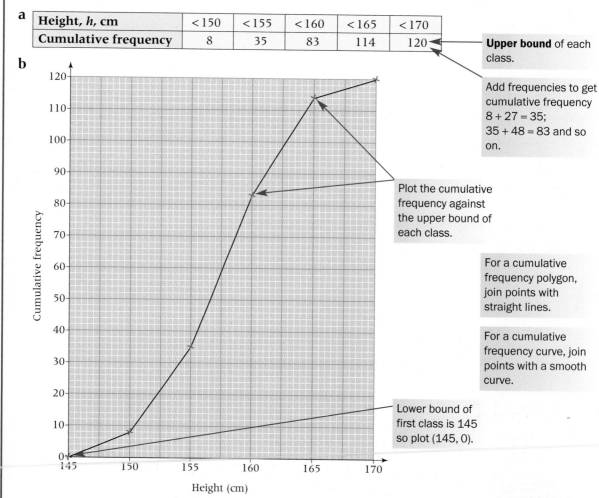

Add frequencies to get cumulative frequency
$8 + 27 = 35$;
$35 + 48 = 83$ and so on.

Plot the cumulative frequency against the upper bound of each class.

For a cumulative frequency polygon, join points with straight lines.

For a cumulative frequency curve, join points with a smooth curve.

Lower bound of first class is 145 so plot (145, 0).

Height (cm)

c Modal class interval $= 155 \leqslant h < 160$

The modal class is the class with the highest frequency.

For each of these data sets

a draw a cumulative frequency table
b draw a cumulative frequency diagram
c write the modal class interval.

> You will need the cumulative frequency diagrams from this exercise in Exercises D5.2 and D5.4.

1 The heights of 100 girls

Height, h, cm	$145 \leqslant h < 150$	$150 \leqslant h < 155$	$155 \leqslant h < 160$	$160 \leqslant h < 165$	$165 \leqslant h < 170$
Frequency	7	25	46	17	5

2 The ages of teachers in a school

Age, A, years	$20 \leqslant A < 30$	$30 \leqslant A < 40$	$40 \leqslant A < 50$	$50 \leqslant A < 60$	$60 \leqslant A < 70$
Frequency	18	37	51	28	16

3 The times taken to complete a crossword puzzle

Time, t, minutes	$0 \leqslant t < 10$	$10 \leqslant t < 20$	$20 \leqslant t < 30$	$30 \leqslant t < 40$	$40 \leqslant t < 50$	$50 \leqslant t < 60$
Frequency	4	11	29	37	27	12

4 The weights of a sample of cats and kittens

Weight, w, grams	$1500 \leqslant w < 2000$	$2000 \leqslant w < 2500$	$2500 \leqslant w < 3000$	$3000 \leqslant w < 3500$	$3500 \leqslant w < 4000$
Frequency	9	22	37	20	12

5 The heights of sunflowers growing in one field

Height, h, cm	$40 \leqslant h < 60$	$60 \leqslant h < 80$	$80 \leqslant h < 100$	$100 \leqslant h < 120$	$120 \leqslant h < 140$	$140 \leqslant h < 160$
Frequency	2	17	28	39	24	10

6 The total spent by 100 shoppers at Tesbury's superstore

Amount, p, £	$0 \leqslant p < 10$	$10 \leqslant p < 20$	$20 \leqslant p < 30$	$30 \leqslant p < 50$	$50 \leqslant p < 70$	$70 \leqslant p < 100$
Frequency	16	14	23	17	15	15

This spread will show you how to:

- Estimate the median and the upper and lower quartiles from a cumulative frequency diagram

Keywords

Cumulative
  frequency
Interquartile
  range
Lower quartile
Median
Upper quartile

- You can estimate the **median** and the **upper** and **lower quartiles** from grouped data.
- You use a cumulative frequency diagram to estimate measures.

Example

The heights of 120 boys are summarised in the cumulative frequency graph.
Use the graph to estimate

a the median
b the **interquartile range**
c the number of boys with height
  i less than 153 cm   ii greater than 163 cm.

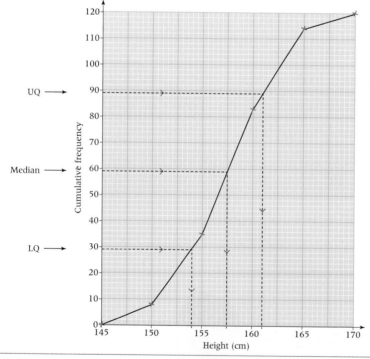

In grouped data you do not know the individual values so you can only make estimates.

To estimate the measures, draw a line from the known values across the graph, then down to the horizontal axis.

Read estimates on the horizontal axis.

a Median = 60th value
      = $157\frac{1}{2}$ cm

Total 120: $\frac{1}{2}$ of 120 = 60th value

b UQ = 90th value
      = 161 cm

$\frac{3}{4}$ of 120 = 90th value

LQ = 30th value
      = 154 cm

$\frac{1}{4}$ of 120 = 30th value

IQR = 161 − 154 = 7 cm

IQR = UQ − LQ

c i 24 boys are less than 153 cm

Read up from 153 to the graph and across to the vertical axis.

ii 19 boys are greater than 163 cm

Read up from 163 and across, then subtract 19 from the total 120.

When the data set is large, use $\frac{1}{2}n$ and not $\frac{1}{2}(n+1)$ to find the median and quartiles.

1 Use the table and graph you have drawn in Exercise D5.1 question **1** to estimate

   a the median

   b the interquartile range

   c the number of girls with height
     i less than 152 cm   **ii** greater than 163 cm.

> You will need some of your answers from this exercise in Exercise D5.4.

2 Use the table and graph you have drawn in Exercise D5.1 question **2** to estimate

   a the median

   b the interquartile range

   c the number of teachers who are aged
     i less than 35   **ii** greater than 55.

3 Use the table and graph you have drawn in Exercise D5.1 question **3** to estimate

   a the median

   b the interquartile range

   c the number of people who took
     i less than 25 minutes
     ii more than 45 minutes to complete the puzzle.

4 Use the table and graph you have drawn in Exercise D5.1 question **4** to estimate

   a the median

   b the interquartile range

   c the number of cats and kittens that weighed
     i less than 2200 g   **ii** more than 3600 g.

5 Use the table and graph you have drawn in Exercise D5.1 question **5** to estimate

   a the median

   b the interquartile range

   c the number of sunflowers that were
     i less than 130 cm   **ii** greater than 90 cm.

6 Use the table and graph you have drawn in Exercise D5.1 question **6** to estimate

   a the median

   b the interquartile range

   c the number of shoppers who spent
     i less than £20
     ii more than £80.

This spread will show you how to:

- Use cumulative frequency diagrams to compare two data sets

Keywords
Cumulative
  frequency
Interquartile
  range
Median

- You can compare two data sets using information from cumulative frequency diagrams.

- You can compare data using a measure of average, such as the **median** and a measure of spread, such as the **interquartile range**.

Example

These cumulative frequency graphs summarise the weights of a sample of 100 men and 100 women.
Make three comparisons between the weights of the men and women.

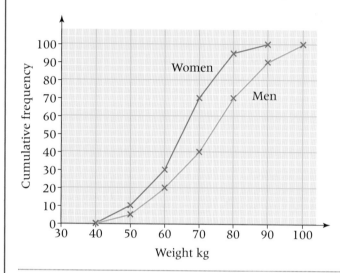

1 Median weight of women = 65 kg
  Median weight of men = 74 kg
  On average, the women are lighter than the men.

2 Range of women's weights = 90 − 40
                           = 50 kg

  Range of men's weights = 100 − 40
                         = 60 kg

3 For the women, upper quartile = 71 kg
                 lower quartile = 56 kg
                          IQR = 15 kg

  For the men, upper quartile = 82 kg
               lower quartile = 62 kg
                        IQR = 20 kg

  The middle half of the women's weights varies less than the middle half of the men's weights.

1   Write three comparisons between the test results of a group of girls and a group of boys.

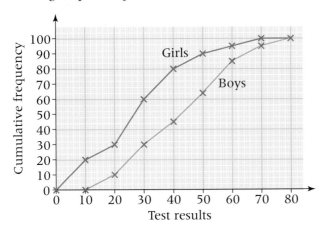

2   Write three comparisons between the heights of samples of sunflowers grown by two farmers.

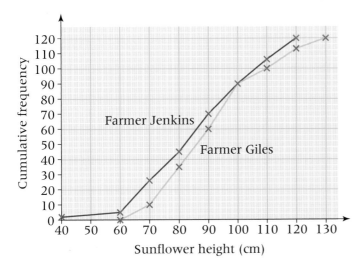

3   Write three comparisons between the mobile phone bills paid by samples of boys and girls.

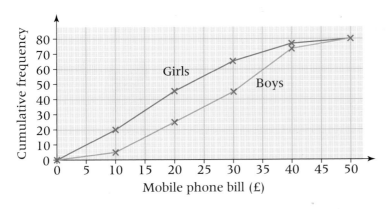

This spread will show you how to:

● Draw box plots

Keywords

Box plot
Cumulative
  frequency
Interquartile
  range
Lower quartile
Median
Range
Upper quartile

● You use a **box plot** to show the range, the median and the IQR of a set of data.

Example

The heights of 120 boys are summarised in this cumulative frequency graph.
Use the graph to draw a box plot.

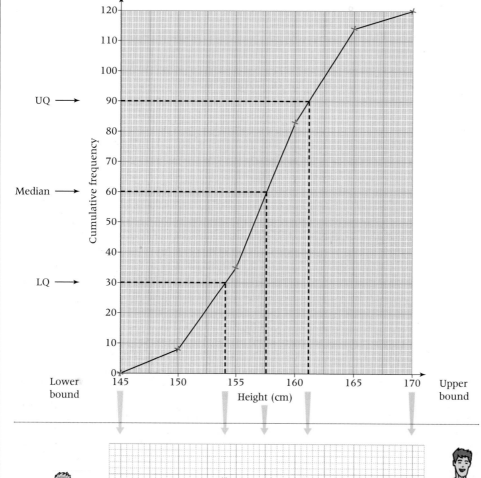

Use the lower bound of the first class and the upper bound of the last class as the lowest value and highest value.

From the graph:
Median = 60th value = 157.5 cm
UQ = 90th value = 161 cm
LQ = 30th value = 154 cm

Box plots are also called box and whisker diagrams.

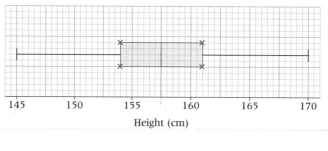

The box shows the IQR.
The whiskers show the range.

1 Use the table, graph and values you found from question **1** in Exercises D5.1 and D5.2 to draw a box plot.

2 Use the table, graph and values you found from question **2** in Exercises D5.1 and D5.2 to draw a box plot.

3 Use the table, graph and values you found from question **3** in Exercises D5.1 and D5.2 to draw a box plot.

4 Use the table, graph and values you found from question **4** in Exercises D5.1 and D5.2 to draw a box plot.

5 Use the table, graph and values you found from question **5** in Exercises D5.1 and D5.2 to draw a box plot.

6 Use the table, graph and values you found from question **6** in Exercises D5.1 and D5.2 to draw a box plot.

7 Use these cumulative frequency diagrams to draw two box plots.

a Ages of students in a school of 400 students

b Waiting times of 80 patients in a doctor's surgery

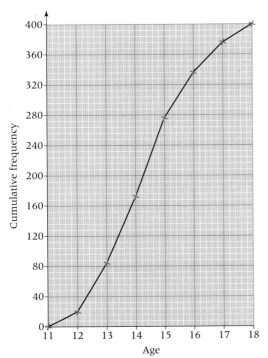

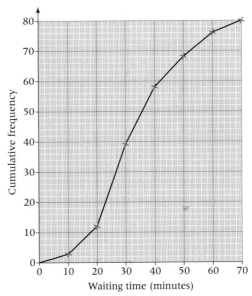

This spread will show you how to:

● Compare two sets of data using box plots

Keywords
Box plot
Interquartile
  range
Median
Skewed
Spread
Symmetrical

● You can compare two or more data sets by using information found in **box plots**.

Example

These box plots summarise the heights of samples of 13- and 14-year-old boys and girls.

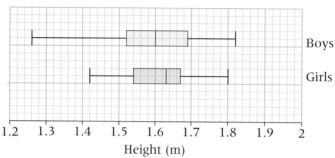

Boys

Girls

Write four comparisons between the heights of the boys and the girls.

Compare like measures such as the medians, IQR etc.

1 Median height of girls = 1.63 m   Median height of boys = 1.60 m
   On average, the girls are taller than the boys.

2 Range of girls' heights = 1.8 – 1.42   Range of boys' heights = 1.82 – 1.26
                   = 0.38 m   = 0.56 m

   The **spread** of boys' heights is greater than the spread of the girls' heights.

3 IQR for girls = 1.67 – 1.54   IQR for boys = 1.69 – 1.51
                = 0.13 m   = 0.18 m

   The IQR for the boys is greater than for the girls so the middle half of the heights is more varied for the boys.

4 The boys' median height is near the centre of the box.
   The boys' heights are symmetrical.
   The girls' median height is nearer to the UQ than the LQ so the girls' heights are negatively skewed.

In **negatively skewed** data, there are more values at the upper end of the range.

In **positively skewed** data there are more values at the lower end of the range.

In **symmetrical** data there are about the same number of values at each end of the range.

1 The box plots summarise the waiting times, to the nearest minute, of a group of patients at the doctor and the dentist.

Write four comparisons between the waiting times.

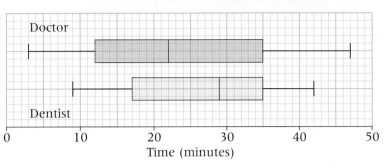

2 The box plots summarise the reaction time, to the nearest tenth of a second, of a group of boys and girls.

Write four comparisons between the reaction times.

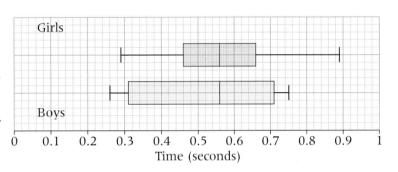

3 The box plots summarise the French and English test results of a group of students.

Write four comparisons between the test results.

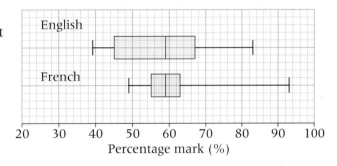

4 The box plots summarise the average length of a phone call, to the nearest minute, made by two groups of girls aged 13 and 17.

Write four comparisons between the average times.

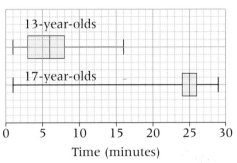

275

D5

Exam review

Key objectives

- Draw and produce, using paper and ICT, cumulative frequency tables and diagrams, box plots and histograms for grouped continuous data

- Interpret a wide range of graphs and diagrams and draw conclusions

- Compare distributions and make inferences using shapes of distributions and measures of average and spread, including median and quartiles

1 40 boys each completed a puzzle. The cumulative frequency graph gives information about the times it took them to complete the puzzle.

a Use the graph to find an estimate for the median time in seconds. (1)

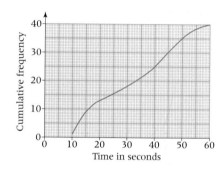

For the boys:
the minimum time to complete the puzzle was 9 seconds and the maximum time to complete the puzzle was 57 seconds.

b Use this information and the cumulative frequency graph to draw a box plot showing information about the boys' times. (3)

The box plot shows information about the times taken by 40 girls to complete the same puzzle:

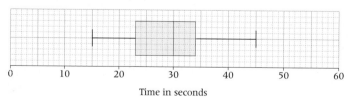

Time in seconds

c Make **two** comparisons between the boys' times and the girls' times. (2)

(Edexcel Ltd., 2004)

276

This unit will show you how to

- Analyse 3-D shapes using 2-D projections and cross-sections, including plans and elevations
- Calculate volumes and surface areas of right prisms
- Know when to round, giving answers to a sensible degree of accuracy
- Solve problems involving surface areas and volumes of prisms
- Convert between volume measures including cm^3 and m^3
- Understand the difference in dimensions of perimeter, area and volume
- Understand and use compound measures, including speed and density

Before you start ...

You should be able to answer these questions.

Review

1 What solid object can be made from each net?

Key stage 3

a 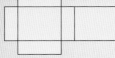   **b**

c

2 Draw a net for a cube.

Key stage 3

3 Work out the area of these shapes.

Unit S1

a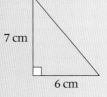

7 cm

6 cm

b

4 cm

c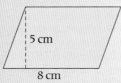

5 cm

8 cm

3-D solids: plans and elevations

This spread will show you how to:

- Analyse 3-D shapes through 2-D projections and cross-sections, including plans and elevations

Keywords
Elevation
Plan
Projection
2-D
3-D

- A **plan** of a solid is the view from directly overhead (bird's eye view).
- An **elevation** is the view from the front or the side of the solid.

Plans and elevations are **projections** of a **3-D** solid onto a **2-D** surface.

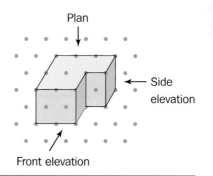

Plan

Side elevation

Front elevation

You draw solids on isometric paper.

Example

For this solid, draw **a** the plan **b** the front elevation.

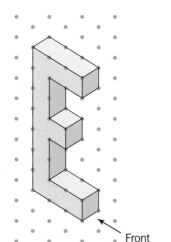

Front

a

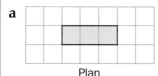

Plan

b

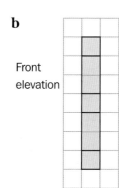

Front elevation

You draw plans and elevations on squared paper.

Example

Here are the plan and front elevation of a prism.
The front elevation shows the cross-section of the prism.

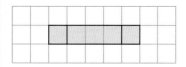

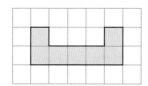

Draw a 3-D sketch of the prism.

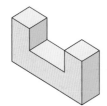

A prism has the same cross-section throughout its length.

You do not need isometric paper for a sketch.

1 For these solids, draw

   i the plan

   ii the front elevation from the direction marked with an arrow.

a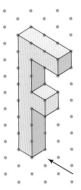

b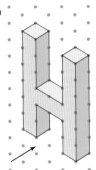

c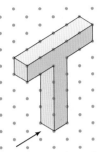

d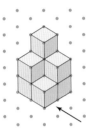

e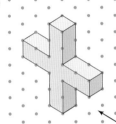

f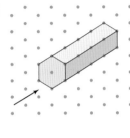

2 The diagrams show the plan and the front elevation of different solids.
Draw a sketch of each solid.

a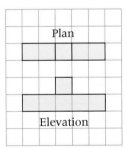

   Plan

   Elevation

b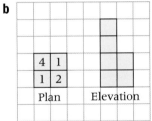

The numbers on the plan tell you the number of cubes in each column.

4	1
1	2

Plan   Elevation

c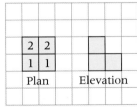

2	2
1	1

Plan   Elevation

d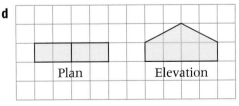

Plan   Elevation

e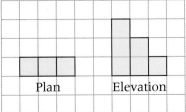

Plan   Elevation

f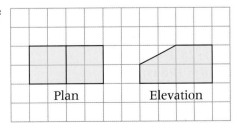

Plan   Elevation

Volume of prisms

This spread will show you how to:

● Calculate volumes of right prisms

● A **prism** is an object with constant **cross-section**.

 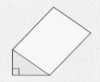

● **Volume** of a prism = area of cross-section × length.
   = **A × l**

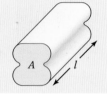

In a right prism there is a right angle between the length and the base.

Example

a Work out the volume of this cuboid.

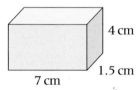

4 cm

1.5 cm

7 cm

b Work out the volume of this prism.

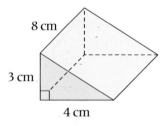

8 cm

3 cm

4 cm

a Area of cross-section =
4 × 1.5 = 6 cm^2
Volume = 6 × 7 = 42 cm^3

b Area of cross-section =
$\frac{1}{2}$ × 4 × 3 = 6 cm^2
Volume = 6 × 8 = 48 cm^3

Area of triangle = $\frac{1}{2}bh$

● A cylinder is a prism with circular cross-section.
● Volume of a **cylinder** = area of circle × height

A

A

h

h

Example

Find the volume of this cylinder.

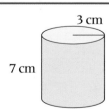

3 cm

7 cm

Do not round intermediate steps of the calculation.

Area of circle = π × 3^2 = 28.274 ... cm^2
Volume = 28.274 ... × 7 = 198 cm^3

Give answers to a sensible degree of accuracy.

1 Find the volume of each cuboid.

a
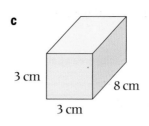
3 cm
5 cm
7 cm

b
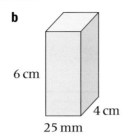
6 cm
4 cm
25 mm

c
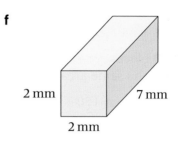
3 cm
8 cm
3 cm

d
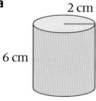
9 cm
2cm
2 cm

e
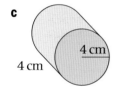
4 cm
4 cm
4 cm

f
2 mm
7 mm
2 mm

2 Find the volume of each cylinder.

a
2 cm
6 cm

b
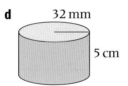
5 cm
8 cm

c
4 cm
4 cm

d
32 mm
5 cm

> Be careful with
> units in part **d**.

3 Find the volume of each prism.

a
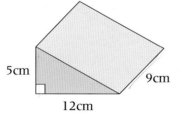
5cm
9cm
12cm

b
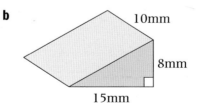
10mm
8mm
15mm

c
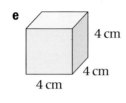
6 cm
2cm
8 cm
15 cm
14 cm

Volume and surface area

This spread will show you how to:

- Solve problems involving surface areas and volumes of prisms

Keywords

Pythagoras'
   theorem
Surface area
Volume

- **Surface area** is the total area of all the faces of a solid.

Example

This cube has surface area 150 cm². Find the volume of the cube.

A cube has six faces.
Each face has the same area.

So area of one face is $150 \div 6 = 25$ cm²

Each face is a square, so length of each side $\sqrt{25} = 5$ cm

Volume of cube = area of cross-section × length
$$= 25 \times 5$$
$$= 125 \text{ cm}^3$$

Each face is a cross-section.

You can use **Pythagoras' theorem** to find the perpendicular height of a triangle.

- **Pythagoras' theorem:** $a^2 + b^2 = c^2$

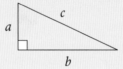

Example

This triangular prism has length 9 cm and its end faces are equilateral triangles with side length 4 cm.
Work out its volume.

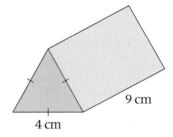

Using Pythagoras' theorem to find the height, h, of the triangle.
$$h^2 = 4^2 - 2^2$$
$$h = \sqrt{12} = 3.464 \ldots$$
Area of cross-section $= \frac{1}{2} \times 4 \times 3.464 \ldots$
$$= 6.928 \ldots$$
Volume of prism: $6.928 \ldots \times 9 = 62.4$ cm³

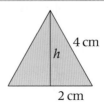

To work out the area of the cross-section you need the perpendicular height of the triangle.

1 Find the volumes of the cubes with these surface areas.

   a Surface area 54 cm^2

   b Surface area 294 cm^2

   c Surface area 96 cm^2

   d Surface area 1.5 m^2

2 Find the volumes of these triangular prisms which have equilateral triangles as cross-sections.

a

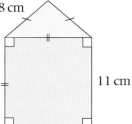

10 cm

6 cm

b

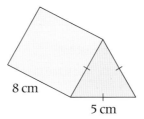

8 cm

5 cm

c 8 cm

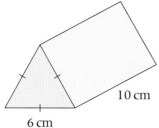

11 cm

d

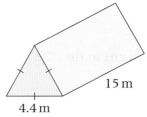

15 m

4.4 m

3 Find the surface areas of the cubes with these volumes.

   a Volume 512 cm^3

   b Volume 1000 cm^3

   c Volume 216 cm^3

   d Volume 1 m^3

4 The volume of a cuboid is 80 cm^3.
The cuboid has square ends.
The sides of the cuboid are whole
numbers of centimetres.

   a Find the possible dimensions of the cuboid.

   b Find

   i the smallest possible surface area

   ii the largest possible surface area of this cuboid.

5 Repeat question **4** for cuboids with volume

   a 24 cm^3

   b 64 cm^3

Measures

This spread will show you how to:

- Convert between volume measures including cm^3 and m^3
- Understand the difference in dimensions of perimeter, area and volume

Keywords
Area
Dimension
Perimeter
Volume

- **Perimeter** is the distance around a shape.
  It is a length measured in mm, cm and m.

- **Area** is the space covered by a two-dimensional shape.
  It is the product of length × length, measured in mm^2, cm^2 and m^2.

- **Volume** is the space inside a three-dimensional solid.
  It is the product of length × length × length, measured in cubic mm^3, cm^3 and m^3.

Example

Change

a 3 700 000 cm to m   **b** 3 700 000 cm^2 to m^2 **c** 3 700 000 cm^3 to m^3

---

a 3 700 000 ÷ 100 = 37 000 m
b 3 700 000 ÷ (100 × 100) = 370 m^2
c 3 700 000 ÷ (100 × 100 × 100) = 3.7 m^3

Larger unit means smaller number, so divide.
Divide twice for squared units.
Divide three times for cubed units.

Example

Change

a 0.042 m to mm   **b** 0.038 m^2 to cm^2 **c** 7 cm^3 to mm^3

---

a 0.042 × 1000 = 42 mm
b 0.038 × (100 × 100) = 380 cm^2
c 7 × (10 × 10 × 10) = 7000 mm^3

Smaller unit means larger number, so multiply.
Multiply twice for squared units.
Multiply three times for cubed units.

You can tell whether an expression represents length, area or volume by considering its dimensions.

- Constants such as π and numbers have no **dimension**.

Length: 1-D
Area = $length^2$: 2-D
Volume = $length^3$: 3-D

Example

In these expressions, r, s and t represent lengths.
For each, decide whether the expression represents length, area, volume or none of these.

$2r + s + 3t$	length
rst	volume
$s^2 t$	volume
$rs - st$	area
$rs + t$	none

Length + length + length = length
Length × length × length = $length^3$
$Length^2$ × length = $length^3$

Length × length = $length^2$ = area
Area − area = area
Area + length = none

1 Change these lengths to the units given.

a 40 000 cm to m **b** 63 000 cm to m **c** 42 m to cm

d 1200 cm to m **e** 80 000 m to km **f** 45 000 mm to m

g 0.05 m to mm **h** 0.0003 km to mm **i** 0.6 m to cm

j 0.007 km to mm **k** 0.00004 km to cm **l** 18.05 km to m

2 Change these areas to the units given.

a 2 600 000 mm^2 to m^2 **b** 700 000 000 cm^2 to m^2

c 0.00045 m^2 to cm^2 **d** 0.12 m^2 to cm^2

e 0.008 m^2 to cm^2 **f** 0.00045 m^2 to cm^2

g 840 000 000 mm^2 to cm^2 **h** 3 000 000 cm^2 to m^2

i 2 m^2 to cm^2 **j** 1 km^2 to m^2

3 Change these volumes to the units given.

a 3 cm^3 to mm^3 **b** 0.0002 m^3 to cm^3

c 0.0048 m^3 to cm^3 **d** 3 000 000 mm^3 to cm^3

e 10 000 000 cm^3 to m^3 **f** 50 000 000 cm^3 to m^3

4 In the following expressions the letters a, b, d, h, r each represent a length.
For each expression, decide whether it represents length, area, volume or none of these.

a ab **b** $\frac{1}{2}dh$ **c** $\frac{1}{3}\pi r^2 - \frac{1}{2}r^2$ **d** abh

e $2\pi r$ **f** $h + r$ **g** $a^2 + b^2$ **h** $\frac{1}{3}d + br$

i $\pi r^2 + ab$ **j** $a^2h + r^3$ **k** $a^2b + 2bh$ **l** $\frac{1}{2}h + d$

5 Jon said the volume of this shape is $\frac{5}{8}\pi rl$.

a Explain why the expression cannot be correct.

b What could the expression $\frac{5}{8}\pi rl$ represent?

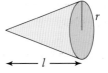

6 The photo shows a window that is rectangular with semi-circular top.
Write an expression for

a the perimeter

b the area.

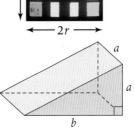

7 The diagram shows a door wedge.
Write an expression for

a the surface area

b the volume.

This spread will show you how to:

- Understand and use compound measures, including speed and density

Keywords
Capacity
Density
Rate of flow
Speed

Compound measures describe one quantity in relation to another. These are examples of compound measures.

- **Speed** = $\dfrac{\text{total distance travelled}}{\text{total time taken}}$ Units such as m/s; km/h

- **Density** = $\dfrac{\text{mass}}{\text{volume}}$ Units such as g/cm³

- **Population density** = $\dfrac{\text{population}}{\text{area}}$ Units such as number of people/km²

Use the triangle to work out which calculation to use.

Cover D (for distance)
You multiply
S(speed) × T(time)

Example

Kerry jogs at an average speed of 5 km/h for $1\frac{1}{2}$ hours. What distance does she jog?

Distance = $5 \times 1\frac{1}{2}$
= 7.5 km

Example

Find the density of a piece of wood with cross-section area 42 cm², length 12 cm and mass 693 g.

Volume = $42 \times 12 = 504$ cm³

Density = $693 \div 504$
= 1.375 g/cm³

Density = $\dfrac{\text{mass}}{\text{volume}}$

- **Capacity** is the volume of liquid that a container can hold.
  Metric units of capacity are litre, centilitre, millilitre.

Mass in grams
Volume in cm³
So Density in g/cm³.

Rate of flow is a compound measure. It is the volume of liquid that passes through a container in a unit of time.

- **Rate of flow** = $\dfrac{\text{volume}}{\text{time}}$ Units such as litres/s

Example

a 12 litres of water flows from a hosepipe in 15 seconds.
What is the rate of flow in litres/s?

b Sand was falling from the back of a lorry at a rate of 0.4 kg/s. It took 20 minutes for all the sand to fall from the lorry. How much sand was the lorry carrying?

a Rate of flow = $\dfrac{\text{volume}}{\text{time}}$
= $12 \div 15$
= 0.8 litres/s

b 20 minutes = 20×60 s
= 1200 s
Rate = $\dfrac{m}{t}$
$0.4 = \dfrac{m}{1200}$
$m = 1200 \times 0.4 = 480$ kg

Here sand is flowing, not a liquid.

Rate of flow = $\dfrac{\text{mass}}{\text{time}}$

Units kg/s
You could use the triangle:

1 Find the rate of flow for pipes A and B in litres/s.

   a Pipe A: 20 litres of water in 8 seconds.

   b Pipe B: 48 litres of water in 30 seconds.

2 Water empties from a tank at a rate of 2 litres/s.
It takes 10 minutes to empty the tank.
How much water was in the tank?

3 An engine uses oil at a rate of 0.3 ml/km.
How much oil will it use on a journey of

   a 100 km   **b** 80 km   **c** 42 km?

4 A car travelled at an average speed of 48 km/h.

   a How far did it travel in

     i 2 hours   **ii** 15 minutes   **iii** 20 minutes?

   b How long did it take to travel

     i 144 km   **ii** 72 km   **iii** 8 km?

5 The table shows information about some journeys Shaun made in one week.
Copy and complete the table.
Remember to show the units.

Distance	Time taken	Average speed
120 km	$1\frac{1}{2}$ hours	
250 miles	4 hours	
4 km		16 km/hour
	24 seconds	5 m/s
300 m	15 seconds	
0.4 km	160 seconds	
3 km		24 m/s
	20 minutes	60 km/h

6 The table shows the densities of different metals.
Use the information in the table to find

   a the mass of 0.8 m³ of zinc

   b the mass of 0.5 m³ of cast iron

   c the mass of 3.2 m³ of gold

   d the volume of 910 g of tin

   e the volume of 220 g of nickel

   f the volume of a brass statue that has mass 17 kg.

Metal	Density
Zinc	7130 kg/m³
Cast iron	6800 kg/m³
Gold	19 320 kg/m³
Tin	7280 kg/m³
Nickel	88 kg/m³
Brass	8500 kg/m³

7 In this question, give your answers in kg/m³.

   a The volume of 24 g of silver is 3 cm³.
Work out the density of silver.

   b The volume of 18 g of titanium is 4 cm³.
Work out the density of titanium.

   c A sheet of aluminium foil has volume 0.4 cm³ and
mass 1.08 g. Work out the density of aluminium foil.

S6

Exam review

Key objectives

- Use 2-D representations of 3-D shapes and analyse 3-D shapes through 2-D projections and cross-sections, including plan and elevation
- Solve problems involving surface areas and volumes
- Understand the difference between formulae for perimeter, area and volume by considering dimensions
- Convert measurements from one unit to another and understand and use compound measures including speed and density

1 The diagram shows a solid object.

  a Sketch

    i   the front elevation

    ii the side elevation

    iii the plan.

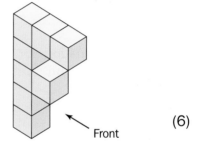
Front

(6)

  b If each cube represents a cm³, what is the volume of the shape? (2)

2 An ice hockey puck is in the shape of a cylinder with a radius of 3.8 cm, and a thickness of 2.5 cm.

3.8 cm
2.5 cm

It is made out of rubber with a density of 1.5 grams per cm³.

Work out the mass of the ice hockey puck.
Give your answer correct to 3 significant figures. (4)

(Edexcel Ltd., 2004)

This unit will show you how to

● Generate points and plot graphs of simple quadratic functions

● Plot graphs of simple cubic and reciprocal functions

● Use graphs to find the solutions or approximate solutions of two simultaneous equations

● Find approximate solutions of equations from their graphs, including one linear and one quadratic and simple cubic equations

Before you start ...

Review

You should be able to answer these questions.

1 Evaluate these expressions, given that $x = 3$.

Unit A5

a x^2 **b** x^3

c $2x^2$ **d** $x^3 + x$

e $3x^2 - x$ **f** $2x^3 + 2x$

g $x^3 + 4x^2$ **h** $2x^3 + 4x^2 - x$

2 Repeat question **1** with $x = -2$.

Unit A5

3 Describe the graphs of these lines.

Unit A4

a $x = 4$ **b** $y = 5$

c $x = -3$ **d** $y = 2x - 1$

4 **a** Copy and completing the table of coordinates for the line $y = 3x + 2$.

Unit A4

x	1	2	3
y			

Draw the graph of this line.

b Repeat part **a** for the line $2x + y = 12$ using this table.

x	−1	0	
y			0

This spread will show you how to:

● Generate points and plot graphs of simple quadratic functions

Keywords
Curve
Minimum
Parabola
Quadratic

● The graph of a **quadratic** equation is a **parabola**, a U-shaped **curve**.

You need at least 6 points to plot a parabola.

Example

Draw the graph of $y = x^2 + 1$.

First make a table of x and y values.

x	-4	-3	-2	-1	0	1	2	3	4
x^2	16	9	4	1	0	1	4	9	16
$y = x^2 + 1$	17	10	5	2	1	2	5	10	17

$(-4)^2 = 16$

Draw the x-axis from -4 to 4 and the y-axis from 0 to 18.
Plot the points, $(-4, 17)$, $(-3, 10)$ and so on, and join them in a smooth curve.

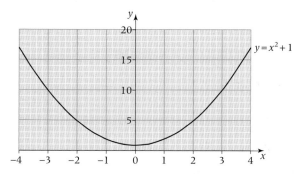

Example

Plot the curve $y = x^2 - x$ for $-2 \leqslant x \leqslant 2$ and find the coordinates of its minimum point.

Make a table of values.

x	-2	-1	0	1	2
x^2	4	1	0	1	4
$-x$	2	1	0	-1	-2
$y = x^2 - x$	6	2	0	0	2

Draw the x-axis from -2 to 2 and the y-axis from -1 to 12.

Plot the points $(-2, 6)$, $(-1, 2)$, ..., $(2, 2)$.

Join the points in a smooth curve.
The **minimum** point is $(\frac{1}{2}, -\frac{1}{4})$.

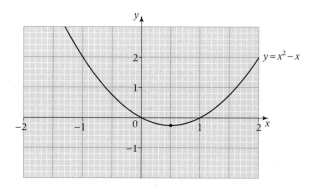

1 **a** Copy and complete this table to show if each graph will be a straight line or a parabola.

 b Add an equation of your own in each column.

Straight line	Parabola

$$y = 3x - 2$$

$$y = x^2 - 2$$

$$3x + 2y = 8$$

$$y = 10 + x^2$$

$$y = x^2 + 2x + 1$$

$$y = x$$

2 **a** Draw axes labelled from −4 to 4 on the x-axis and −5 to +15 on the y-axis.

 b Copy and complete this table for $y = x^2 - 2$.

x	−4	−3	−2	−1	0	1	2	3	4
x^2	16							9	
$y = x^2 - 2$	14							7	

 c Plot the points that you have found in part **b** on your axes from part **a**. Join them to form a smooth parabola.

3 For each equation

 i Make a table with x-values from −4 to 4 and find the corresponding y-values.

 ii Draw an x-axis from −4 to 4 and a suitable y-axis.

 iii Plot the points and join them to form a parabola.

 iv Write the coordinate of the minimum point of each parabola.

 a $y = x^2 + 3$ **b** $y = 2x^2$

 c $y = 3x^2 - 1$ **d** $y = x^2 + x$

> In part **b**, square before you multiply by 2

4 True or false? The point (4, 10) lies on the graph $y = x^2 - 5$. Explain your answer.

5 **a** How do you think the graph $y = 10 - x^2$ will differ from those that you have plotted in questions 2 and 3?

 b Draw and complete a table of coordinates with $-3 \leqslant x \leqslant 3$.

 c Plot your points. Was your prediction correct?

6 **a** Plot the curve $y = x^2 + 5x + 6$ for $-4 \leqslant x \leqslant 4$.

 b Write the coordinates of the points where the graph intersects the x-axis.

 c Use your answer to part **b** to solve $x^2 + 5x + 6 = 0$.

 d Compare your answers to **b** and **c**. What do you notice? Why?

This spread will show you how to:

- Plot graphs of simple cubic and reciprocal functions

Keywords
Cubic
Reciprocal
S-shaped

- A **cubic** equation contains a term in x^3. It has a distinctive curved graph.

Example

Draw the graph of $y = x^3 - 1$ and use it to estimate the value of y when $x = 2.5$.

First make a table of x and y values.

x	−3	−2	−1	0	1	2	3
x^3	−27	−8	−1	0	1	8	27
$y = x^3 - 1$	−28	−9	−2	−1	0	7	26

$(-3)^3 = -27$

Draw the x-axis from −3 to 3 and the y-axis from −30 to 30.
Plot the points, (−3, −28), (−2, −9) and so on, and join them in a smooth curve.

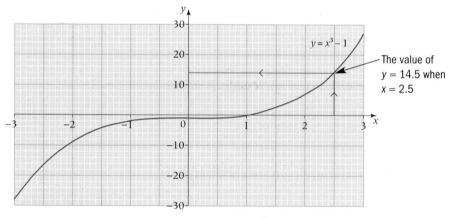

$y = x^3 - 1$

The value of $y = 14.5$ when $x = 2.5$

Cubic graphs have two bends.

The graph of a **cubic** equation is an **S-shaped** curve.

A reciprocal equation contains a term in $\dfrac{1}{x}$.

It has a different-shaped curve.

Example

Draw the graph of $y = \dfrac{1}{x}$.

Use x-axis from −3 to +3.

x	−3	−2	−1	1	2	3
y	$-\frac{1}{3}$	$-\frac{1}{2}$	−1	1	$\frac{1}{2}$	$\frac{1}{3}$

Draw the x-axis from −3 to 3 and the y-axis from −1 to 1.
Plot the points.

Note that you cannot plot $x = 0$ as $\dfrac{1}{0}$ is not a defined function.

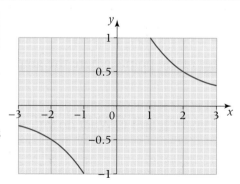

1 Match the graphs with their equations. For the equations that are left over, sketch the shape of their graphs.

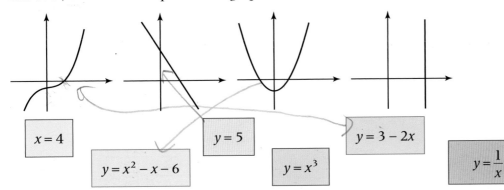

$x = 4$

$y = x^2 - x - 6$

$y = 5$

$y = x^3$

$y = 3 - 2x$

$y = \dfrac{1}{x}$

2 a Draw an x-axis from -3 to 3 and a y-axis from -30 to $+30$.

b Copy and complete the table for $y = x^3 + 1$.

x	-3	-2	-1	0	1	2	3
y	-26						28

when $x = -3$, $(-3)^3 + 1$ which is $-27 + 1$

> When using your calculator, remember brackets.

c Plot the coordinates that you have found in part **b** on your axes from part **a**. Join them to form a smooth, S-shaped curve.

d Use your graph to estimate the value of

  i y when $x = 1.5$ **ii** $0.5^3 + 1$

3 For each equation, copy and complete the table of values. Plot these points on suitable axes and join them to form a smooth curve.

a $y = x^3 - 4$

x	-2	-1	0	1	2	3
x^3					8	
$x^3 - 4$					4	

b $y = \dfrac{2}{x}$

x	-2	-1	1	2	3
$\dfrac{1}{x}$				$\dfrac{1}{2}$	
y				1	

c $y = x^3 + x + 1$

x	-2	-1	0	1	2	3
x^3					8	
$x + 1$					3	
y					11	

d $y = \dfrac{3}{x} + 1$

x	-2	-1	1	2	3
$\dfrac{3}{x}$	$-\dfrac{3}{2}$				
y	$-\dfrac{1}{2}$				

4 a On a pair of axes for $-3 \leqslant x \leqslant 3$ and suitable y values, plot

  i $y = 8 - x^3$ **ii** $y = 2x^2 - 3x$

b Use your graph to find the points where $8 - x^3 = 2x^2 - 3x$.

This spread will show you how to:

● Use graphs to find the solutions or approximate solutions of two simultaneous equations

Keywords
Intersection
Simultaneous
Solution

You can solve **simultaneous** equations on a graph.

Example

Solve $3x - y = 2$ (1) $2x + y = 8$ (2)

Plot their graphs. $3x - y = 2 \Rightarrow y = 3x - 2$

$$2x + y = 8$$

x	1	2	3
y	1	4	7

x	0	4	2
y	8	0	4

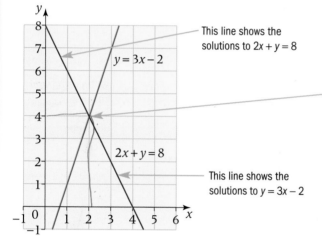

This line shows the solutions to $2x + y = 8$

$y = 3x - 2$

At the point of **intersection** of the two lines, (2, 4) both equations have a **solution**. At the intersection $x = 2$ and $y = 4$

$2x + y = 8$

This line shows the solutions to $y = 3x - 2$

Check:
(1): $3 \times 2 - 4 = 2$
(2): $2 \times 2 + 4 = 8$

The solution is $x = 2$, $y = 4$.

● **The solution to simultaneous equations is where their graphs intersect.**

Example

Three times one number plus twice another is 9. Twice the first number subtract the second is 13. Use a graphical method to find the two numbers.

Let the numbers be x and y. Hence,
$3x + 2y = 9$
    $2x - y = 13$ ...

$3x + 2y = 9$

x	1	3	2
y	3	0	1.5

$2x - y = 13$

x	0	$6\frac{1}{2}$	3
y	−13	0	−7

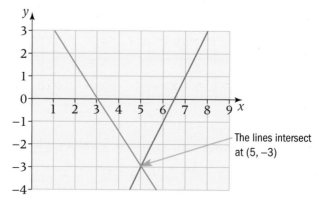

The lines intersect at (5, −3)

Don't just give the coordinates as the solution. Say what each number is.

Hence, the solution is $x = 5$, $y = -3$.
The two numbers are 5 and −3.

1 Use the graph to solve these pairs of simultaneous equations.

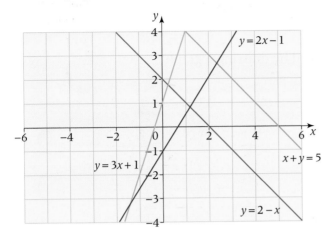

a i $y = 2x - 1$ and $x + y = 5$ **ii** $y = 3x + 1$ and $x + y = 5$

  iii $y = 2 - x$ and $y = 2x - 1$ **iv** $y = 3x + 1$ and $y = 2 - x$

b Using the graph, explain why the simultaneous equations $x + y = 5$ and $y = 2 - x$ have no solution.

2 Plot graphs to solve these simultaneous equations.

  a $y = 2x + 1$ **b** $y = 3x - 2$ **c** $2x + y = 5$
     $x + y = 10$ $x + y = 2$ $x - y = 4$

3 **a** Solve $3x + 2y = 4$ and $x + 4y = 3$ graphically.   See A6.4.

  b Solve $3x + 2y = 4$ and $x + 4y = 3$ algebraically.

  c What should you notice about your answers to part **a** and part **b**?

4 Use a graphical method to solve these problems.

  a Twice one number plus three times another is 4.
    Their difference is 2. What are the numbers?

  b The sum of the ages of James and Isla is 4. The difference between twice Isla's age and treble James' age is 3. How old are they?

5 **a** By plotting graphs if necessary, explain why the simultaneous equations $y = 2x - 1$ and $y = 2x + 4$ have no solution.

  b Is it possible to have a pair of simultaneous equations with more than one solution?

6 Two lines intersect. One has gradient 4 and y-axis intercept 3. The other has gradient 6 and cuts the y-axis at (0, 1).

  a Write the simultaneous equations they represent.

  b Solve the equations algebraically.

  c What is the point of intersection of the lines?

This spread will show you how to:

- Find approximate solutions of equations from their graphs, including one linear and one quadratic and simple cubic equations

Keywords

Intersect
Solution

Quadratic and cubic equations can be solved by drawing a graph.

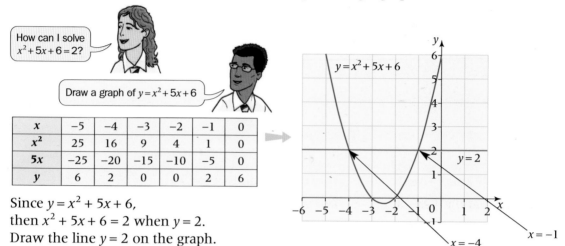

How can I solve $x^2 + 5x + 6 = 2$?

Draw a graph of $y = x^2 + 5x + 6$

x	-5	-4	-3	-2	-1	0
x^2	25	16	9	4	1	0
$5x$	-25	-20	-15	-10	-5	0
y	6	2	0	0	2	6

Since $y = x^2 + 5x + 6$,
then $x^2 + 5x + 6 = 2$ when $y = 2$.
Draw the line $y = 2$ on the graph.

The line $y = 2$ crosses the curve $y = x^2 + 5x + 6$ when $x = -1$ and $x = -4$.

   $x^2 + 5x + 6 = 2$ when $x = -1$ and $x = -4$

The **solution**(s) to $x^2 + 5x + 6 = 2$ are where $y = 2$ and $y = x^2 + 5x + 6$.
These are the points where the line and curve **intersect**.

Example

On the graph of $y = x^3 + x^2 - 6x$, where would you find the solutions to $x^3 + x^2 - 6x = 0$? Find them.

The solution of $x^3 + x^2 - 6x = 0$, is where the curve, $y = x^3 + x^2 - 6x$, and the line, $y = 0$, intersect.

$y = 0$ is the same line as the x-axis.

The graph
$y = x^3 + x^2 - 6x$
intersects the
x-axis at $(-3, 0)$,
$(0, 0)$ and $(2, 0)$,
so the solutions
of $x^3 + x^2 - 6x = 0$
are $x = -3$, $x = 2$
and $x = 0$.

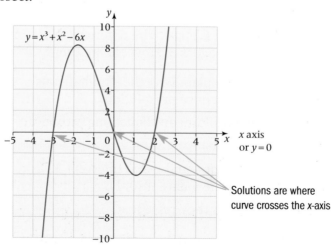

x axis
or $y = 0$

Solutions are where
curve crosses the x-axis

Notice the
characteristic 'S'
shape of the cubic
equation.

You can check by substituting -3, 2 and 0 for x in the equation.

1 Some graphs are drawn on these axes.

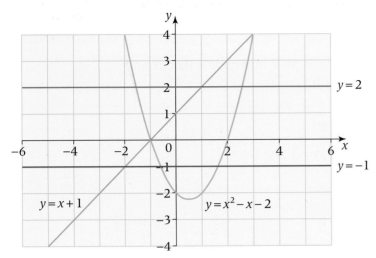

Use the graphs to find the approximate solutions of

a $x^2 - x - 2 = 2$ **b** $x^2 - x - 2 = -1$ **c** $x^2 - x - 2 = x + 1$

2 The graph shows $y = x^2 - 2x - 3$.

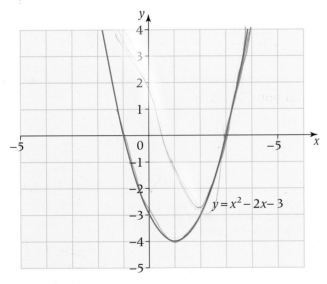

Copy the graph and, by adding lines, use it to find the approximate solutions of

a $x^2 - 2x - 3 = 1$ **b** $x^2 - 2x - 3 = -3$ **c** $x^2 - 2x - 3 = -4$

d $x^2 - 2x - 3 = x - 2$ **e** $x^2 - 2x - 3 = 1 - x$ **f** $x^2 - 2x - 3 = 0$

3 Draw appropriate graphs to find the approximate solutions of

a $x^2 - 2 = 5$ **b** $x^3 + x = 2x - 1$ **c** $2x^2 - x = 0$ **d** $x^3 - x^2 = 2$

This spread will show you how to:

● Find the approximate solutions of simultaneous equations from graphs, when one is linear and one quadratic

Keywords

Linear
Quadratic
Solution

You can solve two simultaneous equations, one of which is **linear** and the other **quadratic**, by drawing a graph.
Look to the points of intersection.

Example

Solve the simultaneous equations $y = 11x - 2$ and $y = 5x^2$.

Make a table of values for each equation.

$y = 11x - 2$

x	-1	0	1
$11x$	-11	0	11
y	-13	-2	9

$y = 5x^2$

x	-3	-2	-1	0	1	2	3
$5x^2$	45	20	5	0	5	20	45

Plot the graphs.

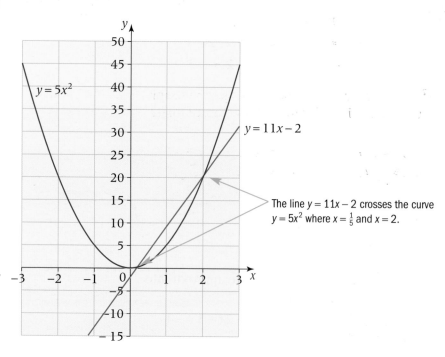

The line $y = 11x - 2$ crosses the curve $y = 5x^2$ where $x = \frac{1}{5}$ and $x = 2$.

The **solutions** of the simultaneous equations
$y = 11x - 2$ and $y = 5x^2$ are
$x = \frac{1}{5}$, $y = \frac{1}{5}$ and $x = 2$, $y = 20$.

1 Use the graphs to find approximate solutions of

   a $x^2 + 2x - 3 = 0$ **b** $x^2 + 2x - 3 = x + 1$

   c $x^2 + 2x - 5 = 0$ **d** $x^2 + x = 0$

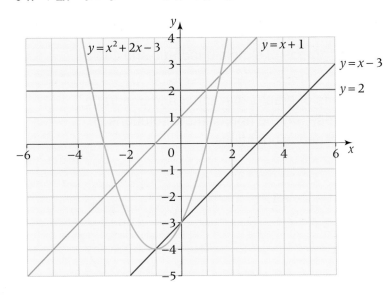

2 Given that the graph of $y = x^2 + 4x - 2$ is already drawn, which one line would you need to draw in order to solve

   a $x^2 + 4x - 2 = 3$ **b** $x^2 + 4x - 2 = 0$ **c** $x^2 + 4x - 2 = 2x + 1$

   d $x^2 + 4x = 6$ **e** $x^2 + 5x = x + 4$ **f** $x^2 + 2 = 6x$?

You do not have to draw them.

3 The graph of $y = x^3 + x^2 - 1$ has been drawn below. Copy the graph and add on graphs of your choice in order to find the approximate solutions of

   a $x^3 + x^2 - 1 = 2$ **b** $x^3 + x^2 = x + 2$ **c** $x^3 + x^2 - 1 = -2$

   d $x^3 + x^2 - x = 0$ **e** $x^3 + x^2 - 2x + 4 = 0$ **f** $x^3 - 1 = 1 - 2x^2$

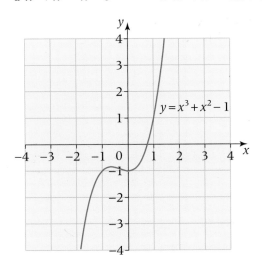

Key objectives

- Construct linear functions and plot the corresponding graphs
- Generate points and plot graphs of simple quadratic and cubic functions
- Find the intersection points of graphs of a linear and quadratic function, knowing that these are the approximate solutions of the corresponding simultaneous equations

1 Here is the graph of $y = x^2 - 2x + 1$.

Copy the graph.

a Draw the graph of $y = 4$ onto the same axes. (2)

b Use your graph to find both solutions of the equation $4 + 2x = x^2 + 1$.

Explain your method. (2)

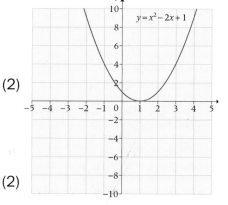

2 a Copy and complete the table of values for $y = 2x + 3$. (2)

x	-2	-1	0	1	2	3
y		1	3			

b Copy the grid and draw the graph of $y = 2x + 3$. (2)

c Use your graph to find

  i the value of y when $x = -1.3$

  ii the value of x when $y = 5.4$ (2)

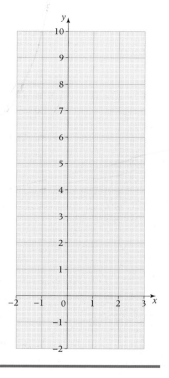

(Edexcel Ltd., 2004)

This unit will show you how to

- Use ratio notation
- Simplify a ratio by cancelling common factors
- Divide a quantity in a given ratio
- Convert between ratios, proportions, percentages, fractions and decimals
- Solve problems involving ratio, proportion and percentages
- Solve problems using reverse percentages
- Check answers by substituting solutions into the initial problem

Before you start ...

You should be able to answer these questions.

Review

1 Gill and Paul share £500 between them.
Gill gets four times as much as Paul.
How much does each person get?

Unit N7

2 In a sale, you can buy three items for the price of two.
What percentage decrease on the original price does this represent?

Unit N7

3 Robyn mixes black and white paint to make two different batches of grey paint.

Unit N4

Batch A contains 3 litres of black paint and 7 litres of white paint.

Batch B contains 5 litres of black paint and 5 litres of white paint.

a Which batch is the darker colour?
Explain your reasoning.

b Robyn wants to make 30 litres of grey paint that is the same shade as Batch A.
How much black paint and white paint will she need?
Explain your answer.

This spread will show you how to:

- Use ratio notation
- Simplify a ratio by cancelling common factors
- Divide a quantity in a given ratio

A **ratio** is used to compare quantities.

- You **simplify** a ratio by cancelling common factors.
  A class contains 16 girls and 12 boys. The ratio of boys to girls is 16 : 12, and you can simplify this ratio to 4 : 3.

- To divide a quantity in a given ratio (for example, dividing 20 m in the ratio 3 : 1)
  - first find the total number of parts: 3 + 1 = 4 ✓
  - now find the size of each part: 20 ÷ 4 = 5 ✓
  - multiply to find each share: 5 × 3 = 15, and 5 × 1 = 5

Example

In a group of 14 people, eight have brown eyes and six have green eyes. Write the ratio of eye colours in its simplest form.

brown : green = 8 : 6 = 4 : 3

Example

Simplify these ratios.

a 1 m to 40 cm

b 6 hours to $2\frac{1}{2}$ days

a 1 m = 100 cm so the ratio of
1 m to 40 cm = 100 cm : 40
$\qquad$ = 5 : 2

b $2\frac{1}{2}$ days = 60 hours so the ratio
6 hours to $2\frac{1}{2}$ days = 6 : 60
$\qquad$ = 1 : 10

Both parts of the ratio must have the same units.

Example

Divide $\qquad$ **a** 120 cm in the ratio 2 : 3 $\qquad$ **b** £108 in the ratio 3 : 5

a $\quad$ 2 + 3 = 5 $\qquad$ Total number of parts = 5
$\quad$ 120 ÷ 5 = 24 $\qquad$ One part = 24 cm
$\quad$ 2 × 24 = 48
$\quad$ 3 × 24 = 72
The two lengths are 48 cm and 72 cm.
To check your answer add the two parts:
$\quad$ 48 + 72 = 120, correct

b 3 + 5 = 8
$\quad$ 108 ÷ 8 = 13.5
$\quad$ 3 × 13.5 = 40.5
$\quad$ 5 × 13.5 = 67.5
The two amounts are £40.50 and £67.50.
Check your answer:
$\quad$ 40.50 + 67.50 = 108, correct

Remember to add £ signs and zeros.

1 Here are the numbers of boys and girls in some Year 11 classes.
   For each class, write down the ratio of boys to girls.

   a Class 11A has 17 boys and 13 girls

   b Class 11B has 11 boys and 19 girls

   c Class 11C has 14 boys and 15 girls

2 The table shows the number of
   students in three different classes who
   own pets. Write the ratio of pet owners
   to non-pet owners in each class.

Class	Pet owners	Non-pet owners
11A	7	23
11B	13	17
11C	16	13

3 Write the ratio of the number of
   vowels to the number of consonants in these words.
   Give your answers in their simplest form.

   a JAGUAR   **b** PANTHER   **c** TIGER   **d** LEOPARD

   e PAPERBACK   **f** PERPETUATE   **g** STAGGERS   **h** MEIOSIS

4 Simplify these ratios.

   a 6 : 4   **b** 12 : 3   **c** 5 : 10   **d** 2 : 8

   e 6 : 9   **f** 12 : 8   **g** 10 : 15   **h** 21 : 14

5 Divide each of these numbers in the ratio 3 : 2. Show your working.

   a 40   **b** 45   **c** 90   **d** 100

   e 250   **f** 2000   **g** 7500   **h** 9250

6 Divide each of these quantities in the ratio 2 : 1. Show your working.

   a 24 hours   **b** 180°   **c** 45 minutes   **d** €360

   e 246 cm   **f** 120 kg   **g** 81 km   **h** 54 miles

7 Divide £360 in these ratios.

   a 1 : 1   **b** 1 : 2   **c** 2 : 1   **d** 3 : 5

8 Karla and Wayne share the tips they receive for working in a café.
   One Saturday, Karla's share was £12.50, and Wayne's was £7.50.

   a Write the ratio of Karla's tips to Wayne's tips in its simplest form.

   b The next week Karla and Wayne shared tips of £22 in the same
   ratio. Find their shares. Show all your working.

9 Mrs Jones wins £500 in a competition. She keeps 40% of the money.

   a How much money does Mrs Jones keep for herself?

   b Mrs Jones shares the rest between her children,
   Annie (8 years old) and Ben (12), in the ratio of their ages.

   Work out how much each child receives. Show your working.

This spread will show you how to:

- Use ratio notation
- Simplify a ratio by cancelling common factors
- Divide a quantity in a given ratio

Keywords
Ratio
Simplify

You can use **ratio** to divide quantities into more than two amounts.

- To divide 55 m in the **ratio** 2 : 3 : 5
  - first add: 2 + 3 + 5 = 10
  - then divide: 55 m ÷ 10 = 5.5 m.
  - then multiply: 2 × 5.5 m = 11 m, 3 × 5.5 m = 16.5 m, and 5 × 5.5 m = 27.5 m

Check:
11 + 16.5 + 27.5
= 55

Example

Alan and Betty share a bingo prize of £48 in the ratio of their ages. Alan is 32, and Betty is 26. How much does each get?

Ratio of ages = 32 : 26 = 16 : 13   **Simplify** the ratio

$\quad$ 16 + 13 = 29, so one part = £48 ÷ 29   Add the parts

Alan gets $\dfrac{£48}{29} \times 16 = £26.48$ Round to nearest penny

Betty gets $\dfrac{£48}{29} \times 13 = £21.52$ Check:
£26.48 + £21.52
= £48

Example

Roni weighs 56 kg, Mike weighs 64 kg and Steffi weighs 72 kg. Write the ratio of their weights in its simplest form.

Roni's weight : Mike's weight : Steffi's weight = 56 : 64 : 72 = 7 : 8 : 9   Cancel by 8

Example

The Smith, the Brown and the Jones families go on holiday to Cumbria. They stay in a farmhouse which they rent for £2000. They agree to share the rent according to the number of people in each family. There are 3 in the Smith family, 6 in the Brown family and 8 in the Jones family. How much does each family pay?

Ratio of family size = 3 : 6 : 8

$\quad$ 3 + 6 + 8 = 17, so one part = £2000 ÷ 17

The Smiths pay $\dfrac{£2000}{17} \times 3 = £352.94$

The Browns pay $\dfrac{£2000}{17} \times 6 = £705.88$

The Joneses pay $\dfrac{£2000}{17} \times 8 = £941.18$ Check: £352.94 + £705.88 + £941.18 = £2000

1 Write these ratios in their simplest form.

   a 2 : 4   **b** 2 : 4 : 4   **c** 9 : 3 : 6

   d 10 : 20 : 8   **e** 8 : 20 : 12   **f** 15 : 35 : 10

2 Split £240 in these ratios.

   a 1 : 2   **b** 2 : 3   **c** 5 : 3

   d 1 : 2 : 3   **e** 5 : 1 : 2   **f** 7 : 2 : 3

3 Divide 250 m in these ratios, giving your answers correct to the nearest centimetre.

   a 1 : 4   **b** 2 : 3   **c** 2 : 5

   d 6 : 1   **e** 6 : 7   **f** 9 : 5

4 Peter, Bob and Yasmin share prize money of £7500 between them in the ratio 9 : 5 : 11. How much do they each receive?

5 Ann, Charles and Edward divide prize money of £850 between them in the ratio 4 : 3 : 2. Find the amount that each receives, giving your answers to the nearest penny.

6 Copy and complete the table to show how each quantity can be divided in the ratio given. Give your answers to a suitable degree of accuracy.

Quantity	Ratio	Share 1	Share 2	Share 3
200 km	2:3:5			
38 kg	1:2:3			
450 cm	2:3:8			
£720	4:5:10			
95 litres	1:6:7			

7 A fruit drink contains mango juice, pineapple juice and passion fruit juice, in the ratio 4 : 3 : 2 by volume. Calculate the volume of each type of juice in a 1-litre pack of the fruit drink.

8 Mrs Williams won a £500 Premium Bond prize, and decided to share it among her three children in the ratio of their ages. The children are aged 5, 7 and 9.

   a Calculate the amount that each child receives, giving your answers correct to the nearest penny.

   b Exactly one year later, Mrs Williams wins another £500 prize. Again, she decides to divide it among her children in the ratio of their ages. Calculate the amount that each child receives this time.

9 Robert and Kathleen are business partners. At start up, Robert invested £150 000 and Kathleen invested £200 000. Each year they share the profits in the ratio of their investments. In 2004 the profit amounted to £29 540. Work out each partner's share.

This spread will show you how to:

- Solve problems involving ratio and proportion

Keywords

Fraction
Proportion
Ratio

- **Ratios** compare one number with another.
  The ratio of boys to girls is 3 : 2.

- **Proportions** tell you what fraction of the whole amount something is.
  $\frac{3}{5}$ of the students are boys.

- A ratio can be used to find a proportion.
  A drink is made by mixing squash and water in the ratio 1 : 4, so one part out of every five is squash. The proportion of squash in the drink is $\frac{1}{5}$.

Example

A concrete mixer contains 5 kg of cement and 20 kg of sand. Find

a the ratio of sand to cement

b the proportion of sand in the mixture.

a Ratio sand : cement = 20 : 5 = 4 : 1

b Proportion of sand = $\frac{20}{25} = \frac{4}{5}$

Write the weight of sand as a fraction of the total.

Example

Pink paint is made from one-third red paint and two-thirds white paint. Write the ratio of red paint to white paint.

Ratio red : white = $\frac{1}{3} : \frac{2}{3}$

= 1 : 2

Multiply by 3 to simplify.

Example

The ratio of hardback books to paperbacks on a bookshelf is 2 : 7. What proportion of the books are paperbacks?

Proportion of paperbacks = $\frac{7}{2+7}$

= $\frac{7}{9}$

Write the fraction $\frac{\text{paperbacks}}{\text{total}}$

Example

The ratio of boys to girls in a class is 2 : 3.
Alex is working out the proportion of the class that are boys.
He says it's $\frac{2}{3}$.
What is wrong with Alex's reasoning?

The answer can't be right – there are more girls than boys in the class.
The proportion of boys = $\frac{2}{2+3} = \frac{2}{5}$

Remember to add the parts when you find a proportion.

1 Orange paint is made from 2 parts red to 3 parts yellow paint.

   a Write the ratio of red paint to yellow paint in the mixture.

   b Find the proportion of red paint in the mixture, giving your
   answer as a fraction.

   The answer is
   not $\frac{2}{3}$!

2 A bowl contains 200 grams of flour and 100 grams of sugar.

   a Write the ratio of sugar to flour.

   b Find the proportion of sugar in the mixture.

3 A string of decorative lights has 7 red bulbs, 21 green bulbs and
   14 blue bulbs.

   a Write the ratio of the number of bulbs of each colour in its simplest
   form.

   b Calculate the proportion of

   i red bulbs   **ii** green bulbs   **iii** blue bulbs.

4 The table shows the ratio of blue paint to yellow paint in mixtures to
   make different shades of green paint. For each one, calculate the
   proportion of the mixture that is blue paint.

a 1:1	**b** 1:2	**c** 2:3	**d** 3:5	**e** 4:3	**f** 5:2

5 Write the proportion of yellow paint in each of the mixtures given in
   question **4**.

6 A fruit drink contains $\frac{2}{5}$ water and $\frac{3}{5}$ fruit juice. Find the ratio of
   water to fruit juice in the drink.

7 A swimming club had a total income of £2000. It spent 30% of this
   income on pool hire, and the rest on instructors' fees.

   a Calculate the amount spent on pool hire.

   b Work out the ratio of the amount spent on pool hire to the
   amount spent on fees.

8 Tom has £2200. He gives $\frac{1}{4}$ to his son and $\frac{2}{5}$ to his daughter.
   How much does Tom keep for himself? You must show all
   your working.

9 Andrew records how he spends his time over a 24-hour period.
   He spends $\frac{1}{3}$ of the time sleeping and $\frac{1}{6}$ of the time travelling.
   What proportion of his time does Andrew spend on activities other
   than sleeping and travelling? Show your working.

10 Mrs Smith inherits £16 000. She divides the money between her
   three children John, Sarah and Mark in the ratio 6 : 7 : 8,
   respectively. How much does Sarah receive? What proportion of the
   money does Mark receive?

N8.4 Ratio, proportion and percentages

This spread will show you how to:

● Convert between ratios, proportions, percentages, fractions and decimals

Keywords
Approximation
Fraction
Percentage
Proportion
Ratio

You can describe **proportions** using **percentages**.

● Proportions expressed as percentages are easy to compare.

'The proportion of sugar is $\frac{1}{2}$ in recipe A and $\frac{2}{5}$ in recipe B'

is the same as

'Recipe A contains 50% sugar and recipe B contains 40% sugar'.

Sometimes it is impossible to express a proportion exactly as a percentage. For example, $\frac{1}{7}$ is an exact proportion, but 14% (or even 14.29%) is only an **approximation** to it.

Example

The metal alloy, constantan, is made from nickel and copper in a **ratio** of 2 : 3. What percentage of the alloy consists of nickel?

$2 + 3 = 5$

Proportion of nickel $= \frac{2}{5}$

Percentage of nickel $= \frac{2}{5} = \frac{2 \times 20}{5 \times 20} = \frac{40}{100} = 40\%$

Add the parts of the ratio.

Example

Here is Emma's fruitcake recipe.

Fruitcake (1kg)

125 g of butter
150 g of flour
100 g of sugar
450 g of dried fruit
170 g of eggs
5 g of spices

What percentage of the cake is butter?

Proportion of butter $= \frac{125}{1000} = \frac{1}{8}$

Percentage of butter $= \frac{125}{1000} \times 100\%$

$= 12.5\%$

Find the proportion of butter first.

Example

The ratio of the number of wins, draws and losses for a football team one season is 5 : 4 : 3. What percentage of the games are wins?

$5 + 4 + 3 = 12$

Proportion of wins $= \frac{5}{12}$

$\approx 42\%$ to nearest whole number

1 A tub of fruit yogurt contains fresh yogurt and fruit in the ratio 9 : 1 by weight. Find the percentage of fruit in the contents of the tub.

2 **a** Write the ratio of the number of vowels to the number of consonants in the word CATERPILLAR.

 b Find the percentage of the letters in the word CATERPILLAR that are vowels. Show your working.

3 The compositions of three different alloys of copper are shown in this table. Find the percentage of copper in each alloy.

Alloy	Composition
Nickeline	4 parts copper, 1 part nickel
US nickel coinage	3 parts copper, 1 part nickel
Medal bronze	93 parts copper, 7 parts tin

4 Scott and Maxine buy a present for their father. They share the cost in the ratio 3 : 2. What percentage of the cost does Scott pay?

5 The ratio of boys to girls in a class is 5 : 4. What percentage of the class is girls? Show your working.

6 The ABC mobile telephone company keeps records of calls made on its network. The ratio of the number of calls that are successfully connected to those that are missed for any reason is 4 : 1.

 a Find the percentage of calls on the network that are successfully connected.

 b The 123 mobile telephone company also keeps records of calls on their network. They say that 75% of calls are connected successfully.
 Find the ratio of successful to missed calls on the 123 network.

7 Dan and Phil share £3800 between them in the ratio 3 : 7, respectively.

 a Calculate the amount received by each person.

 b Find the percentage of the total amount of money that was received by Dan.

8 Leon and Frieda divide $500 in the ratio 61 : 33, respectively.

 a Calculate the amount received by each person.

 b Find the percentage of the total amount that Frieda receives.

9 German silver is made from copper, zinc and nickel in the ratio 16 : 5 : 3.

 a Calculate the mass of each metal in 500 g of German silver.

 b Work out the percentage of each metal in German silver.

This spread will show you how to:

• Solve problems using reverse percentages

Keywords

Decrease
Increase
Original
Percentage

In a reverse **percentage** problem, you are given an amount after a percentage change, and you have to find the **original** amount.

You have already solved some reverse percentage problems in N7.5.

Example

In a sale, a pair of shoes cost £38.25 after a 15% **decrease**.

Find the original price of the shoes.

Sale price is $(100 - 15)\% = 85\%$ of original price
$$85\% = 0.85$$
Original price $\times 0.85 =$ sale price
$$\text{Original price} = \frac{£38.25}{0.85} = £45$$

First work out the reverse percentage.

0.85 is the decimal equivalent of 85%.

Example

A table costs £88, including 17.5% VAT. Find the cost (to the nearest pound) before VAT was added.

£88 is 117.5% of cost before VAT.
$$117.5\% = 1.175$$
Cost before VAT $\times 1.175 = £88$

Cost before VAT $= \dfrac{£88}{1.175} = £75$ to nearest pound.

Find the decimal equivalent.

Example

The price of unleaded petrol is increased by 4% to 93.7p per litre. Find the price before the **increase**.

Increased price is 104% of old price.
$$100\% = 1.04$$
Old price $\times 1.04 =$ increased price

Old price $= \dfrac{93.7}{1.04} = 90.1$p per litre to nearest 0.1p

Example

A car costs £15 000, including VAT at 17.5%. Veejay is working out the cost without VAT. He thinks: 'I'll just work out 17.5% of £15 000 and take that away, to get £12 375'.
What is wrong with Veejays's reasoning?

Original price   +   VAT   =   £15 000
     100%   17.5%   =   117.5%

£15 000 ÷ 1.175 = £12 766 (to nearest pound)

1 Find the result of these following percentage increases.
Do these mentally.

  a £100 is increased by 25%   **b** £50 is increased by 20%

  c 40 m is increased by 15%   **d** 36 cm is increased by 50%

2 Calculate mentally the result of these percentage decreases.

  a £60 is decreased by 25%   **b** 80 cm is decreased by 75%

  c 720 cm is decreased by 50%   **d** 120 mm is decreased by 15%

3 Write a decimal number that you could multiply a quantity by
to find the results of these percentage changes.

  a An increase of 20%   **b** A decrease of 15%

  c An increase of 6%   **d** A decrease of 5%

  e A decrease of 6%   **f** A decrease of 17%

4 Calculate the result of these percentage increases and decreases, by
multiplying by the correct decimal number. Show your method.
You may use a calculator.

  a £64 decreased by 7%   **b** 45 kg increased by 14%

  c 10.4 seconds decreased by 3%   **d** 120 m increased by 65%

  e €240 increased by 8.5%   **f** $340 decreased by 11.5%

5 Calculate the original cost of these items,
before the percentage changes shown.
Show your method. You may use a calculator.

  a A hat that costs £46.50 after a 7% price cut.

  b A skirt that costs £32.80 after a price rise of 6%.

6 A computer costs £658, including VAT at 17.5%.
Find the price, before VAT was added.

7 To decrease an amount by 8%, multiply it by 0.92.
For a further decrease of 8%, multiply by 0.92 again, and so on.
Use this idea to calculate

  a the final price of an item with an original price of £380, which is
  given two successive price cuts of 8%.

  b the final price of an item with an original price of £2400,
  which is given three successive price cuts of 10%.

Exam review

Key objectives

- Divide a quantity in a given ratio
- Solve word problems about ratio and proportion
- Reverse percentage problems

1 Dave is looking for a pair of trainers.
He finds the same type of trainers in two different shops.

Shop A is selling the trainers with a 30% reduction.
The new price of the trainers in Shop A is £30.

The original price of the trainers in Shop B was £42.
Shop B is now selling the trainers at $\frac{3}{4}$ their original price.

a Which shop has the lowest original price for the trainers? (3)

b Which shop is selling the trainers for the lowest price now? (3)

2 In 2002, Shorebridge Chess Club's total income came
from a council grant and members' fees.

Council grant   £50
Members' fees   240 at £5 each

a i Work out the total income of the club for the year 2002.

  ii Find the council grant as a fraction of the club's total
income for the year 2002. Give your answer in its
simplest form. (3)

In 2001, the club's total income was £1000. The club
spent 60% of its total income on a hall. It spent a further
£250 on prizes.

b Work out the ratio
the amount spent on the hall : the amount spent on prizes.
Give your answer in its simplest form. (3)

(Edexcel Ltd., 2003)

This unit will show you how to

- Enlarge objects, given a centre of enlargement and scale factor, including fractional scale factors
- Recognise that enlargements preserve angle but not length
- Understand how enlargement affects perimeter
- Describe an enlargement by giving the scale factor and centre of enlargement
- Understand, identify and use scale factors
- Understand similarity of 2-D shapes, using this to find missing lengths and angles
- Use common sense to check answers to geometric problems

Before you start ...

You should be able to answer these questions.

1 Draw a grid with axes from −3 to +5. Draw these points on the grid.

A (1, 3)   B (3, −2)   C (4, 5)

D (5, 2)   E (−3, 3)   F (0, 4)

G (−1, 0)

2 Write these ratios in their simplest form.

a 4 : 12   **b** 5 : 30   **c** 6 : 9

d 16 : 8   **e** 15 : 6

3 Two parallel lines have lengths 8 cm and 3 cm.

Write the ratio of the lengths of these lines.

4 Solve these equations.

a $\frac{x}{2} = 7$ **b** $\frac{x}{5} = 3$

c $\frac{x}{6} = \frac{5}{3}$ **d** $\frac{x}{4} = \frac{9}{2}$

Review

Key stage 3

Unit N8

Unit N8

Unit A2

Enlargement

This spread will show you how to:

- Enlarge objects, given a centre of enlargement and scale factor, including fractional scale factors
- Recognise that enlargements preserve angle but not length
- Understand how enlargement affects perimeter

Keywords

Centre of enlargement
Enlargement
Scale factor
Similar

To enlarge a shape, you multiply all the lengths by the same scale factor.

- In an **enlargement**
  - corresponding angles are the same
  - corresponding lengths are in the same ratio.

You draw an enlargement from a centre.

In the diagram, △PQR is enlarged by **scale factor** 2 from the centre (0, 1).

The image, P′Q′R′, is **similar** to the object, PQR.

All the distances are × 2 so CQ′ = 2CQ, CP′ = 2CP, CR′ = 2CR

Lengths on the image are 2 × corresponding lengths on the object so P′R′ = 2PR and so on.

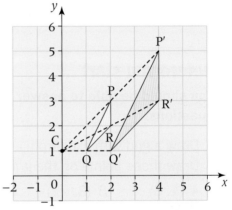

- Perimeter of image = scale factor × perimeter of object.

Example

a Enlarge the white triangle by scale factor 3, centre (2, 1).
b How much larger is the perimeter of the image than the perimeter of the object?

a

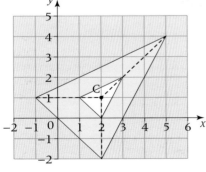

Lengths on the image are 3 × the corresponding lengths on the object.

The distance from C to a point on the image is 3 × the distance from C to the corresponding point on the object.

b Perimeter of image = 3 × perimeter of object.

1 **a** Copy this diagram, but extend both axes to 16.

   b Enlarge triangle T by scale factor 2, centre (0, 0). Label the image U.

   c Enlarge triangle T by scale factor 3, centre (0, 0). Label the image V.

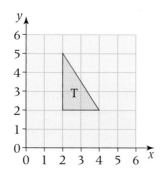

2 **a** Copy this diagram, but extend both axes from −7 to 7.

   b Enlarge kite K by scale factor 2, centre (1, 3). Label the image L.

   c Enlarge kite K by scale factor 4, centre (1, 3). Label the image M.

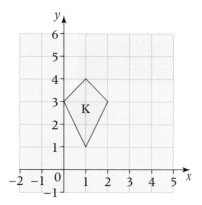

3 **a** Draw a grid with an x-axis from −6 to 6 and a y-axis from −3 to 15. Plot the points (1, 3) (1, 5) (3, 7) (4, 6). Join them to make quadrilateral Q.

   b Enlarge quadrilateral Q by scale factor 2, centre (4, 8). Label the image R.

   c Enlarge quadrilateral Q by scale factor 2, centre (0, 6). Label the image S.

   d How much larger is the perimeter of R than the perimeter of Q?

4 **a** Draw a grid with x- and y-axes from 0 to 13. Plot the points (1, 3) (4, 4) (2, 1). Join them to make triangle A.

   b Enlarge triangle A by scale factor 3, centre (0, 0). Label the image B.

   c Enlarge triangle A by scale factor 2, centre (2, 1). Label the image C.

   d How much larger is the perimeter of B than the perimeter of A?

DID YOU KNOW?
The E. coli. bacteria in this picture has been enlarged by a scale factor of 5000.

315

Fractional scale factors

This spread will show you how to:

- Enlarge objects, given a centre of enlargement and scale factor, including fractional scale factors

Keywords

Centre of enlargement
Enlargement
Scale factor

A map or a scale drawing is an enlargement by a fractional scale factor.

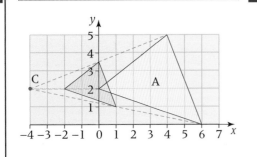

This map is an enlargement of St Michael's Mount by scale factor $\frac{1}{25000}$.

- Enlargement by a scale factor less than 1 produces a smaller image.

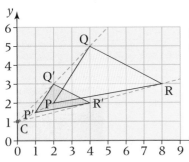

Enlargement by scale factor $\frac{1}{2}$, centre $(0, 1)$

All the distances are multiplied by $\frac{1}{2}$ so $CQ' = \frac{1}{2}CQ$, $CP' = \frac{1}{2}CP$, $CR' = \frac{1}{2}CR$.

Lengths on the image are half the corresponding lengths on the object so $P'R' = \frac{1}{2}PR$ and so on.

Enlargement by scale factor $\frac{1}{2}$ is the inverse of enlargement by scale factor 2.

Example

Enlarge triangle A by scale factor $\frac{1}{2}$, centre $(-4, 2)$.

Example

Enlarge triangle B by scale factor $\frac{1}{3}$ centre $(-2, -2)$.

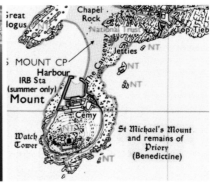

Lengths on the image are $\frac{1}{3}$ corresponding lengths on the object.

1 a Copy this diagram.

   b Enlarge triangle T by scale factor $\frac{1}{2}$, centre (0, 0). Label the image U.

   c Enlarge triangle T by scale factor $\frac{1}{2}$, centre (2, 2). Label the image V.

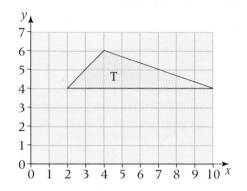

2 a Copy this diagram.

   b Enlarge kite K by scale factor $\frac{1}{3}$, centre (0, 0). Label the image L.

   c Enlarge kite K by scale factor $\frac{1}{3}$, centre (3, 3). Label the image M.

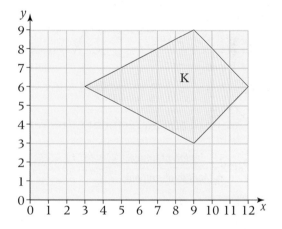

3 a Draw a grid with an x-axis from −3 to 5 and a y-axis from −2 to 10.
Plot the points (1, 3) (1, 6) (3, 9) (4, 6).
Join them to make quadrilateral Q.

   b Enlarge quadrilateral Q by scale factor $\frac{1}{3}$, centre (4, 3).
Label the image R.

   c Enlarge quadrilateral Q by scale factor $\frac{1}{2}$, centre (−2, −2).
Label the image S.

4 a Draw a grid with x-axis from −6 to 6 and y-axis from 0 to 11.
Plot the points (−6, 6) (0, 6) (3, 3).
Join them to make triangle A.

   b Enlarge triangle A by scale factor $\frac{1}{3}$, centre (0, 0).
Label the image B.

   c Enlarge triangle A by scale factor $\frac{1}{2}$, centre (0, 10).
Label the image C.

Describing an enlargement

This spread will show you how to:

- Describe an enlargement by giving the scale factor and centre of enlargement
- Understand, identify and use scale factors

Keywords

Centre of
    enlargement
Enlargement
Scale factor

To describe an **enlargement** you give the **scale factor** and the **centre of enlargement**.

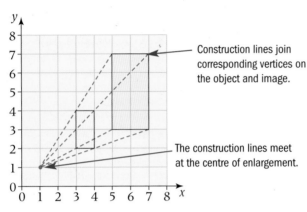

Construction lines join corresponding vertices on the object and image.

This is an enlargement of scale factor $\frac{1}{2}$, centre (1, 1).

The construction lines meet at the centre of enlargement.

The scale factor of an enlargement is the ratio of corresponding sides.

- **Scale factor** = $\dfrac{\text{length of image}}{\text{length of original}}$

You can write this as a ratio
length of image : length of object

Example

Describe fully the single transformation that maps triangle PQR onto P′Q′R′.

The construction lines meet at $(-3, -2)$

Scale factor = $\dfrac{\text{P′R′}}{\text{PR}} = \dfrac{6}{2} = 3$.

The transformation that maps PQR onto P′Q′R′ is an enlargement, centre $(-3, -2)$, scale factor 3.

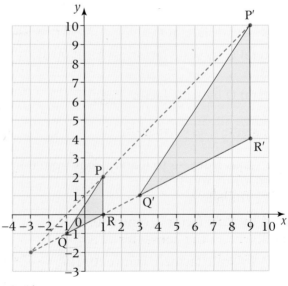

1 Describe fully the single transformation that maps

   a triangle P onto triangle Q

   b triangle Q onto triangle P.

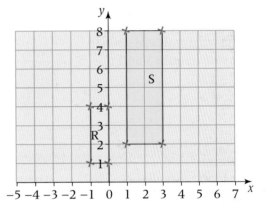

2 a Describe fully the single transformation that maps

   i rectangle R onto rectangle S

   ii rectangle S onto rectangle R.

   b The perimeter of rectangle R is 8 units.

     What is the perimeter of rectangle S?

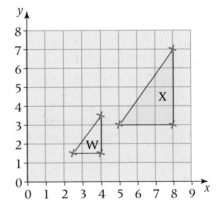

3 Describe fully the single transformation that maps

   a triangle W onto triangle X

   b triangle X onto triangle W.

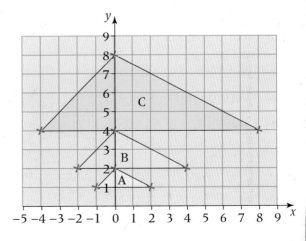

4 a Describe fully the single transformation that maps

   i triangle B onto triangle A

   ii triangle B onto triangle C

   iii triangle A onto triangle C.

   b How many times bigger is the perimeter of triangle C than triangle A?

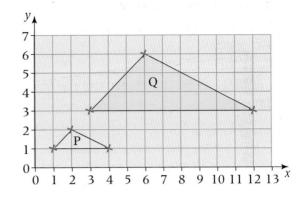

This spread will show you how to:

- Understand similarity of 2-D shapes, using this to find missing lengths and angles

Keywords

Enlargement
Ratio
Similar

In an **enlargement** the object and the image are mathematically **similar**.

- **In similar shapes**
  - **corresponding pairs of angles are equal**
  - **corresponding pairs of sides are in the same ratio.**

Any two circles are similar.
Any two squares are similar.

You can use the **ratio** between similar shapes to find missing side lengths.

Example

These quadrilaterals are similar.
Find the side lengths *s* and *t*.

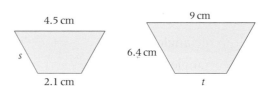

Side 4.5 cm corresponds to side 9 cm.
As a fraction,
the ratio larger : smaller is $\frac{9}{4.5} = 2$ the ratio smaller : larger is $\frac{4.5}{9} = \frac{1}{2}$

Side *s* corresponds to side 6.4 cm.   Side *t* corresponds to side 2.1 cm.

$\frac{s}{6.4} = \frac{1}{2}$

$\frac{t}{2.1} = 2$

$s = \frac{1}{2} \times 6.4 = 3.2$ cm

$t = 2.1 \times 2 = 4.2$ cm

Choose the ratio that gives a fraction with the side you want on top.

You need to be able to identify corresponding sides when one shape is upside down.

Example

These two pentagons are similar.
Find the side lengths *x* and *y*.

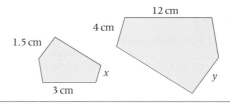

Side 3 cm corresponds to side 12 cm.
As a fraction,
the ratio larger to smaller is $\frac{3}{12} = \frac{1}{4}$ the ratio smaller to larger is $\frac{12}{3} = 4$

Side *x* corresponds to side 4 cm.   Side *y* corresponds to side 1.5 cm.

Side *x* is smaller. $4 \times \frac{1}{4} = 1$ Side *y* is larger. $1.5 \times 4 = 6$
$x = 1$ cm $y = 6$ cm

1 These two trapeziums are similar.
Find the lengths a and b

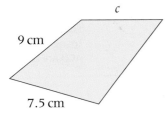
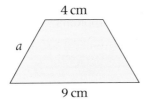

2 These two quadrilaterals are similar.
Find the lengths c and d

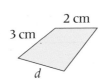

3 These two pentagons are similar.
Find the lengths e and f.

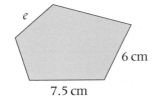

4 These two parallelograms are similar.
Find the perimeter of the smaller parallelogram.

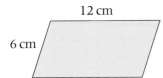

5 These two isosceles triangles are similar.

a Write the ratio
larger : smaller
for these triangles.

b Find the length x.

c Find the perimeter of each triangle.

d Write the ratio of their perimeters.
What do you notice?

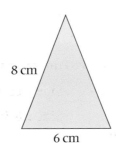
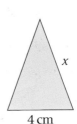

This spread will show you how to:

- Understand similarity of 2-D shapes, using this to find missing lengths and angles
- Use common sense to check answers to geometric problems

Keywords
Ratio
Similar

- In similar triangles
  - corresponding pairs of angles are equal
  - corresponding pairs of sides are in the same ratio.

If a line is drawn parallel to one side of a triangle, the smaller triangle and the larger triangle are similar.

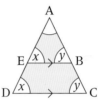

Corresponding angles are equal (using properties of angles in parallel lines), so the triangles ABE and ACD are similar.

Example

Find the length CD in this triangle.

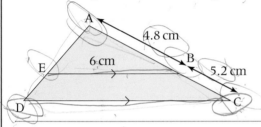

Triangle ACD is similar to triangle ABE

$$\frac{CD}{BE} = \frac{AC}{AB}$$
$$\frac{CD}{6} = \frac{10}{4.8}$$
$$CD = \frac{6 \times 10}{4.8} = 12.5 \text{ cm}$$

CD corresponds to BE, and AC corresponds to AB

Check that your answer makes sense. Did you expect a longer or shorter length?

Example

Find the length QS in this triangle.

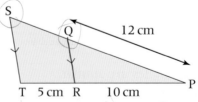

Triangle PQR is similar to triangle PST.

$$\frac{PS}{PQ} = \frac{PT}{PR}$$
$$\frac{PS}{12} = \frac{15}{10}$$
$$PS = \frac{12 \times 15}{10}$$

PS = 18 cm
QS = PS − PQ
    = 18 − 12 = 6 cm

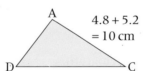
$$\frac{PS}{PQ} = \frac{PT}{PR} = \frac{PS}{12} = \frac{15}{12} = \frac{12 \times 15}{10}$$
$$QS = PS - PQ$$
$$18 - 12 = 6$$

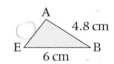

PQ corresponds to PS, and PR corresponds to PT

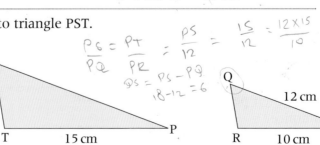

1 In the diagram, BE is parallel to CD. Find the lengths

a AB

b AD

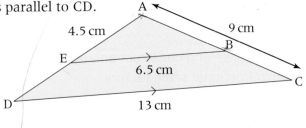

Start by sketching the two triangles.

$AB = AE$

$\dfrac{AB}{4.5} = \dfrac{4.5}{9}$

$\dfrac{9 \times 4.5}{6.5} =$

$EB - AC = AB$

$9 = AB$

2 In the diagram, QT is parallel to RS. Find the lengths

a PT

b QR

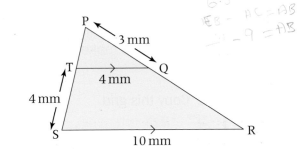

3 In the diagram, VZ is parallel to WY. Find the lengths

a WY

b VW

c VX

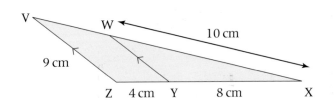

4 In the diagram, KL is parallel to NM.

a Find the lengths MN and JN.

b Find the perimeter of the trapezium KLMN.

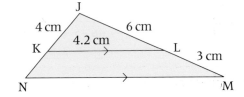

5 EB is parallel to DC. Work out the perimeter of

a triangle ABE

b trapezium EBCD.

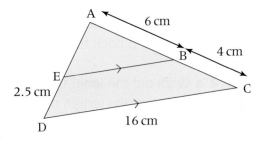

Exam review

Key objectives

- Recognise, visualise and construct enlargements of objects using positive fractional scale factors

- Identify the scale factor of an enlargement as the ratio of the lengths of any two corresponding line segments and understand the implications of enlargement for perimeter

- Understand similarity of triangles and other plane figures and use this to make geometric inferences

1 Copy this grid.

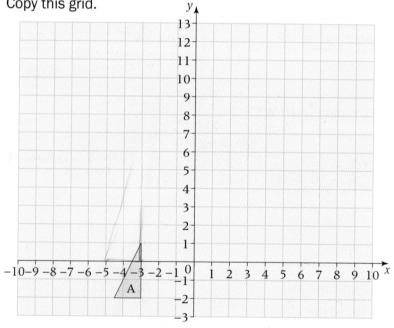

a Enlarge triangle *A* by a scale factor 4, centre (−5.5, −3). (2)
Label the new triangle *B*.

b Describe the single transformation that maps triangle *B* back to triangle *A*. (3)

2 a Work out the length of *AD*. (2)
  b Work out the length of *BC*. (2)

(Edexcel Ltd., 2000)

D6

This unit will show you how to

- Understand and use the probability scale
- Find probabilities of mutually exclusive events
- Calculate theoretical probabilities
- Calculate probabilities of independent events
- Draw and use tree diagrams for the outcomes of several events
- Use a tree diagram to calculate the probability of an event that can happen in more than one way

Before you start ...

You should be able to answer these questions.

Review

1 Work out.

a $1 - 0.45$ **b** $1 - 0.96$

c $1 - 0.28$ **d** $1 - 0.375$

Unit N2

2 Work out.

a $0.2 + 0.4$ **b** $0.3 + 0.04$

c $0.65 + 0.25$ **d** $0.7 + 0.05$

Unit N2

3 Work out.

a 0.5×0.36 **b** 0.25×0.68

c 0.64×0.3 **d** 0.16×0.75

Unit N2

4 Work out.

a $1 - \frac{5}{6}$ **b** $1 - \frac{1}{5}$

c $1 - \frac{7}{9}$ **d** $2 - 1\frac{1}{4}$

Unit N3

5 Work out.

a $\frac{1}{5} + \frac{2}{3}$ **b** $\frac{3}{4} + \frac{1}{6}$

Unit N3

6 Work out.

a $\frac{2}{3} \times \frac{5}{6}$ **b** $\frac{2}{9} \times \frac{4}{5}$

Unit N3

This spread will show you how to:

- Understand and use the probability scale
- Find probabilities of mutually exclusive events
- Calculate theoretical probabilities

Keywords
Event
Expected
 number
Mutually
 exclusive
Outcome

You write a probability as a decimal, fraction or percentage.

- **Probability takes a value from 0 to 1.**
- **Probability of an outcome** $= \dfrac{\text{number of favourable outcomes}}{\text{total number of outcomes}}$.

For an **event** the total probability for all possible outcomes = 1.

- **For an event A, P(not A) = 1 – P(A)**

P(A) means the probability of outcome A.

Two or more events are **mutually exclusive** if they cannot happen at the same time. You use the addition rule to find the probability of mutually exclusive events.

Sometimes called the OR rule.

- **For two mutually excusive events A and B, P(A or B) = P(A) + P(B).**
- **Expected number** = total number of outcomes × probability of outcome.

Example

A spinner has 8 sides, of which 4 show squares, 3 show triangles, and 1 shows a circle.

a The spinner is spun once. Find the probability that the spinner

  i   shows a triangle

  ii does not show a triangle

  iii does not show a circle

  iv shows a square or a circle

  v shows a triangle or a square.

b The spinner is spun 400 times. Find the number of times the spinner is expected to show

  i a circle   **ii** a triangle.

---

a i   P(triangle) $= \frac{3}{8}$

  ii P(not triangle) $= 1 - \frac{3}{8} = \frac{5}{8}$

  iii P(not circle) $= 1 - \frac{1}{8} = \frac{7}{8}$

  iv P(circle or square) = P(circle) + P(square) $= \frac{1}{8} + \frac{4}{8} = \frac{5}{8}$

  v P(triangle or square) = P(triangle) + P(square) $= \frac{3}{8} + \frac{4}{8} = \frac{7}{8}$

b i   Expected number of circles = 400 P(circle) $= 400 \times \frac{1}{8} = 50$

  ii Expected number of triangles = 400 × P(triangle) $= 400 \times \frac{3}{8} = 150$

1 A spinner has ten equal sides, of which 4 show squares, 3 show pentagons, 2 show hexagons and 1 shows a circle. The spinner is spun once. Find the probability that the spinner

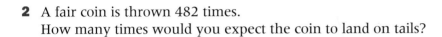

    a shows a square   **b** does not show a square

    c shows a square or pentagon   **d** shows a pentagon or circle

    e does not show a hexagon   **f** does not show a circle.

2 A fair coin is thrown 482 times.
How many times would you expect the coin to land on tails?

3 An ordinary fair dice is rolled 612 times.
How many times would you expect it to land on 6?

4 In a class of 28 students, 15 are girls and 13 are boys. 22 students wear glasses. One student is chosen at random.
What is the probability that this student

    a is a boy   **b** is a girl

    c wears glasses   **d** does not wear glasses?

5 A spinner has 16 equal sides. The table shows the colours of the sides and the shapes drawn on them. The spinner is spun once.

	Circle	Triangle	Square
Red	3	5	1
Black	4	1	2

    a Find the probability that the spinner

        i   shows red   **ii** shows a triangle

        iii shows a circle or triangle   **iv** does not show a circle

        v shows a red triangle   **vi** shows a black square.

    b Copy and complete this sentence

> 'The probability of the spinner landing on the red square is the same as the probability of the spinner landing on _____ _____'

6 A fair eight-sided dice has equal sides. Two sides show 2s, three sides show 3s and three sides show 5s.

    a The dice is thrown once. Find the probability that the dice shows

        i   an even number   **ii** an odd number

        iii a prime number   **iv** a multiple of 4.

    b The dice is thrown 256 times.
       How many times would you expect the dice to land on 2?

This spread will show you how to:

● Calculate probabilities of independent events

● Two or more events are **independent** if the outcome of one event has no effect on the outcome of the other.

When a dice is rolled and a coin is thrown, the number rolled on the dice has no effect on whether the coin lands on a head or a tail. The events are independent.

When two or more coins are thrown, a head or tail on the first coin has no effect on what shows on any of the other coins. All the events are independent.

You use the multiplication rule to find the probability of two or more independent events.

● For two independent events A and B, P(A and B) = P(A) × P(B)

Sometimes called the AND rule.

A spinner has 12 equal sides: five green, four blue, two red and one white.

A fair coin is thrown and the spinner is spun.

a What is the probability of getting
i a head and green   **ii** a tail and white
iii a tail and red   **iv** a head and red?

b Why are the answers to **iii** and **iv** the same?

a **i** $P(H) = \frac{1}{2}$ $P(\text{green}) = \frac{5}{12}$
    $P(\text{head and green}) = P(H) \times P(\text{green}) = \frac{1}{2} \times \frac{5}{12} = \frac{5}{24}$

ii $P(T) = \frac{1}{2}$ $P(\text{white}) = \frac{1}{12}$
    $P(\text{head and white}) = P(H) \times P(\text{white}) = \frac{1}{2} \times \frac{1}{12} = \frac{1}{24}$

iii $P(T) = \frac{1}{2}$ $P(\text{red}) = \frac{2}{12}$
    $P(\text{head and red}) = P(H) \times P(\text{red}) = \frac{1}{2} \times \frac{2}{12} = \frac{2}{24}$

iv $P(H) = \frac{1}{2}$ $P(\text{red}) = \frac{2}{12}$
    $P(\text{head and red}) = P(H) \times P(\text{red}) = \frac{1}{2} \times \frac{2}{12} = \frac{2}{24}$

$P(H) = P(T)$

b The answers to **iii** and **iv** are the same since the coin is fair.

A dice is rolled twice.

Find the probability that on the first roll the dice shows a four and on the second roll the dice shows an odd number.

$P(4) = \frac{1}{6}$ $P(\text{odd number}) = \frac{3}{6}$ $P(4, \text{odd number}) = \frac{1}{6} \times \frac{3}{6} = \frac{3}{36}$

1 A fair coin is thrown and an ordinary dice is rolled.

   a Copy and complete the table to list all the outcomes.
One has been done for you.

Coin/Dice	1	2	3	4	5	6
Head		H2				
Tail						

   b Find the probability of getting

     i a head and a 2   **ii** a tail and a 4   **iii** a tail and a 5.

2 A red dice and a blue dice are rolled.

   a Draw a table to show all the possible outcomes.

   b Find the probability of getting

     i   6 on the red dice and 6 on the blue dice

     ii 3 on the red dice and 5 on the blue dice

     iii 3 on the red dice and an odd number on the blue dice

     iv 5 or greater on the red dice and 1 on the blue dice.

3 An ordinary fair dice is rolled twice. Find the probability that the
dice shows an even number on the first roll and a number greater
than 4 on the second roll.

4 A spinner has ten equal sides: 4 show squares, 3 show pentagons,
2 show hexagons and 1 shows a circle. A fair coin is thrown and the
spinner is spun. Find the probability of getting

   a a square and a head   **b** a circle and a head

   c a pentagon and a tail   **d** a hexagon and a tail

   e a circle and a tail.

5 A spinner has ten equal sides. 4 show squares, 3 show pentagons,
2 show hexagons and 1 shows a circle. The spinner is spun twice.
Find the probability of getting

   a a circle on the first spin and a circle on the second spin

   b a circle on the first spin and a hexagon on the second spin

   c a triangle on the first spin and a square on the second spin

   d a square on the first spin and a circle on the second spin

   e a hexagon on first spin and a pentagon on second spin.

6 A 10 pence coin and a 2 pence coin are spun.

   a Draw a table to show all the possible outcomes.

   b Find the probability that the 10p shows tails and the 2p shows heads.

7 One coin is spun twice. Find the probability that both times the coin
shows heads.

This spread will show you how to:

- Draw and use tree diagrams for the outcomes of several events

Keywords
Random
Replaced
Tree diagram

- You can use a **tree diagram** to show the possible outcomes of two events.

- In a tree diagram,
  - write the outcomes at the end of each branch
  - write the probability on each branch
  - the probabilities on each set of branches should add to 1.

Example

A bag contains 7 yellow and 3 blue marbles.

A marble is chosen at **random** from the bag, its colour is noted and then it is **replaced** in the bag.

The bag is shaken and then a second marble is chosen at random.

Draw a tree diagram to show all the possible outcomes.

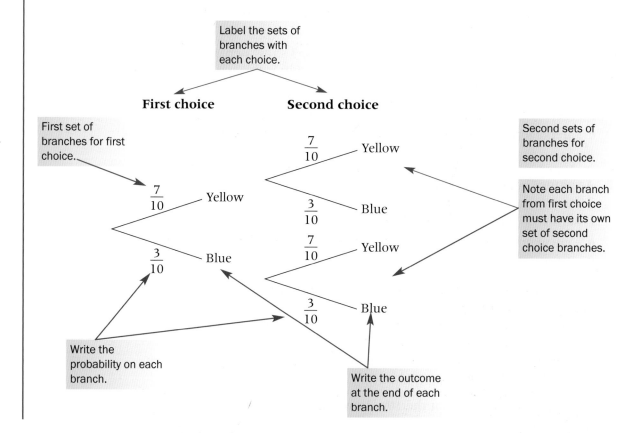

Label the sets of branches with each choice.

First choice   **Second choice**

First set of branches for first choice.

Second sets of branches for second choice.

Note each branch from first choice must have its own set of second choice branches.

$\frac{7}{10}$ — Yellow

$\frac{7}{10}$ — Yellow

$\frac{3}{10}$ — Blue

$\frac{3}{10}$ — Blue

$\frac{7}{10}$ — Yellow

$\frac{3}{10}$ — Blue

Write the probability on each branch.

Write the outcome at the end of each branch.

1 A bag contains 6 purple counters and 11 orange counters. A counter is chosen at random from the bag, its colour noted and then it is replaced in the bag. The bag is shaken and a second counter is chosen at random. Copy and complete the tree diagram to show all the possible outcomes.

You will need the tree diagrams drawn for questions **2**, **3** and **4** for Exercise D6.4.

First choice   **Second choice**

```
                          ____ Purple
        ____ Purple  <
                          ____ Orange

    <
                          ____ Purple
        ____ Orange  <
                          ____ Orange
```

2 A bag contains 4 green and 5 red marbles. A marble is chosen at random from the bag, its colour noted and then it is replaced in the bag. The bag is shaken and a second marble is chosen at random. Draw a tree diagram to show all the possible outcomes.

3 A bag contains 3 white counters and 8 black counters. A counter is chosen at random from the bag, its colour noted and then it is replaced in the bag. The bag is shaken and then a second counter is chosen at random. Draw a tree diagram to show all the possible outcomes.

4 A 10 pence and a 2 pence coin are thrown. Draw a tree diagram to show all the possible outcomes.

5 A bag contains 2 white and 5 black counters. A counter is chosen from the bag and a fair coin is thrown. Draw a tree diagram to show all the possible outcomes.

6 Dave owns 13 CDs. Three of the CDs are by his favourite group, Cloudplay. Dave chooses one of the CDs at random, notes whether it is a Cloudplay CD, and replaces it. He then chooses another one of the 13 CDs at random. Copy and complete the probability tree diagram that has been started.

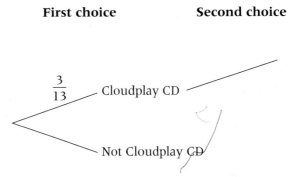

First choice   **Second choice**

$\frac{3}{13}$ Cloudplay CD

Not Cloudplay CD

This spread will show you how to:

● Draw and use tree diagrams for the outcomes of several events

● You can use a tree diagram to find the probability of an outcome of an event.

● To find the probability of an outcome, multiply the probabilities along the branches leading to that outcome.

Example

A bag contains 7 yellow and 3 blue marbles.

A marble is chosen at random from the bag, its colour noted and it is replaced in the bag.

The bag is shaken and then a second marble is chosen at random.

The tree diagram shows all the possible outcomes.

First choice   **Second choice**

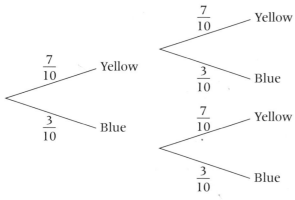

Use the tree diagram to find the probability of choosing

a two yellow marbles
b two blue marbles
c a yellow marble then a blue marble, in that order.

a $P(YY) = \frac{7}{10} \times \frac{7}{10} = \frac{49}{100}$

$\qquad = \frac{49}{100}$

b $P(BB) = \frac{3}{10} \times \frac{3}{10} = \frac{9}{100}$

$\qquad = \frac{9}{100}$

c $P(YB) = \frac{7}{10} \times \frac{3}{10} = \frac{21}{100}$

$\qquad = \frac{21}{100}$

1 A bag contains 3 green and 4 white marbles. A marble is chosen at random from the bag, its colour noted and then it is replaced in the bag. The bag is shaken and a second marble is chosen at random. The tree diagram shows all the possible outcomes. Use the tree diagram to find the probability of choosing

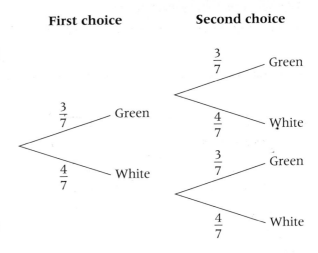

First choice **Second choice**

a two green marbles

b two white marbles

c a white marble then a green marble in that order.

2 A bag contains 4 green and 5 red marbles. A marble is chosen at random from the bag, its colour noted and then it is replaced in the bag. The bag is shaken and a second marble is chosen at random. Use the tree diagram from Exercise D6.3 question **2** to find the probability of choosing

a two green marbles   **b** two red marbles.

3 A bag contains 3 white counters and 8 black counters. A counter is chosen at random from the bag, its colour noted and then it is replaced in the bag. The bag is shaken and a second counter is chosen at random. Use the tree diagram from Exercise D6.3 question **3** to find the probability of choosing

a two white counters   **b** two black counters.

4 A 10 pence and a 2 pence coin are thrown. Use the tree diagram from Exercise D6.3 question **4** to find the probability of getting

a two heads   **b** two tails   **c** a head on the 10p and a tail on the 2p.

5 Bag A contains 5 red counters and 7 black counters. Bag B contains 2 yellow counters and 8 red counters. A counter is chosen from each bag.

a Copy and complete the tree diagram to show the possible outcomes.

b Find the probability of choosing

  i a black and a yellow counter

  ii two red counters.

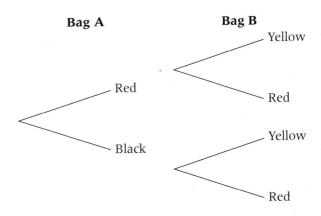

Bag A   **Bag B**

This spread will show you how to:

- Use a tree diagram to calculate the probability of an event that can happen in more than one way

Keywords
Event
Independent
Mutually
    exclusive
Outcome

You can use a tree diagram to find the probability of an event that can happen in more than one way.

- **To find probabilities when an event can happen in different ways**
  - multiply the probabilities along the branches
  - add the probabilities for the different ways of getting the chosen event.

Example

Pamela makes two pottery vases.

Each vase is made independently.

The probability that a vase cracks while it is in the kiln is 0.2.

a Find the probability that the vase does not crack while it is in the kiln.

b Draw a tree diagram to show all the possible outcomes for two vases.

c Find the probability that one of the vases will crack while it is in the kiln.

a P(does not crack) = 1 − P(cracks)
P(cracks) = 0.2
So, P(does not crack) = 1 − 0.2 = 0.8

The outcomes 'cracks' and 'does not crack' are **mutually exclusive**.

b

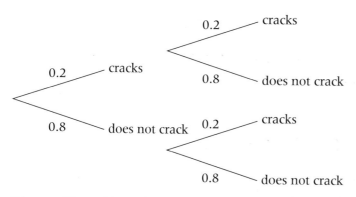

First Vase   Second Vase

0.2 — cracks
0.2 — cracks
0.8 — does not crack
0.8 — does not crack
0.2 — cracks
0.8 — does not crack

The state of the first vase has no effect on the state of the second vase – the events are **independent**.

c P(one will crack) = P(cracks, does not crack) + P(does not crack, cracks)
= (0.2 × 0.8) + (0.8 × 0.2)
= 0.16 + 0.16
= 0.32

Two ways of getting one cracked vase.

Exercise D6.5

1 Batteries are placed in two toy cars, a red car and a blue car. The probability that a battery lasts for more than 20 hours is 0.8.

  a Find the probability that a battery does not last for more than 20 hours.

  b Draw a tree diagram to show the outcomes for the batteries in the two cars.

  c Find the probability that the batteries last more than 20 hours

    i in both cars   **ii** in only one car

    iii in at least one car.

2 A bag contains 6 white and 2 black counters. A counter is chosen at random from the bag, its colour noted and then it is replaced. A second counter is chosen at random.

  a Draw a tree diagram to show all the outcomes of choosing two counters.

  b Find the probability of choosing

    i two black counters   **ii** one counter of each colour

    iii at least one black counter.

3 Two delicate glasses are placed in a dishwasher. The probability that a glass breaks in the dishwasher is $\frac{1}{20}$.

  a Find the probability that a glass does not break in the dishwasher.

  b Draw a tree diagram to show all the outcomes for the two glasses.

  c Find the probability that while in the dishwasher

    i neither glass breaks   **ii** only one glass breaks

    iii at least one glass breaks.

4 A spinner has five equal sectors. Three are coloured orange and two are coloured black. The spinner is spun twice.

  a Draw a tree diagram to show all the outcomes of two spins on the spinner.

  b Find the probability that on two spins the spinner lands

    i on black both times   **ii** on black at least one time

    iii once on each colour.

5 Josh makes two model aeroplanes, a grey plane and an orange plane. He flies both of them. The probability that one crashes is 0.1.

  a Find the probability that a model aeroplane does not crash.

  b Draw a tree diagram to show all the outcomes for the two model aeroplanes.

  c Find the probability that both model aeroplanes will crash.

  d Find the probability that only one of the aeroplanes will crash.

Key objectives

- Identify different mutually exclusive outcomes and know that the sum of the probabilities of all these outcomes is 1
- Know when to add or multiply two probabilities
- Use tree diagrams to represent outcomes of compound events, recognising when events are independent

1 a Anna chooses a ball at random from the bag. Calculate the probability that she picks

   i   a red ball

   ii a ball that is not red

   iii a red ball or a white ball   (3)

  b Write the probability that Anna picks a purple ball. Explain your answer.   (2)

2 Julie does a statistical experiment. She throws a dice 600 times. She scores six 200 times.

  a Is the dice fair? Explain your answer.   (1)

Julie then throws a fair red dice once and a fair blue dice once.

  b Copy and complete the probability tree diagram to show the outcomes.   (1)
     Label clearly the branches of the probability tree diagram. The probability tree diagram has been started in the space below.

          Red Dice   **Blue Dice**

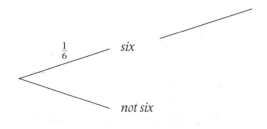

(Edexcel Ltd., 2003)

This unit will show you how to

- Draw and interpret distance–time graphs
- Understand and use compound measures, including speed
- Draw and interpret graphs modelling real-life situations
- Label axes correctly, choosing appropriate quantities
- Form linear functions, using the corresponding graphs to solve real-life problems
- Interpret the gradient and *y*-intercept of graphs modelling real-life situations
- Draw and use scatter diagrams and lines of best fit

Before you start ...

You should be able to answer these questions.

Review

1 Give the speed in miles per hour for each journey.

   a 20 miles is travelled in 1 hour.
   b 40 miles is travelled in $\frac{1}{2}$ hour.
   c 30 miles is travelled in 20 minutes.
   d 120 miles is travelled in 5 hours.

Unit S6

2 Give the key characteristics of each straight line graph.

Unit A4

Equation	Gradient	y-axis intercept	Direction
$y = 3x + 4$			
$y = 10 - 4x$			
$2y = 8x + 10$			
$2y - 4x = 15$			
$y = 7$			

3 Give the gradient of each line segment.

Unit A4

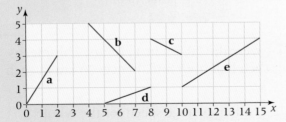

4 Find the gradient of the line segment joining

Unit A4

   a (3, 4) to (4, 9)   **b** (5, 8) to (9, 13)

   c (−2, 7) to (−3, −4)

This spread will show you how to:

● Draw and interpret distance–time graphs

● Understand and use compound measures, including speed

● You can represent a journey on a distance–time graph.

● Time is always plotted on the horizontal axis.

● Distance is plotted on the vertical axis.

This graph shows Dan's journey on his bike.

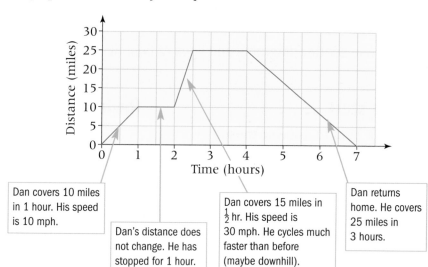

Dan covers 10 miles in 1 hour. His speed is 10 mph.

Dan's distance does not change. He has stopped for 1 hour.

Dan covers 15 miles in $\frac{1}{2}$ hr. His speed is 30 mph. He cycles much faster than before (maybe downhill).

Dan returns home. He covers 25 miles in 3 hours.

Example

Janine leaves home at 1 p.m. and cycles to her friend's house, 30 km away, at a speed of 20 km/h. She stays for 2 hours, then cycles home, arriving at 6 o'clock.

Draw a distance-time graph to represent the journey and determine her speed on the way home and her average speed for the entire journey.

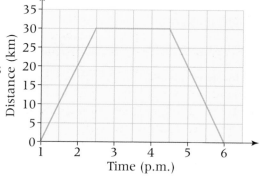

Janine returning home, is shown by the graph going down, back to the x-axis

On the journey home, Janine covers the 30 km in $1\frac{1}{2}$ hours. This means that she has covered 10 km in each half hour and, hence, 20 km in each hour. Her speed is 20 km/h.

Since she covers 60 km in 5 hours, her average speed for the whole journey, including the stop, is 12 km/h.

average speed =
$\dfrac{\text{total } \textbf{distance}}{\text{total } \textbf{time}}$

1 The distance–time graph shows the journey of a car between Birmingham and Stoke-on-Trent.

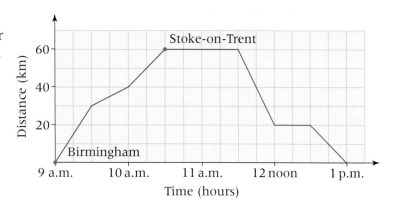

  a How far is it from Birmingham to Stoke-on-Trent?

  b For how long did the car stop?

  c What was the speed of the car for the first part of the journey?

  d Between which two times was the car travelling fastest?

  e What was the average speed of the car for the whole journey?

2 Three students have drawn distance–time graphs. Two have made mistakes. Which two students have made a mistake and what mistake is it?

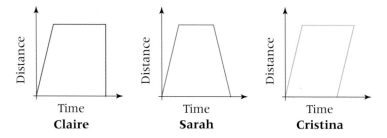

3 Construct a distance–time graph to show each journey.

  a A car travels between Bristol and London. On the outward journey, it travels the 120 miles to London in $2\frac{1}{4}$ hours. The driver remains in London for $1\frac{1}{2}$ hours. The car travels half way back to Bristol at 40 miles per hour, as the motorway is busy, then the remaining distance at 80 miles per hour.

  b Two brothers both went to see each other on the same day. Henry left his home at 2 p.m. to go and see Leo, who lives 5 miles away. Henry walked at an average speed of 4 miles per hour, but he stopped half way for a 15 minute rest. At 2:30 p.m., Leo set out on his bicycle from his home in order to go and visit Henry. He cycled straight there in $\frac{1}{2}$ hour.

  > Draw the two journeys on one graph.

4 The following graph represents the journey of a car. Construct a graph of speed (km/h) against time (h) for this journey.

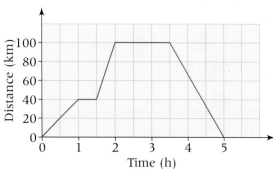

Other real-life graphs

This spread will show you how to:

● Draw and interpret graphs modelling real-life situations

Keywords

Model

You can use graphs to **model** the depth of water flowing in or out of a container at a constant rate.

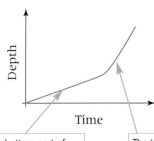

Imagine water filling this container.

The bottom part of the container has the same diameter, so fills at a steady rate

The top part starts wide then narrows. As the container gets narrower, it fills faster.

● In a real-life graph involving time, time is usually represented by the *x*-axis.

Sketch a graph to show what happens as

(1) Sam fills a bath with both taps running

(2) He realises it is too hot so turns off the hot tap

(3) He turns off the cold tap

(4) He gets in

(5) Has a long soak

(6) Gets out

(7) Pulls out the plug.

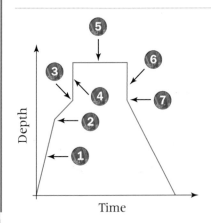

Label the axes with the quantities they represent. A scale is not needed for a sketch graph.

1 Match the four sketch graphs with the containers.

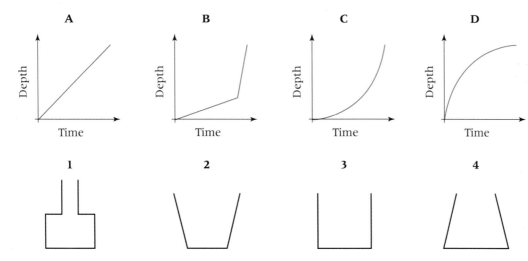

2 The sketch graph shows Andrew's height from 3 to 23 years of age. Explain what the graph shows at each stage and explain why this might be.

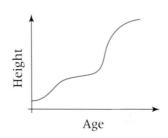

3 Sketch a graph of depth against time as this container is filled.

4 Construct sketch graphs to represent these situations.

   a A woman is pregnant and puts on weight. When her baby arrives, she finds it difficult to lose any of the weight for six months, but then joins an exercise class and eats healthily. She is back to her natural weight a year and a half after becoming pregnant. (Graph is weight against time.)

   b A frozen chicken is taken out of the freezer and left to defrost. Two hours later it is put in the microwave briefly to speed up and finish off the process. The chicken is then put in the oven to roast for Sunday lunch. (Graph is temperature against time.)

5 A skateboarder at a skate park uses a ramp, as shown. For one go on the ramp, construct a sketch graph of

   a speed against time

   b acceleration against time.

This spread will show you how to:

- Draw and interpret graphs modelling real-life situations
- Form linear functions, using the corresponding graphs to solve real-life problems

Keywords

Formula
Gradient
Graph
Intercept

- You can use a graph to represent a real-life situation.

Suppose you have a mobile phone. You pay £10 line rental each month, then 50p for every minute you spend making calls.

Work in £
50p = £0.5

Time on calls in minutes (x)	0	1	2	3	4
Price of calls in pounds	0	0.50	1.00	1.50	2.00
Line Rental (£)	10	10	10	10	10
Total cost in pounds (y)	10	10.50	11	11.50	12

To get the total cost, you add £10 (the line rental) to the call cost.
The call cost is the number of minutes on the phone multiplied by 0.50.
Hence,

Total cost = 0.50 × time on phone + 10

$$y = 0.5x + 10$$

This formula is linear and its graph is a straight line. The y-axis intercept is 10 and the gradient is 0.5.

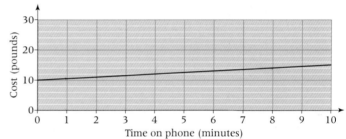

Plot a graph to represent the total cost of hiring a party venue, if the owner charges £100 hire fee and £5 per guest.
Use the graph to estimate the number of guests if the total bill is £285.

If x is the number of guests and y is the cost (£s), then

$$y = 5x + 100$$

Draw a horizontal line from £285 to the graph. Draw a vertical line from the graph to the horizontal axis.

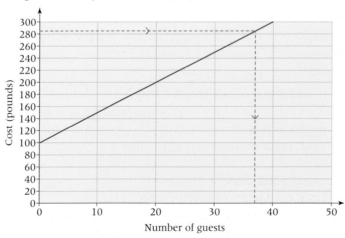

Label axes with the quantities they represent and the units in which they are measured.

For cost of £285, number of guests = 37.

1 **a** This graph is a conversion graph for miles to kilometres and vice versa.
Use the graph to convert

   i 20 miles into kilometres

   ii 60 kilometres into miles.

b If Dan ran 30 miles and Charlie ran 50 kilometres, who ran further?

c By finding the gradient of the line, give a formula to connect the number of miles (x) with the number of kilometres (y).

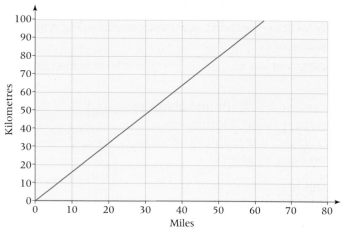

2 Pauline and her family are going on holiday and exchange £400 spending money into euros (€) before they go. At the bank, the exchange rate is £1 = €1.45.

£1 = €1.45

a Construct a graph that the family can take on holiday to convert any amount of their spending money from pounds to euros or vice versa.

b Use the graph to find

   i the cost, in euros, of a side trip which is advertised for £95

   ii the cost, in pounds, of a meal in a restaurant that comes to €85.

3 A campsite charges £15 per night per tent, plus an extra £3 per person.

a Construct a table of charges and, hence, a graph to show the cost of the campsite depending upon how many people stay in the tent. (The largest tent available is one that sleeps 15 people.)

b Use your graph to calculate how many people stayed in the tent if the total cost was £36.

c Explain why the total cost could never be £50.

d Suggest an equation for your graph, stating clearly the meaning of any letters you use.

e Use your equation to work out the cost of pitching a new Supertent that sleeps 27 people.

4 Two competing electricity companies use these formulae to work out customers' bills.
The number of units of electricity used is x.
The price of the electricity is £y.

POWER UP!
$y = 3x + 5$

SPARKS ARE US!
$y = 2x + 15$

Using graphs, compare the pricing policies of the two companies and advise householders from which company they should buy their electricity.

This spread will show you how to:

- Form linear functions, using the corresponding graphs to solve real-life problems
- Draw and use scatter diagrams and lines of best fit

Keywords
Gradient
Graph
Intercept

You can use a straight line graph to represent real-life information.
For example, this graph shows a hire car company's charges.

You can find the gradient and the y-axis intercept.

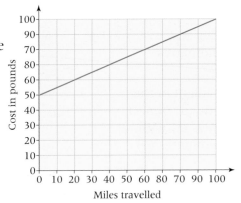

The line intercepts the y-axis at (0, 50).

For every 10 squares you move across, you travel 5 squares up:

Gradient = $\frac{\text{rise}}{\text{run}} = \frac{5}{10}$ or $\frac{1}{2}$

The equation of the graph is $y = \frac{1}{2}x + 50$.

- Graphs help to give you more information.
  - The intercept of 50 tells you that you are charged £50 for a car.
  - The gradient of $\frac{1}{2}$ tells you that for every 2 miles you travel, you are charged an extra £1.

Example

Interpret the line of best fit on this scatter diagram of 18 students' heights and weights.

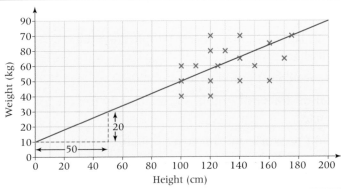

The gradient, $m = \frac{20}{50} = \frac{2}{5}$
The y-axis intercept, $c = 10$
The equation of the graph is $y = \frac{2}{5}x + 10$.

The gradient tells you that for every 5 cm you grow you gain 2 kg.
The intercept tells you that at 0 cm height, you weigh 10 kg – in other words, it does not always make sense to interpret the intercept on a straight line graph for a big age range.

1 Match each graph with an equation.

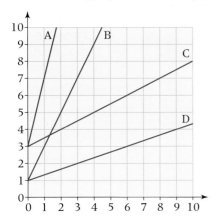

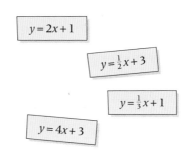

$y = 2x + 1$

$y = \frac{1}{2}x + 3$

$y = \frac{1}{3}x + 1$

$y = 4x + 3$

2 For each graph, find its equation in the form $y = mx + c$ and interpret the meaning of m and c, deciding if it is sensible to interpret c.

a

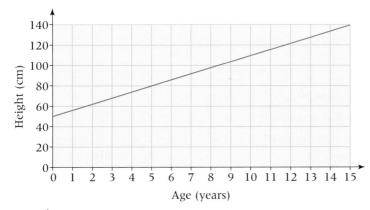

b

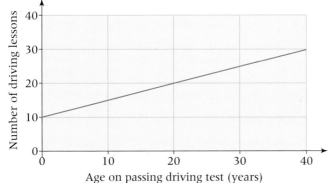

3 Interpret these equations representing real-life situations and discuss their limitations.

a $y = 0.3x + 2$ ... x is age in years, y is amount of pocket money in £.

b $y = 1\frac{4}{5}x + 32$ ... x is temperature in degrees Celsius (°C) and y is temperature in degrees Fahrenheit (°F).

Using quadratic graphs

This spread will show you how to:
- Plot graphs of simple quadratic functions

Keywords

Maximum
Model
Parabola
Quadratic

- You can model some real-life situations with **quadratic** graphs.

This **parabola** shows the height of a javelin as it is thrown.

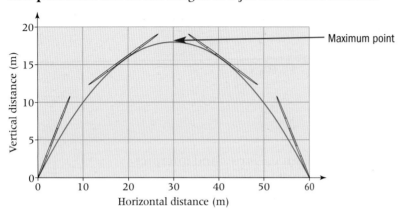

The **maximum** point on the graph shows that the javelin reaches a height of 18 m when it is 30 m away from the throwing line.

Example

One part of a roller coaster is modelled using the equation

$y = \dfrac{x^2 - 20x}{10} + 10$ where x is the horizontal distance and y is the vertical distance from the start of the section.

a Complete this table of values and use it to plot the graph that shows the path of the roller coaster.

x	0	5	10	15	20
x^2	0	25			
$-20x$	0	-100			
$\dfrac{x^2 - 20x}{10}$	0	-7.5			
y	10	2.5			

b How long is this part of the roller coaster ride?

a

x	0	5	10	15	20
x^2	0	25	100	225	400
$-20x$	0	-100	-200	-300	-400
$\dfrac{x^2 - 20x}{10}$	0	-7.5	-10	-7.5	0
y	10	2.5	0	2.5	10

b 20 m

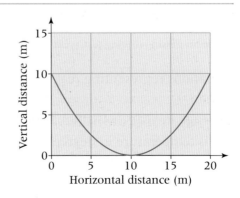

1 The graph $y = 2x^2 + x - 4$ is shown. What are the coordinates of the minimum point of the graph?

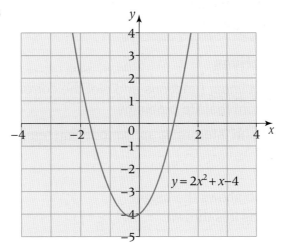

$y = 2x^2 + x - 4$

2 a Copy and complete the table of values for the graph $y = x^2 + x + 1$.

x	−3	−2	−1	0	1	2	3
x^2	9						
y	7				3		

b Plot the points for x and y and join them to form a smooth parabola.

c What is the approximate minimum value of $x^2 + x + 1$ and for what value of x does it occur?

3 A ball is thrown into the air.
The formula, $y = 20x - 4x^2$, shows its height, y metres, above the ground x seconds after it is thrown.

a Copy and complete the table of values to show the height of the ball during its first five seconds.

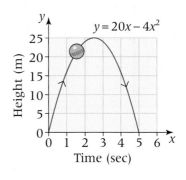

$y = 20x - 4x^2$

Time (x)	0	1	2	3	4	5
$20x$						
$4x^2$						
Height (y)						

b Use the table to plot a graph to show the ball's height against time.

c Use your graph to find

  i the maximum height reached by the ball and the time at which it reaches this height

  ii two times when the ball is 12 metres above the ground

  iii the interval of time when the ball is above 15 metres.

Key objectives

- Construct linear functions and plot the corresponding graphs arising from real-life problems
- Discuss and interpret graphs modelling real situations

1 This graph shows the height of a ball as it is thrown.

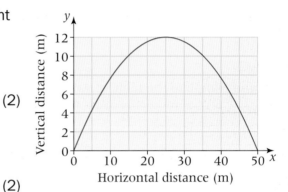

   a What is the maximum height reached by the ball? (2)

   b How far away is the ball from its starting position when it lands? (2)

2 Here is part of a distance–time graph of Siân's journey from her house to the shops and back.

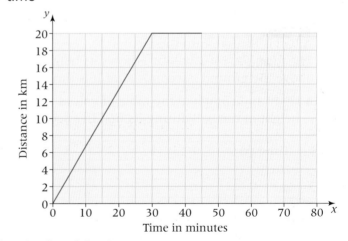

   a Work out Siân's speed for the first 30 minutes of her journey. Give your answer in km/h. (2)

Siân spends 15 minutes at the shops.
She then travels back to her house at 60 km/h.

   b Copy and complete the distance–time graph. (2)

(Edexcel Ltd., 2003)

This unit will show you how to

- Understand, recall and use trigonometry in right-angled triangles
- Understand and use similarity and ratio to find missing angles and lengths in triangles
- Use the trigonometric functions of a scientific calculator
- Understand, recall and use Pythagoras' theorem and trigonometry in 2-D problems
- Use bearings, maps and scale drawings in problem-solving

Before you start ...

You should be able to answer these questions.

Review

1 Rearrange these equations to make x the subject.

Unit A2

a $y = \frac{x}{6}$ **b** $y = \frac{x}{5}$

c $y = \frac{x}{10}$ **d** $y = \frac{2}{x}$

e $y = \frac{5}{x}$ **f** $y = \frac{8}{x}$

2 Use Pythagoras' theorem to find the missing side in these triangles.

Unit S4

a
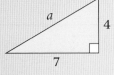

b
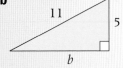

c
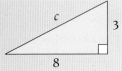

d
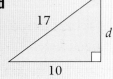

This spread will show you how to:

- Understand, recall and use trigonometry in right-angled triangles
- Understand and use similarity and ratio to find missing angles and lengths in triangles

Keywords
Adjacent
Opposite
Right-angled
   triangle
Tangent

These two right-angled triangles with angle 30° are similar.

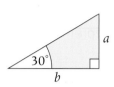

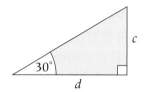

In similar triangles, corresponding pairs of angles are equal.

$\dfrac{c}{a} = \dfrac{d}{b}$ which you can rearrange to $\dfrac{c}{d} = \dfrac{a}{b}$

The ratio holds for all right-angled triangles with an angle 30°.

This result is true for all similar right-angled triangles.

Corresponding pairs of sides are in the same ratio.

$\dfrac{e}{f} = \dfrac{g}{h}$

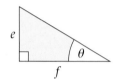

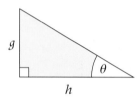

θ, the Greek letter, theta, is often used for angles.

For a right-angled triangle with angle θ, the ratio $\dfrac{\text{opposite side}}{\text{adjacent side}}$ is constant.

This ratio is called the **tangent** ratio.

Opposite side is opposite the angle θ.
Adjacent side is adjacent (next) to angle θ.

- **Tan** $\theta = \dfrac{\text{opposite side}}{\text{adjacent side}}$
- **You can use the tangent ratio in any right-angled triangle.**

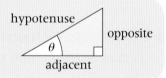

Example

Find the missing sides.

a

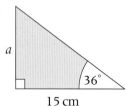

b

a $\tan 36° = \dfrac{a}{15}$

$15 \times \tan 36° = a$

$a = 10.9$ cm (1 dp)

b Third angle $= 180° - (90° + 65°)$

$= 25°$

$\tan 25° = \dfrac{b}{20}$

$b = 20 \tan 25°$

$= 9.3$ cm (1 dp)

Use the tan button on your calculator.

$\boxed{\text{tan}}\ \boxed{3}\ \boxed{6}\ \boxed{=}$

Make sure it is in degree mode.

1 Find the missing side in each triangle.
Give your answers to 3 significant figures.

In some triangles you will need to find the third angle to use in your calculations.

a
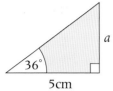

b
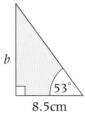

c
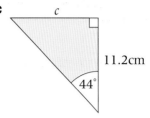

d
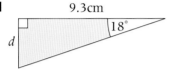

e
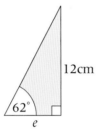

f
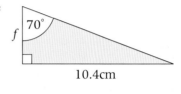

g
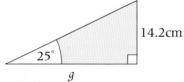

h
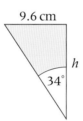

i
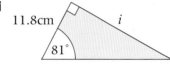

j
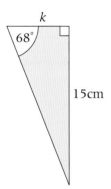

k
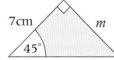

l
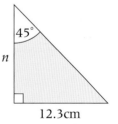

2 In question **1**, what sort of triangles are **k** and **l**?
How could you find the missing sides *m* and *n* without using the tangent ratio?

351

Sine and cosine ratios

This spread will show you how to:

- Understand, recall and use trigonometry in right-angled triangles
- Use the trigonometric functions of a scientific calculator

Keywords
Cosine
Hypotenuse
Right-angled
 triangle
Sine
Tangent

The tangent ratio is:

$$\text{Tan } \theta = \frac{\text{opposite side}}{\text{adjacent side}}$$

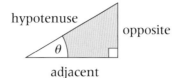

The **hypotenuse** is the longest side, opposite the right angle.

There are two other ratios you can use in right-angled triangles.

- **Sine ratio**

$$\sin \theta = \frac{\text{opposite side}}{\text{hypotenuse}}$$

- **Cosine ratio**

$$\cos \theta = \frac{\text{adjacent side}}{\text{hypotenuse}}$$

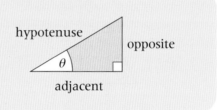

Label the sides you want to find and the side you know.
Remember that opposite and adjacent refer to the angle.

Example

Find the missing sides.

a

14 cm, a, 32°

a $\sin 32° = \frac{a}{14}$

$14 \times \sin 32° = a$

$a = 7.42$ cm (3 sf)

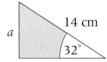
opp and hyp so use sine

Use the $\boxed{\sin}$ key on your calculator.

b

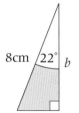

8cm, 22°, b

b $\cos 22° = \frac{b}{8}$

$8 \times \cos 22° = b$

$b = 7.42$ cm (3 sf)

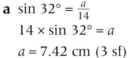
adj and hyp so use cosine

Use the $\boxed{\cos}$ key on your calculator.

c

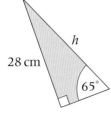

h, 28 cm, 65°

c $\sin 65° = \frac{28}{h}$

$h = \frac{28}{\sin 65}$

$h = 30.9$ cm (3 sf)

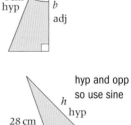
hyp and opp so use sine

To find the hypotenuse, you will always need to divide by either sin or cos.

1 Find the missing side in each of these right-angled triangles.
Give your answer to 3 significant figures.

a
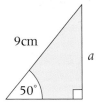
9cm
a
50°

b
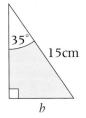
35°
15cm
b

c

10.6cm
c
48°

d
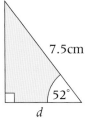
7.5cm
52°
d

e
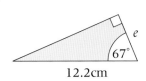
e
67°
12.2cm

f

f
80°
6.9cm

g
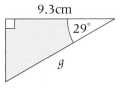
9.3cm
29°
g

h
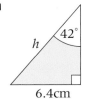
42°
h
6.4cm

i
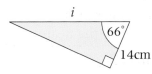
i
66°
14cm

j
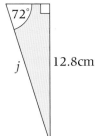
72°
j
12.8cm

k
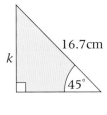
16.7cm
k
45°

l
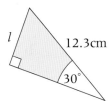
l
12.3cm
30°

m
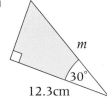
m
30°
12.3cm

n
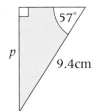
57°
p
9.4cm

o
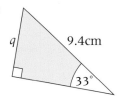
q
9.4cm
33°

This spread will show you how to:

- Understand, recall and use trigonometry in right-angled triangles
- Use the trigonometric functions of a scientific calculator

Keywords
Cosine
Right-angled
  triangle
Sine
Tangent

You can use the **sine**, **cosine** and **tangent** ratios in a right-angled triangle.

$$\sin \theta = \frac{\text{opp}}{\text{hyp}} \qquad \cos \theta = \frac{\text{adj}}{\text{hyp}} \qquad \tan \theta = \frac{\text{opp}}{\text{adj}}$$

- You can use the inverse functions $\sin^{-1}$, $\cos^{-1}$ and $\tan^{-1}$ to find the angle if you know two sides.

Always start you calculation by labelling opposite side and adjacent side in relation to the angle.

Find the $\boxed{\sin^{-1}}$, $\boxed{\cos^{-1}}$ and $\boxed{\tan^{-1}}$ keys on your calculator. They may be 2nd functions, or you may need to use the $\boxed{\text{INV}}$ key.

Example

Find the missing angles.
Give your answers to the nearest degree.

a

a You have adjacent and hypotenuse, so use cosine.

$$\cos x = \frac{6}{10.5}$$
$$x = \cos^{-1} \frac{6}{10.5}$$
$$x = 55° \text{ (to the nearest degree)}$$

On your calculator, use the brackets keys for the fraction.

b

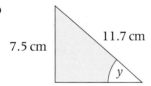

b You have opposite and hypotenuse, so use sine.

$$\sin y = \frac{7.5}{11.7}$$
$$y = \sin^{-1} \frac{7.5}{11.7}$$
$$y = 40° \text{ (to the nearest degree)}$$

c

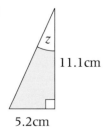

c You have opposite and adjacent, so use tan.

$$\tan z = \frac{5.2}{11.1}$$
$$z = \tan^{-1} \frac{5.2}{11.1}$$
$$z = 25° \text{ (to the nearest degree)}$$

Tan can be a fraction >1. sin and cos are always fractions <1.

1 Find the missing angle in each triangle.
Give your answers to 3 significant figures.

a

9cm
4cm
a

b
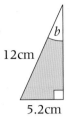
12cm
b
5.2cm

c
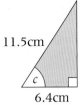
11.5cm
c
6.4cm

d

17.2cm
d
6.6cm

e
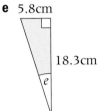
5.8cm
18.3cm
e

f
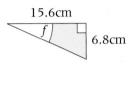
15.6cm
f
6.8cm

g
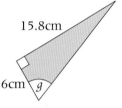
15.8cm
6cm
g

h
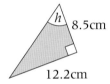
h
8.5cm
12.2cm

i
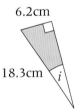
6.2cm
18.3cm
i

j
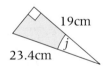
19cm
j
23.4cm

k
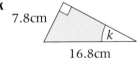
7.8cm
k
16.8cm

l
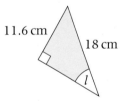
11.6 cm
18 cm
l

m
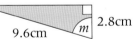
9.6cm
m
2.8cm

n
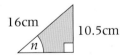
16cm
10.5cm
n

o
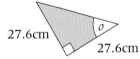
27.6cm
o
27.6cm

p
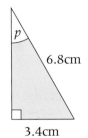
p
6.8cm
3.4cm

q
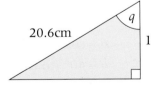
20.6cm
q
10.3cm

355

This spread will show you how to:

● Understand, recall and use Pythagoras' theorem and trigonometry in 2-D problems

Keywords
Cosine
Pythagoras'
 theorem
Sine
Tangent

Pythagoras'
theorem is
$a^2 + b^2 = c^2$

You use Pythagoras' theorem in a right-angled triangle when you know two sides and want to find the third.

Example

Use Pythagoras' theorem to find the missing sides in these triangles.

a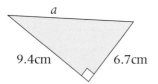

b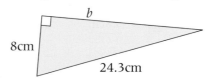

a $a^2 = 9.4^2 + 6.7^2 = 133.25$
$a = \sqrt{133.25} = 11.5$ cm (1 dp)

b $b^2 + 8^2 = 24.3^2$
$b^2 = 24.3^2 - 8^2 = 526.49$
$b = \sqrt{526.49} = 22.9$ cm (1 dp)

You can use **sine**, **cosine** and **tangent** ratios in a right-angled triangle:

● When you know a side and an angle and want to find another side
● When you know two sides and want to find an angle.

● $\sin \theta = \frac{\text{opp}}{\text{hyp}}$ $\cos \theta = \frac{\text{adj}}{\text{hyp}}$ $\tan \theta = \frac{\text{opp}}{\text{adj}}$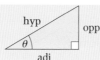

$\textbf{S}\text{in } \theta =$
$\dfrac{\textbf{O}\text{pp}}{\textbf{H}\text{yp}}$

$\textbf{C}\text{os } \theta =$
$\dfrac{\textbf{A}\text{dj}}{\textbf{H}\text{yp}}$

$\textbf{T}\text{an } \theta =$
$\dfrac{\textbf{O}\text{pp}}{\textbf{A}\text{dj}}$

Example

a Calculate the length FG.
b Calculate the size of angle GEH.

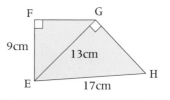

a Using Pythagoras in triangle EFG:
$9^2 + FG^2 = 13^2$

$FG^2 = 169 - 81 = 88$

$FG = \sqrt{88} = 9.4$ cm (to 1 dp)

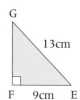

b In triangle GEH:
$\cos \theta = \frac{13}{17}$

$\theta = \cos^{-1}\frac{13}{17}$

$\theta = 40°$ (to the nearest degree)

You have adjacent and hypotenuse, so use cosine.

1 Find the missing lengths.

a

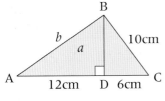

Use BCD, then ABD

b

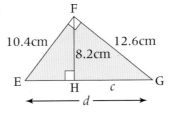

Use EFG then FGH

c

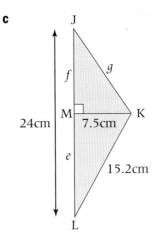

2 Find the missing angles.

a

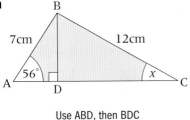

Use ABD, then BDC

b

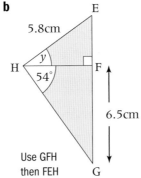

Use GFH then FEH

c

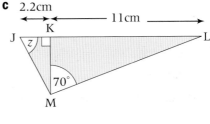

3 ABC and ACD are right-angled triangles.

   a Find AC.

   b Find angle CAD.

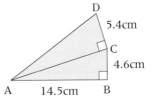

4 JKL and JLM are right-angled triangles.

   a Find JL.

   b Find angle JML.

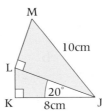

5 PQR and PRS are right-angled triangles.

   a Find PR.

   b Find RQ.

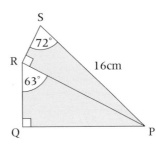

This spread will show you how to:

- Understand, recall and use Pythagoras' theorem and trigonometry in 2-D problems
- Use bearings, maps and scale drawings in problem-solving

Keywords

Bearing
Cos
Pythagoras' theorem
Sin
Tan

In more complex problems involving lengths and angles, it helps to sketch the situation.

Example

PQRS is a parallelogram. PQ = 9.4 cm. QR = 7.8 cm. Angle PQR = 47°. Find the area of the parallelogram.

Draw a diagram.

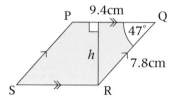

First find the vertical height, h, between PQ and RS.

h is at right-angles to PQ and RS.

$\sin 47° = \frac{h}{7.8}$

$h = 7.8 \times \sin 47° = 5.704 \ldots$

Area of PQRS $= h \times b = 5.704 \ldots \times 9.4 = 53.6$ cm^2 (to 3 sf)

Do not round intermediate values in the calculation.

Example

A boat sails from a harbour on a bearing of 072° to a buoy 12 km away. Then it changes direction and sails 20 km to a lighthouse due east of the harbour.

a On what bearing does the boat sail from the buoy to the lighthouse?
b How far is it from the lighthouse back to the harbour?

Draw a diagram.

H is the harbour.
B is the buoy.
L is the lighthouse.

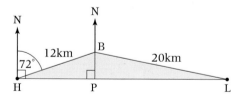

Draw in BP to divide HBL into two right-angled triangles.

a $\angle BHP = 90° - 72° = 18°$

$\sin 18° = \frac{BP}{12}$

$BP = 12 \times \sin 18° = 3.708 \ldots$

$\cos PBL = \frac{3.708 \ldots}{20}$

$\angle PBL = \cos^{-1}\left(\frac{3.708 \ldots}{20}\right)$

$= 79.314 \ldots°$

Bearing of L from B
$= 180° - 79.314 \ldots°$
$= 101°$ (nearest degree)

b $BP = 3.708\ldots$

$PL^2 = 20^2 - 3.708 \ldots^2$ Pythagoras' theorem

$PL = \sqrt{386.249\ldots}$

$= 19.653 \ldots$

$PH^2 = 12^2 - 3.708 \ldots^2$

$PH = \sqrt{140.291\ldots}$

$= 11.412 \ldots$

$HP = PL + PH$

$= 11.412 \ldots + 19.653 \ldots$

$= 31.1$ km (to 3 sf)

1 Find the area of a parallelogram with side lengths 5 cm and 11 cm and smaller angle 64°.

2 A rhombus has side lengths 9 cm and smaller angle 52°.
Find the area of the rhombus.

3 Find the area of a rhombus with sides 7 cm and smaller angle 40°.

4 A chord AB with length 10 cm is drawn inside a circle with centre O and radius 7 cm.
Find the angle AOB.

5 A chord PQ is drawn inside a circle with centre O and radius 8 cm such that angle POQ = 80°.
Find the length of the chord PQ.

6 A chord ST is drawn inside a circle with centre O and radius 9 cm such that angle SOT = 110°.
Find the length of the chord ST.

7 An isosceles triangle has side lengths 9 cm, 9 cm and 6 cm.
Find the angle between the two equal sides.

8 Jenny walks 4 km on a bearing 052°. She changes direction and walks a further 5 km to finish due east of her starting point.
Find how far Jenny is from her starting point.

9 Liz leaves home and cycles 16 km on a bearing 215° to a lake.
She changes direction and cycles 12 km to a wood which is due south of her home.

a On what bearing does she cycle from the lake to the wood?

b How far does she have to cycle home?

10 A flag pole TP, with T at the top, is held upright by two ropes, TX and TY, fixed on horizontal ground at X and Y.
Angle PXT = 23°. Angle PYT = 36°. TX = 10 m.
Find the length of TY.

11 Ali and Pete are estimating the height of a phone mast.
Ali stands 15m from the mast and measures the angle of elevation to the top as 60°.
Pete stands 25m from the mast and measures the angle of elevation to the top as 46°.
Can they both be correct? Discuss.

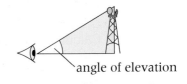

angle of elevation

Key objectives

- Understand, recall and use Pythagoras' theorem in 2-D problems
- Understand, recall and use trigonometrical relationships in right-angled triangles, and use these to solve problems

1 In a treasure hunt the contestants are given these instructions.

> Treasure hunt!
>
> Walk 20 paces north.
> Next walk 40 paces on a bearing of 120°.
> The treasure lies 30 paces on a bearing of 110°.

a Draw a diagram using a scale of 1 cm = 1 pace to show the path of the hunt. (2)

b i Use Pythagoras' theorem and trigonometry to calculate how far the treasure is from the starting point. (2)

ii What bearing do the contestants take to return the treasure to the starting point? (2)

2 DE = 6 m.
EG = 10 m.
FG = 8 m.
Angle DEG = 90°.
Angle EFG = 90°.

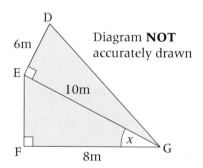

Diagram **NOT** accurately drawn

a Calculate the length of DG.
Give your answer correct to 3 significant figures. (3)

b Calculate the size of the angle marked *x*.
Give your answer correct to one decimal place. (3)

(Edexcel Ltd., 2004)

GCSE formulae

In your Edexcel GCSE examination you will be given a formula sheet like the one on this page.

You should use it as an aid to memory. It will be useful to become familiar with the information on this sheet.

Volume of a prism = area of cross section × length

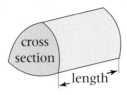

Volume of sphere = $\frac{4}{3}\pi r^3$

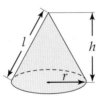

Surface area of sphere = $4\pi r^2$

Volume of cone = $\frac{1}{3}\pi r^2 h$

Curved surface area of cone = $\pi r l$

In any triangle ABC

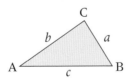

Sine rule $\dfrac{a}{\sin A} = \dfrac{b}{\sin B} = \dfrac{c}{\sin C}$

Cosine rule $a^2 = b^2 + c^2 - 2bc \cos A$

The Quadratic Equation

The solutions of $ax^2 + bx + c = 0$ where $a = 0$, are given by

$$x = \frac{-b \pm \sqrt{(b^2 - 4ac)}}{2a}$$

Answers

N1 Before you start ...

1 a Four thousand   b Four hundred
   c Four tenths   d Four thousandths

2 a

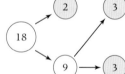

   −2.5
   −5 −4 −3 −2 −1 0 1 2 3 4 5

   b −3, −2.4, −1.8, 0, +1.5, +5

3 ai   1, 2, 3, 6
   aii 1, 2, 3, 4, 6, 12
   aiii 1, 2, 4, 7, 14, 28
   aiv 1, 2, 3, 4, 6, 9, 12, 18, 36
   b   2, 3, 5, 7, 11, 13, 17, 19, 23, 29, 31, 37, 41,
        43, 47

N1.1

1 a 0.1, 0.3, 1, 1.3, 2, 3.1
   b 6.07, 7.06, 27.6, 77.2, 607

2 a 682.8, 862.6, 6000.8, 6008, 8000.6
   b 47.9, 49.7, 74.9, 79.4, 94.7, 97.4

3 a 167   b 248   c 7.16
   d 10.95   e 2430   f 2813

4 a 21.4   b 6.73   c 410.6
   d 20.07   e 0.6025   f 8.6

5 a 4.52   b 5.5   c 16.8   d 16.8

6 a 7.03, 7.08, 7.3, 7.38, 7.8, 7.83
   b 2.18, 2.4, 4.18, 4.2, 8.24, 8.4

7 a 18.7, 18.16, 17.6, 17.16, 16.7, 16.18
   b 13.2, 13.145, 2.5, 2.38, 1.1, 1.06

8 ai 3050   aii 3000   bi 1760   bii 1800
   ci 290   cii 300   di 50   dii 100
   ei 40   eii 0   fi 740   fii 700

9 a 3000   b 1000   c 0
   d 25 000   e 16 000   f 168 000

10 ai 39.1   aii 39.11   bi 7.1   bii 7.07
   ci 5.9   cii 5.92   di 512.7   dii 512.72
   ei 4.3   eii 4.26   fi 12.0   fii 12.01
   gi 0.8   gii 0.83   hi 26.9   hii 26.88

11 ai 0.1   aii 0.07   aiii 0.070
   bi 15.9   bii 15.92   biii 15.918
   ci 128.0   cii 128.00 ciii 127.998
   di 887.2   dii 887.17 diii 887.172
   ei 55.1   eii 55.14   eiii 55.145
   fi 0.0   fii 0.01   fiii 0.007

12 a 1306   b 2.085   c 1085   d 2.487
   e 0.0008   f 6.19   g 0.04513 h 0.0045

N1.2

1 a −2   b −1   c −2   d −2   e −5   f −5
2 a +1   b +3   c +4   d +9   e +12   f +5
3 a −5   b −2   c −4   d +2   e +4   f +7
4 a −2   b +   c +, −   d −, −
5 a +22   b −12   c −2   d −14

e +12   f +12   g +19   h −15
i +11   j −61   k +344   l +49
6 a −10.8   b +6.3   c +13.5   d −0.7
   e −38.2   f +112.7
7 a £5.49   b £172.38
8 a −13 °C b 15 °C   c 41 °C

N1.3

1 a −2   b −1   c −2   d −2   e −5   f −5
1 a −15   b −18   c −21   d −56   e −36   f −12
2 a +16   b +16   c +15   d +42   e +56   f +81
3 a −25   b −32   c −72   d −20   e +30   f +49
   g +16   h −20   i −18   j +26   k −42   l −48
4 a −3   b −4   c −7   d −2   e +19   f +5
5 a −2   b −5   c +5   d +4   e −22   f −1
   g +40   h +4   i +5   j −17   k −3   l +27
6 a +   b −6   c +45   d −12
7 a +450   b −150   c +63   d −25
   e −0.73   f +0.092
8 a −49   b +63   c +3.77   d −619.7
   e +140.9   f +0.09
9 a −36   b +0.1   c −0.98   d −0.0087
   e +0.0073 f −0.00006
10 a −55   b +0.48   c +5.266 d −156
   e +0.0082 f +0.50005
11 a +0.18   b −27   c −7   d −380

N1.4

1 a 12   b 8   c 4   d 6
2 a $77 = 7 \times 11$ b $51 = 3 \times 17$
   c $65 = 5 \times 13$ d $91 = 7 \times 13$
   e $119 = 7 \times 17$ f $221 = 13 \times 17$
3

4 a $2^2 \times 3^2$ b $2^3 \times 3 \times 5$
   c 2×17 d 5^2
   e $2^4 \times 3$ f $2 \times 3^2 \times 5$
   g 3^3 h $2^2 \times 3 \times 5$
5 a $2^2 \times 263$ b $2^9 \times 5$
   c $2 \times 3^2 \times 5 \times 7$ d $3 \times 5^2 \times 11$
   e $5 \times 11 \times 13$ f $7 \times 11 \times 13$
   g 3×73 h 17^2
   i $2^3 \times 5 \times 71$ j $5 \times 7^2 \times 11$
   k $7 \times 13 \times 19$ l $2 \times 3^2 \times 11 \times 17$
   m $2^2 \times 11 \times 13 \times 17$ n $2 \times 5 \times 7 \times 13^2$
   o $2^2 \times 23 \times 31$ p $3^3 \times 13 \times 29$
6 a $2 \times 3 \times 35$, $2 \times 5 \times 21$, $2 \times 7 \times 15$, $3 \times 7 \times 10$,
      $3 \times 5 \times 14$, $5 \times 6 \times 7$
   b $1 \times 1 \times 121$, $2 \times 3 \times 35$, $2 \times 5 \times 21$, $2 \times 7 \times 15$,
      $3 \times 7 \times 10$, $3 \times 5 \times 14$, $5 \times 6 \times 7$

N1.5

1 **a** $6^2 = 36$, so 6 joins to itself.
  b 1, 2, 3, 4, 6, 8, 12, 16, 24, 48

  c 12
2 **a** 1   **b** 1   **c** 3   **d** 2   **e** 8   **f** 10
3 **a** Multiples of 12 = 12, 24, 36, 48, ...
    Multiples of 9 = 9, 18, 27, 36, 45, ...
  b 36
4 **a** 20   **b** 36   **c** 30   **d** 60   **e** 70   **f** 40
5 8
6 **a** 5600   **b** 4432   **c** 12 720
  d 14 168   **e** 5105   **f** 10 220

N1 Exam review

1 **ai** −25.9   **aii** −25.9   **aiii** 25.9
  bi −53.2   **bii** −0.532   **biii** 5.32
2 **ai** $2^2 \times 3 \times 5$ **aii** $2^5 \times 3$
  b 12   **c** 480

S1 Before you start ...

1 **ai** 16 cm   **aii** 160 mm   **aiii** 12 cm^2
  bi 18 cm   **bii** 180 mm   **biii** 20.25 cm^2
  ci 19.1 cm   **cii** 191 mm   **ciii** 14 cm^2
2 54 cm^2

S1.1

1 **a** 28 cm^2 **b** 22.26 cm^2 **c** 26.1 cm^2
  d 7440 mm^2 **e** 10 290 mm^2
2 **a** 7.5 cm^2 **b** 11.76 cm^2 **c** 126 mm^2
  d 6 cm^2 **e** 14 cm^2 **f** 7.2 cm^2
  g 10.8 cm^2
3 **ai** 40 cm   **aii** 72 cm^2
  bi 33 cm   **bii** 32 cm^2
  ci 36 cm   **cii** 44 cm^2
4 479 cm^2

S1.2

1 **a** 15 cm^2 **b** 33.48 cm^2 **c** 45.6 cm^2
  d 31.5 cm^2 **e** 13.34 cm^2
2 **a** 12 cm^2 **b** 20 cm^2 **c** 20.5 cm^2
  d 600 mm^2 **e** 1250 mm^2
3 205.5 cm^2

S1.3

1 **a** 25.1 cm   **b** 239 mm   **c** 50.3 cm
  d 47.1 cm   **e** 75.4 mm   **f** 132 mm
  g 82.9 cm
2 **a** 75.4 mm   **b** 145 cm   **c** 330 mm
  d 3.77 cm   **e** 22.6 cm   **f** 393 cm
3 **a** 50.3 cm^2 **b** 4540 mm^2 **c** 201 cm^2
  d 177 cm^2 **e** 452 mm^2 **f** 1390 mm^2
  g 547 cm^2
4 **a** 452 mm^2 **b** 1660 cm^2 **c** 8660 mm^2
  d 1.13 cm^2 **e** 40.7 cm^2 **f** 12 300 cm^2
5 18.8 m

6 7.0 cm to 1 dp
7 **a** 13.9 cm^2 **b** 3.79 cm^2

S1.4

1 **a** 39.3 cm^2 **b** 81.4 cm^2 **c** 905 mm^2
  d 127 cm^2 **e** 402 mm^2 **f** 373 cm^2
  g 422 cm^2 **h** 103 cm^2
2 **a** 226 cm^2 **b** 8.31 cm^2 **c** 65.3 m^2
  d 195 cm^2 **e** 142 mm^2 **f** 4.51 cm^2
3 **a** 25.7 cm   **b** 37.0 cm   **c** 123 mm
  d 46.3 cm   **e** 82.3 mm   **f** 79.2 cm
  g 84.3 cm   **h** 33.5 cm
4 **a** 61.7 cm   **b** 11.8 cm   **c** 33.2 m
  d 57.3 cm   **e** 48.8 mm   **f** 8.71 cm
5 **a** 6.28 m^2
  b 20 flowers; he has 0.28 m^2 left
6 113 m
7 2510 cm^2

S1.5

1 **a** 142 cm^2 **b** 98 cm^2 **c** 114 cm^2
  d 65.6 cm^2 **e** 96 cm^2 **f** 102 mm^2
2 **a** 44.0 cm^2 **b** 165 cm^2 **c** 75.4 cm^2
  d 66.4 cm^2
3 **a** 330 cm^2 **b** 468 cm^2

S1 Exam review

1 32 cm
2 88.4 cm^2

A1 Before you start ...

1 **a** 45   **b** 52   **c** −26   **d** 196
  e 7   **f** 13   **g** −50   **h** 30
2 **a** 15   **b** 21   **c** 15   **d** 15
3 **a** 1, 2, 3, 4, 6, 8, 12, 24
  b 1, 2, 3, 4, 6, 9, 12, 18, 36
  c 1, 2, 4, 5, 10, 20, 25, 50, 100
  d 1, 11, 121
4 **a** 3   **b** 4   **c** 10   **d** 6
  e 4   **f** 25   **g** 33   **h** 7

A1.1

1 **a** $5w$ **b** $\frac{6}{k}$ **c** y^2 **d** $6ab$ **e** $8k^3$
2 **a** 20   **b** 4   **c** 36   **d** 22   **e** 108
3 **a** $11a + 6b$ **b** $2t + 26$ **c** $x - 12y$
  d $p^2 + 14p$ **e** $20xy$ **f** $7ab$
4 Abdul
5 **a** $28mn$ **b** $12m^2$ **c** $10p$
  d 2   **e** $24abc$ **f** $6k^3$
  g $4b$ **h** $9c$

6

$4p + 7q$	$4p + 7q$	**$4p + 2q$**
$6mn$	**$10mn$**	$6mn$
$2d$	**2**	$2d$
$2n - 8$	**$2n$**	$2n - 8$

7 **ai** $8p + 16$ **aii** $32p$ **b** $3x, 2y$

8

$9b-2a$	$12a^2$	$4b$
$2ab$	$5p^3+7p^2+10p$	$13abc$
$5m-4$	$60m^3$	$\dfrac{2}{a^2}$

A1.2

1 a $4n+20$ **b** $6b-42$
   c a^2+3a **d** $ab-ac$
   e $8x+12y-16z$ **f** $2h^2+18h$
2 a $-3k-27$ **b** $-2h+10$
   c $-w+4$ **d** $-t+p$
   e $-k^2-7k$ **f** $-18m+9k-36$
   g $-x^2+x+8$ **h** $-2x^2-6$
   i $-3+3x$
3 a $10c+62$ **b** $23x+67$
   c $2x^2+10x$ **d** $17t^2+32t$
   e $7x-45$ **f** $2x-11$
   g $2m-26$ **h** $-11g+33$
   i $p+14$ **j** $2q-7$
4 $-5x^2+44x$
5 a $3(2x-1)$ **b** $6x-3$
   c $6x-3=15$, which gives $6x-18=0$
6 a For example, $8(3x+2)$
   b For example, $2(2x+3)+5(4x+2)$
7 $y^2(y+7)$, y^3+7y^2

A1.3

1 a x^2+5x+6 **b** $p^2+11p+30$
   c w^2+5w+4 **d** $c^2+10c+25$
   e x^2+2x-8 **f** $y^2+5y-14$
   g $t^2+4t-12$ **h** $x^2-7x+10$
   i $y^2-14y+40$ **j** w^2-3w+2
   k $p^2-10p+25$ **l** $q^2-24q+144$
2 a $6x^2+17x+7$ **b** $10p^2+19p+6$
   c $6y^2+11y+4$ **d** $4y^2+24y+36$
   e $10t^2+12t-16$ **f** $15w^2+42w-9$
   g $6x^2-6y^2$ **h** $9m^2-24m+16$
   i $6p^2-pq-40q^2$ **j** $4m^2-12mn+9n^2$
3 a $(x-3)(x+6)=x^2+3x-18$
   b $(2m-3)^2=4m^2-12m+9$
4 a $a^2+2ab+b^2$ **b** 16
   c For example,
      $2.5^2+2\times2.5\times3.5+3.5^2=6^2$
5 a $(3x-1)(2x+3)=75$, which gives
      $6x^2+7x-78=0$
   b $(3x-4)(5x+2)=2\times75$, which gives
      $15x^2=14x^2+158$

A1.4

1 a $2(x+2)$ **b** $3(y-2)$
   c $12(p+3q)$ **d** $5(5w-1)$
   e $x(6y+w)$ **f** $b(a-2c)$
   g $q(pr+rt-sw)$ **h** $x(5y-1)$
   i $2x(y+3)$ **j** $2a(2b-3a)$
   k $5p(5p-2)$ **l** $7x(1+2y)$
   m $2a(c+2a-4)$ **n** $5m(3n-1+2m^2)$
   o $6p(p^3-2)$

2 Correct factorisations are: Clare $5x(1+2y)$,
    Ben $3p(2q+1)$, Vicky $7p(3+2q)$
3 a $23(x+y)$ **b** $(a-b)(a-b+5)$
   c $(q+r)(6-(q+r)^2)$ **d** $7(pt-w)$
4 a $(a+b)(x+y)$ **b** $(c+b)(d+m)$
   c $(a+b)(a+2)$ **d** $(c-m)(d+e)$
5 a $4(x-1)$ **b** $20(b+2)$
6 a 6 **b** 16.5 **c** 58.6 **d** 33.2
7 $4(3x+2)-2(2x-1)=8x+10=2(4x+5)$

A1.5

1 a $(x+2)(x+4)$ **b** $(x+3)(x+7)$
   c $(x+4)(x+7)$ **d** $(x+3)(x+8)$
   e $(x-2)(x-6)$ **f** $(x-3)(x-6)$
   g $(x-9)(x-4)$ **h** $(x+4)(x-3)$
   i $(x-7)(x+5)$ **j** $(x+9)(x-3)$
   k $(x-16)(x+2)$ **l** $(x+20)(x-2)$
2 a $x^2+9x-22$ **b** $(x+11)(x-2)$
3 a $(x+12)(x-6)$ **b** $(x-12)(x+2)$
   c $(x-15)(x-5)$ **d** $(x+16)(x-4)$
   e $(x-8)(x+8)$ **f** $(x-4)(x-25)$
4 a $(x+2)(x+19)$ **b** $x(5x+5+y)$
   c $(x+11)^2$ **d** $(x+9)(x-2)$
   e $(p+3)(p+11)$ **f** $x(2x+3y)$
5 $(x+5)(x+4)=12$
    $x^2+9x+20=12$
    $x^2+9x+8=0$
    $(x+1)(x+8)=0$
6 a $2(x+4)(x+7)$ **b** $x(x-8)(x+3)$
   c $x(x-4)(x+4)$
7 $(2.3+1.7)^2=4^2=16$

A1 Exam review

1 a $2x(2+x)$ **b** $x^2+3x-10$ **c** $12a^2b^5$
2 a $4x+8$ **b** 15.5

N2 Before you start …

1 a 120 **b** 138 **c** 90 **d** 265
2 a 4438 **b** 1977 **c** 857 **d** $14\,224$
3 a 147 **b** 1515 **c** $66\,560$ **d** $51\,450$
4 a $11\,700$ **b** $78\,408$ **c** 205 **d** $67\,564$

N2.1

1 a 30 **b** 30 **c** 50 **d** 210
   e 780 **f** $23\,780$
2 a 6 **b** 4 **c** 22 **d** 39
   e 18 **f** 454
3 a 200 **b** 200 **c** 100 **d** 700
   e 1400 **f** $134\,600$
4 a 2000 **b** $13\,000$ **c** 8000 **d** $11\,000$
   e $78\,000$ **f** $156\,000$
5 a 0.3 **b** 0.7 **c** 0.3 **d** 0.2
   e 4.6 **f** 105.4
6 a 0.32 **b** 0.46 **c** 15.30 **d** 104.68
   e 16.45 **f** 0.00
7 a 480 **b** 1200 **c** 490 **d** $14\,000$
   e 530 **f** $15\,000$

8 a 0.36   **b** 0.42   **c** 0.057   **d** 0.0047
   e 1.4   **f** 0.0000042
9 a 200   **b** 2000   **c** 5   **d** 10
   e 0.0005   **f** 100 000
10 a 0.62   **b** 0.57   **c** 0.56   **d** 380
   e 550   **f** 7 300 000
11 a 400 ÷ 20   **b** 40 × 40   **c** 1000 ÷ 90
   d 4000 + 10 000   **e** 100 + (2000 ÷ 50)
12 a 20   **b** 1600   **c** 11   **d** 14 000   **e** 140
13 a 16.9047619, estimate is slightly high
   b 1677, estimate is close
   c 11.44565217, estimate is close
   d 16 040, estimate is slightly low
   e 153.3846154, estimate is close

N2.2

1 a 0.8   **b** 0.5   **c** 0.4   **d** 1.1
   e 0.4   **f** 1
2 a 5.8   **b** 6.5   **c** 7.4   **d** 4.1
   e 11.4   **f** 10
3 a 10   **b** 10.1   **c** 10.3   **d** 11.1
   e 16.4   **f** 7.5
4 a 7.77   **b** 5.25   **c** 3.6   **d** 3.9
   e 2.13   **f** 13.04
5 a 4.15   **b** 6.85   **c** 4.58   **d** 10.42
   e 12.14   **f** 2.04
6 a 0.7   **b** 0.6   **c** 0.3   **d** 3.4
   e 10.2   **f** 14.1
7 a 0.8   **b** 0.7   **c** 0.8   **d** 4.1
   e 10.7   **f** 14.4
8 a 0.74   **b** 1.05   **c** 1.51   **d** 7.14
   e 1.08   **f** 0.64
9 a 1.62   **b** 3.51   **c** 0.45   **d** 4.37
   e 14.52   **f** 11.66
10 a 1.72   **b** 3.61   **c** 0.65   **d** 4.67
   e 15.02   **f** 12.36

N2.3

1 a 12.4   **b** 12   **c** 13.1   **d** 22.5
   e 23.1   **f** 26.1
2 See Q1
3 a 13   **b** 1.871   **c** 201.321   **d** 45
   e 38.97   **f** 21.69
4 a 140.33   **b** 242.3   **c** 98.807   **d** 203.59
   e 161.002   **f** 102
5 See Q4
6 a 2.2   **b** 9.1   **c** 15   **d** 4.9
   e 8.4   **f** 2.9
7 See Q6
8 a 2.6   **b** 7.86   **c** 5.9   **d** 92.54
   e 0.97   **f** 24.27
9 a 13.896   **b** 19.45   **c** 359.79   **d** 7.683
   e 0.326   **f** 11.42
10 See Q9
11 a 4.1   **b** 40.2   **c** 11.288   **d** 5.968
   e 0.892   **f** 0.469

N2.4

1 a 4.8   **b** 4.8   **c** 0.48   **d** 48   **e** 48
2 a 9.1   **b** 9.1   **c** 0.91   **d** 91   **e** 91
3 a 42   **b** 16   **c** 7   **d** 104   **e** 2   **f** 28
4 a 4.2   **b** 1.6   **c** 0.7   **d** 10.4   **e** 2   **f** 0.28
5 a $9 \times 7 = 63$, $63 \div 7 = 9$, $63 \div 9 = 7$
   b $8 \times 6 = 48$, $48 \div 6 = 8$, $48 \div 8 = 6$
   c $7 \times 13 = 91$, $91 \div 13 = 7$, $91 \div 7 = 13$
   d $18 \times 15 = 270$, $270 \div 15 = 18$, $270 \div 18 = 15$
   e $3.5 \times 5 = 17.5$, $17.5 \div 3.5 = 5$, $17.5 \div 5 = 3.5$
   f $3.9 \times 2.4 = 9.36$, $9.36 \div 2.4 = 3.9$,
     $9.36 \div 3.9 = 2.4$
7 a 4.5   **b** 4500   **c** 45   **d** 0.45
   e 5   **f** 90   **g** 500
8 a 345.8   **b** 345.8   **c** 38   **d** 345.8
   e 9.1   **f** 0.0091
9 a 10 185   **b** 101.85   **c** 0.10185   **d** 0.35
   e 3.5   **f** 0.00291
10 a 29.61   **b** 0.2961   **c** 6300

N2.5

1 a 98   **b** 152   **c** 273
   d 323   **e** 308
2 See Q1
3 a 9.8   **b** 15.2   **c** 2.73
   d 0.0323   **e** 0.308
4 a 80   **b** 12   **c** 12
   d 21   **e** 22
5 See Q4
6 a 0.8   **b** 1.2   **c** 1.2
   d 0.21   **e** 0.22
7 a 24.91   **b** 4.284   **c** 105.84
   d 130.8985   **e** 42.9442
8 See Q7
9 a 3.87   **b** 0.775   **c** 0.916
   d 7.53   **e** 18.13
10 See Q9
11 a 3.45   **b** 4.15   **c** 7.74
   d 4.08   **e** 2.35
12 See Q11
13 a 5.26   **b** 28.88   **c** 1384.29
   d 175.56   **e** 28.65
14 See Q13

N2 Exam review

1 a 2.7   **b** 20   **c** 49   **d** 4.8
2 a 119.31   **b** 119 310   **c** 1.23

A2 Before you start …

1 a 9   **b** 4   **c** 30   **d** 10
2 a I think of a number and multiply it by 6.
   b I think of a number and multiply it by itself.
   c I think of a number, multiply it by 2 and
     then subtract 3.
   d I think of a number, subtract 4 from it and
     then multiply by 4.

e I think of a number and divide it by 7.

f I think of a number, square it and then multiply by 2.

g I think of a number, multiply it by 2 and then square.

h I think of a number, multiply it by 3 and then subtract the result from 10.

3 a $3x + 27$ **b** $8x - 4$ **c** $6 - 12y$

   d $x^2 - 7x$

4 a $5 > 3$ **b** $-9 < 1$ **c** $-2 > -5$

   d $0.9 > 0.85$ **e** $-\frac{1}{4} > -\frac{1}{2}$

A2.1

1

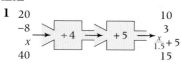

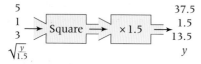

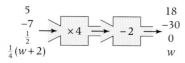

2 The starting numbers are different, so 10% of the final answer is different from 10% of the starting value.

3 a 5.5   **b** 5   **c** −3   **d** 40   **e** 100

4 Yes, −5, as 25 has two possible square roots

5 ai Add 3

   aii Multiply by 2, subtract 1

   aiii Add 5, divide by 2

   aiv Square, multiply by 3

   bi 14   **bii** 9   **biii** 43   **biv** 4 or −4

6 a −2   **bi** 7   **bii** 10   **biii** $-\frac{3}{16}$

7 a Many possibilities, e.g. ×2, ×2, ×2, +20, −7

   b Depends on **7a**

A2.2

1 a 5   **b** 6   **c** 10   **d** −2   **e** 5 or −5

   f 13   **g** $8\frac{1}{3}$ **h** $\sqrt[3]{16}$ **i** 100

2

¹1	3		²4
0		³4	0
⁴5	⁵1		
		⁶3	2	5

3 sides = 1 and 14

4 51

5 a 5   **b** 1

6 a $2x - 10$ **b** $5x - 20 = 180$ **c** 40°, 70°, 70°

A2.3

1 a −1   **b** 2   **c** 3   **d** $-\frac{13}{16}$

2 a $6a - 2 = 2a + 6$ **b** $3b - 14 = b$

   c $1 - 4c = 2 - 8c$ **d** $5d + 7d - 3 = 10 - d$

3 a $8x - 2 = 2x + 10, x = 2$ **b** $5x + 3 = 24 - 2x, x = 3$

   c $11 - 2x = 14 - 3x, x = 3$

4 a 40°, 60°, 80°

   b Square: 8 × 8, rectangle: 10 × 6

   c Mark: 160 cm, Miranda: 144 cm

A2.4

1 a $\frac{1}{3}$ **b** $\frac{1}{3}$ **c** $1\frac{1}{3}$ **d** $\frac{7}{8}$

   e −5   **f** $\frac{5}{11}$ **g** $-2\frac{1}{3}$ **h** $\frac{1}{3}$

2 a 5   **b** 10   **c** 28   **d** −1

3 a −8   **b** $\frac{3}{4}$ **c** $\frac{2}{11}$ **d** $-1\frac{1}{4}$

4 a −7   **b** 2   **c** −4   **d** 2

5 a $\frac{16}{x} = 10, x = 1.6$ **b** $\frac{12}{x + 4} = 7, x = -2\frac{2}{7}$

   c $\frac{11}{x - 3} = \frac{8}{x}, x = -8$

6 a 36   **b** 14   **c** 80

7 $x = 37\frac{2}{3}$

   Values for Set 1: $74\frac{1}{3}$, 115, $192\frac{1}{3}$, $263\frac{2}{3}$, 222, $-65\frac{1}{3}$

   Values for Set 2: 106, $196\frac{1}{3}$, $-24\frac{2}{3}$, 228, $162\frac{2}{3}$

A2.5

1 a $x \leqslant 3$ **b** $2 \leqslant x \leqslant 8$ **c** $-5 < x < 12$

2 a $x \leqslant 2$ **b** $x > -1$ **c** $x \geqslant -1$

   d $1 \leqslant x < 5$ **e** $-7 < x \leqslant 0$

3 True

4 a $x \leqslant 7$ **b** $x > 11$ **c** $p \leqslant -16$

   d $x > -3$ **e** $y \leqslant \frac{2}{3}$ **f** $y < -3$

   g $x \leqslant 2$ **h** $x > -5$ **i** $x \leqslant 10$

   j $p \leqslant -3$ **k** $x > -18$ **l** $x \geqslant \frac{1}{2}$

5 a $6(x - 2) > 12 + 2(x - 2), x > 5$ **b** 6

A2 Exam review

1 a $x \geqslant -3$ **b** $-3 < x \leqslant 2$

$$-5\,-4\,-3\,-2\,-1\ \ 0\ \ 1\ \ 2\ \ 3\ \ 4\ \ 5$$

2 a $y = 3\frac{1}{2}$ **b** $x = 7$

D1 Before you start ...

1 a A census is a survey where every member of a population is questioned.

   b A sample looks at a fraction of the population.

2 a Primary: Nicola, Secondary: Maddy

   b Primary data is data collected by yourself. Secondary data has already been collected by someone else.

3 ai There are two groups containing 4 hours.
  aii There are no groups containing 2 hours.
  b Use inequalities for the groups, e.g. $0 \leqslant t < 2$, $2 \leqslant t < 4$, 4 or more.

D1.1

1 a Assumes you visit cinema. Answers may differ at different times of year so could say on average. No answer choices given.
  b On average how many times do you go to the cinema in one month? Once or less often, 2 or 3 times a month, 4 times or more often.
  c On average how much do you spend when you go to the cinema? Less than £5, £5 to £10, more than £10.

2 ai Leading.
  aii What is your favourite sport? Tennis, swimming, football, rugby, cricket, other.
  b On average, how many times a week do you play sport? Never, 1 or 2 times, 3 or 4 times, more than 4 times.

3 a May not like either/not enough choices.
  b What is your favourite flavour of crisps? Plain, cheese and onion, salt and vinegar, smokey bacon, ketchup, other.
  ci 'lots' is vague; options do not cover all possible answers; needs a time frame.
  cii How many times have you visited the tuck shop in the last month? Never, once or twice, three to five times, more than five times.

4 ai Does not cover all possible answers.
  aii How far would you travel to see your favourite band? Less than 1 mile, 1 mile to 5 miles, between 5 and 10 miles, 10 miles or more.
  b How much would you pay for a ticket to see your favourite band? Less than £5, £5 to £10, £10.01 to £15, more than £15.

D1.2

1 Obviously visit cinema so not typical of population.
2 People in an athletics club will probably play sport more often than those who aren't.
3 a It is not representative of the people who use the school tuck shop.
  b It is not representative of the whole school.
  c Pick names out of a hat, or take every 20th person on a list of all the people in the school.
4 a His friends are not representative of the whole population, they might particularly like (or dislike) travelling to see bands.
  b People listening to MP3 players may be more interested in music than is typical.
5 All girls/all friends so may have same taste in music/small sample.

6 Cars passing at similar time/small sample.
7 a People at bus stops are more likely to take the bus to work.
  b Biased against people who are ex-directory, don't have a landline or aren't in when phoned.

D1.3

1 Two-way table with number of visits per week and amount spent.
2 Two-way table with number of times per week play sport and gender.
3 Two-way table with crisp flavour and Year groups.
4 Two-way table with distances and money.
5 Two-way table with favourite bands and numbers of CDs.
6 Two-way table with colour and make of car.
7 Two-way table with mode of travel and time taken.
8 a 500
  bi 20.8% **bii** 24% **biii** 39.2%

D1.4

1 ai	7	**aii**	6	**aiii** 5.82	**aiv**	6	**av**	2
bi	75	**bii**	63	**biii** 60.1	**biv**	63	**bv**	27
ci	8	**cii**	96	**ciii** 95.6	**civ**	96	**cv**	2
di	71	**dii**	22, 37	**diii** 40.4	**div**	37	**dv**	38
ei	26	**eii**	88, 89	**eiii** 84.2	**eiv**	87	**ev**	7
fi	72	**fii**	27	**fiii** 46.9	**fiv**	34	**fv**	37
gi	8	**gii**	105	**giii** 105.2	**giv**	105	**gv**	3

2 Range is unduly affected by one extreme value, which IQR ignores.
3 Mode is the lowest value.
4 a 1, 6, 8, 2, 8, 5, 6, 9, 3, 5, 7, 4, 4, 5, 5
  bi 8 **bii** 5 **biii** 5.2 **biv** 5 **bv** 3
  c Range and IQR stay same.
  d Answers are as for Q1 less 100.
  e Adding 100 does not affect spread of values but does affect averages.
5 ai 200, 200 **aii** 100, 100 **aiii** 100, 100
  aiv 100, 100 **av** 2, 2
  b All measures the same although sets of numbers are different.

D1.5

1 80.4 minutes
2 7.325 hours
3 43 lessons
4 78.6%
5 7.33
6 8.29

D1 Exam review

1 $\dfrac{30a + 20b}{50}$
2 Two-way table with crisp flavour and gender.

N3 Before you start …

1 **a** 6 squares shaded **b** 8 squares shaded
  c 9 squares shaded **d** 10 squares shaded
  e 7 squares shaded

2 **a** $\frac{1}{2}$ **b** $\frac{3}{4}$ **c** $\frac{4}{5}$ **d** $\frac{19}{20}$ **e** $\frac{3}{4}$

3 **a** 0.75 **b** 0.4 **c** 0.7 **d** 0.45 **e** 0.17

4 **a** $\frac{1}{2}$ **b** $\frac{1}{4}$ **c** $\frac{3}{10}$ **d** $\frac{4}{5}$ **e** $\frac{9}{20}$

5 **a** 50% **b** 25% **c** 10% **d** 20% **e** 5%

N3.1

1 **a** 4 **b** 10 **c** 24 **d** 12
  e 20 **f** 35 **g** 60 **h** 60

2 **a** 6 squares, 3 shaded
  b 6 squares, 4 shaded
  c 15 squares, 9 shaded
  d 20 squares, 15 shaded

3 **a** $\frac{20}{60}$ **b** $\frac{15}{60}$ **c** $\frac{40}{60}$ **d** $\frac{24}{60}$

4 **a** $\frac{18}{24}$ **b** $\frac{8}{24}$ **c** $\frac{9}{24}$ **d** $\frac{10}{24}$

5 **a** $\frac{20}{30}$ **b** $\frac{18}{42}$ **c** $\frac{35}{45}$ **d** $\frac{25}{40}$

6 **a** $\frac{2}{10}$ and $\frac{3}{10}$, $\frac{3}{10}$ **b** $\frac{8}{12}$ and $\frac{9}{12}$, $\frac{3}{4}$
  c $\frac{6}{15}$ and $\frac{5}{15}$, $\frac{2}{5}$ **d** $\frac{21}{30}$ and $\frac{20}{30}$, $\frac{7}{10}$

7 **a** Two sets of 10 squares; 2 and 3 shaded
  b Two sets of 12 squares; 8 and 9 shaded
  c Two sets of 15 squares; 6 and 5 shaded
  d Two sets of 30 squares; 21 and 20 shaded

8 **a** 30 **b** 12 **c** 24 **d** 28

9 **a** $\frac{6}{12}$, $\frac{8}{12}$, $\frac{9}{12}$ **b** $\frac{4}{20}$, $\frac{15}{20}$, $\frac{7}{20}$
  c $\frac{3}{24}$, $\frac{14}{24}$, $\frac{16}{24}$ **d** $\frac{56}{84}$, $\frac{64}{84}$, $\frac{24}{84}$

10 **a** $\frac{2}{15}$, $\frac{1}{5}$, $\frac{2}{3}$ **b** $\frac{1}{4}$, $\frac{7}{20}$, $\frac{2}{5}$
  c $\frac{5}{14}$, $\frac{3}{8}$, $\frac{3}{7}$ **d** $\frac{2}{7}$, $\frac{2}{3}$, $\frac{5}{6}$

11 **a** $\frac{1}{2}$, $\frac{2}{5}$, $\frac{3}{10}$, $\frac{1}{4}$ **b** $\frac{4}{5}$, $\frac{1}{4}$, $\frac{3}{20}$, $\frac{1}{10}$
  c $\frac{3}{4}$, $\frac{17}{40}$, $\frac{2}{5}$, $\frac{3}{8}$ **d** $\frac{5}{6}$, $\frac{5}{8}$, $\frac{7}{12}$, $\frac{11}{24}$

N3.2

1 **a** 3 squares, 1 and 2 shaded different colours
  b 5 squares, 1 and 3 shaded different colours

2 **a** $\frac{3}{5}$ **b** 1 **c** $\frac{5}{7}$ **d** 1

3 **a** 5 squares, 1 and 2 shaded different colours
  b 4 squares, 1 and 3 shaded different colours
  c 7 squares, 2 and 3 shaded different colours
  d 8 squares, 3 and 5 shaded different colours

4 **a** $\frac{1}{3}$ **b** $\frac{3}{5}$ **c** $\frac{2}{3}$ **d** $\frac{3}{5}$

5 **a** 6 squares, 2 and 3 shaded different colours
  b 10 squares, 6 and 3 shaded different colours

6 **a** $\frac{3}{10}$ **b** $\frac{5}{6}$ **c** $\frac{11}{20}$ **d** $\frac{3}{8}$

7 **a** 10 squares, 2 and 1 shaded different colours
  b 6 squares, 4 and 1 shaded different colours
  c 20 squares, 8 and 3 shaded different colours
  d 8 squares, 1 and 2 shaded different colours

8 **a** $\frac{1}{8}$ **b** $\frac{1}{2}$ **c** $\frac{3}{4}$ **d** $\frac{5}{16}$

9 **a** $\frac{5}{21}$ **b** $\frac{2}{15}$ **c** $\frac{1}{18}$ **d** $\frac{7}{20}$

10 **a** $1\frac{1}{10}$ **b** $1\frac{7}{12}$

11 **a** $1\frac{1}{6}$ **b** $1\frac{3}{10}$ **c** $1\frac{2}{15}$ **d** $1\frac{1}{14}$

12 **a** $1\frac{1}{4}$ **b** $1\frac{4}{5}$ **c** $1\frac{5}{8}$ **d** $4\frac{1}{4}$

13 **a** $\frac{7}{4}$ **b** $\frac{23}{16}$ **c** $\frac{14}{9}$ **d** $\frac{18}{7}$

14 **a** $3\frac{14}{15}$ **b** $4\frac{7}{12}$ **c** $7\frac{13}{14}$ **d** $7\frac{55}{63}$
  e $1\frac{7}{20}$ **f** $\frac{3}{4}$ **g** $1\frac{19}{20}$ **h** $4\frac{13}{14}$

N3.3

1 **a** $\frac{1}{4}$ **b** $\frac{1}{6}$ **c** $\frac{1}{10}$ **d** $\frac{1}{12}$

2 **a** 5 **b** 9 **c** 2 **d** 3

3 **a** $8 \times \frac{1}{5}$ **b** $6 \times \frac{1}{4}$ **c** $9 \times \frac{1}{5}$ **d** $17 \times \frac{1}{3}$

4 **a** $\frac{1}{4}$ **b** 2 **c** 3 **d** $\frac{1}{5}$

5 **a**

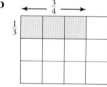

  b

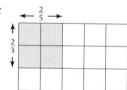

  c

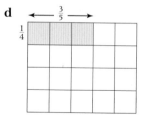

  d

6 **a** $\frac{3}{20}$ **b** $\frac{4}{27}$ **c** $\frac{1}{14}$ **d** $\frac{1}{4}$
  e $\frac{20}{63}$ **f** $\frac{1}{12}$ **g** $\frac{12}{65}$ **h** $\frac{1}{15}$

7 **a** $\frac{3}{4}$ **b** $\frac{3}{4}$ **c** $\frac{4}{5}$ **d** $\frac{4}{5}$
  e $\frac{3}{7}$ **f** $1\frac{4}{5}$ **g** $3\frac{3}{7}$ **h** $2\frac{3}{4}$

8 **a** $\frac{5}{32}$ **b** $\frac{1}{8}$ **c** $\frac{2}{21}$ **d** $\frac{1}{48}$
  e 20 **f** 9 **g** $12\frac{1}{2}$ **h** $25\frac{2}{3}$

9 **a** $\frac{1}{3}$ **b** $\frac{1}{4}$ **c** $\frac{1}{6}$ **d** $\frac{1}{4}$
  e $\frac{8}{9}$ **f** $\frac{45}{56}$ **g** $\frac{2}{9}$ **h** $\frac{8}{21}$

10 a $1\frac{1}{8}$ **b** $1\frac{1}{10}$ **c** 3 **d** 3

e $4\frac{19}{24}$ **f** $1\frac{3}{14}$ **g** $1\frac{21}{22}$ **h** $4\frac{4}{27}$

N3.4

1 a 0.5 **b** 0.75 **c** 0.4
d 0.1 **e** 0.2 **f** 0.25

2 a 50% **b** 75% **c** 40%
d 10% **e** 20% **f** 25%

3 a 62.5% **b** 80% **c** 87.5%
d 60% **e** 37.5% **f** 12.5%

4 See Q3

5 a 0.0625 **b** 0.28 **c** 0.056
d 0.075 **e** 0.4375 **f** 0.03125

6 a 6.25% **b** 28% **c** 5.6%
d 7.5% **e** 43.75% **f** 3.125%

7 a $0.\dot{3}$ **b** $0.1\dot{6}$ **c** $0.\dot{6}$
d $0.\dot{1}428\,5\dot{7}$ **e** $0.\dot{1}$ **f** $0.8\dot{3}$

8 a $33.\dot{3}$% **b** $16.\dot{6}$% **c** $66.\dot{6}$%
d $14.\dot{2}85\,71\dot{4}$% **e** $11.\dot{1}$% **f** $83.\dot{3}$%

9 See Q7 and Q8

10 a $0.\dot{4}2857\dot{1}$ **b** 0.1875 **c** 0.2125
d $0.\dot{5}$ **e** 0.16 **f** $0.\dot{7}1428\dot{5}$

11 a Terminating (only prime factor of denominator is 5)
b Terminating (only prime factors of denominator are 2 and 5)
c Recurring (denominator has a prime factor of 11)
d Recurring (denominator has prime factors of 3 and 7)
e Terminating (only prime factor of denominator is 5)
f Terminating (only prime factor of denominator is 2)

12 The decimal does recur. The restricted number of digits on the calculator display, and the rounding of the final digit, obscure the recurring pattern.

N3.5

1 a 0.43 **b** 0.86 **c** 0.94
d 0.455 **e** 0.0375 **f** 1.05

2 a $\frac{1}{2}$ **b** $\frac{1}{4}$ **c** $\frac{1}{5}$
d $\frac{1}{8}$ **e** $\frac{3}{4}$ **f** $\frac{9}{10}$

3 a $\frac{51}{100}$ **b** $\frac{43}{100}$ **c** $\frac{413}{1000}$
d $\frac{719}{1000}$ **e** $\frac{91}{100}$ **f** $\frac{871}{1000}$

4 a $\frac{49}{100}$ **b** $\frac{53}{100}$ **c** $\frac{73}{100}$
d $\frac{81}{100}$ **e** $\frac{37}{100}$ **f** $\frac{19}{100}$

5 a $\frac{8}{25}$ **b** $\frac{11}{20}$ **c** $\frac{11}{25}$
d $\frac{31}{200}$ **e** $\frac{16}{25}$ **f** $\frac{53}{200}$

6 a $\frac{11}{20}$ **b** $\frac{31}{50}$ **c** $\frac{21}{25}$

d $\frac{13}{20}$ **e** $\frac{18}{25}$ **f** $\frac{37}{200}$

7 $\frac{8}{11}$, 11 not multiple of 2 or 5

8 a $0.\dot{1}$ **b** $0.\dot{5}$ **c** $0.7\dot{5}$
d $0.\dot{3}4\dot{6}$ **e** $0.7\dot{6}\dot{5}$

9 a $\frac{2}{9}$ **b** $\frac{2}{3}$ **c** $\frac{25}{99}$
d $\frac{3}{11}$ **e** $\frac{545}{999}$ **f** $\frac{605}{999}$

10 a $\frac{47}{90}$ **b** $\frac{1}{18}$ **c** $\frac{249}{550}$
d $\frac{752}{9000}$ **e** $\frac{2}{3}$ **f** $\frac{8197}{99\,900}$

11 $\frac{13\,717\,421}{1\,111\,111\,111}$

N3 Exam review

1 ai $\frac{3}{5}$ **aii** $\frac{13}{25}$
bi $\frac{97}{100}$ **bii** $\frac{13}{100}$ **biii** $\frac{9}{14}$

2 a 0.067, 0.56, 0.6, 0.605, 0.65
b −10, −6, −4, 2, 5
c $\frac{2}{5}, \frac{1}{2}, \frac{2}{3}, \frac{3}{4}$

S2 Before you start ...

1 $a = 56°$, $b = 112°$
2 $c = 235°$, $d = 160°$
3 $e = 72°$, $f = 63°$
4 $g = 118°$, $h = 75°$

S2.1

1 ai

aii

bi

bii

ci

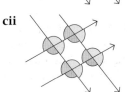

cii

2 a $a = 17°$, angles on a straight line; $b = 17°$, alternate angles; $c = 163°$, corresponding angles
  b $d = 125°$, alternate angles; $e = 105°$, vertically opposite angles
  c $f = 134°$, vertically opposite angles; $g = 134°$, corresponding angles; $h = 24°$, angles in a triangle
3 a $180°$
  b x (bottom left) and z (bottom right)
  c angles in the triangle are x, y, z (part **b**) and add up to $180°$ (part **a**)
4 a $a = 25°$, $b = 155°$, $c = 25°$, $d = 155°$
  b Parallelogram
5 $a = 80°$, $b = 110°$, $c = 70°$, $d = 30°$, $e = 150°$

S2.2

1

Shape	△	□	⬠	⬡	⯃	◯
Number of sides	3	4	5	6	8	10
Number of triangles the shape splits into	1	2	3	4	6	8
Sum of the interior angles in the shape	180°	360°	540°	720°	1080°	1440°
Size of one interior angle	60°	90°	108°	120°	135°	144°
Size of one exterior angle	120°	90°	72°	60°	45°	36°

2 a $x = 80°$ **b** $x = 77°$, $2x = 154°$
  c $x = 55°$, $3x = 165°$
3 a $x = 125°$, $y = 50°$ **b** $x = 110°$, $y = 65°$
4 ai $120°$ **aii** $108°$ **aiii** $118°$
  b The sum of the two opposite angles
5 Drawing a diagonal splits the quadrilateral into two triangles, so the sum of the interior angles is $2 \times 180° = 360°$

S2.3

1 a $a = 63°$, angle at centre is double the angle at the circumference
  b $b = 46°$, angle at centre is double the angle at the circumference
  c $c = 118°$, angle at centre is double the angle at the circumference
  d $d = 32°$, angles in same arc
  e $e = 78°$, angles in same arc
  f $f = 103°$, angle at centre is double the angle at the circumference
  g $g = 122°$, angles in same arc
  h $h = 21°$, angles in same arc
  i $i = 140°$, angle at centre is double the angle at the circumference
  j $j = k = l = 38°$, angles in same arc
  k $m = 47°$, $n = 62°$, angles in same arcs
  l $p = 75°$, $q = 25°$, angles in same arcs
  m $r = 90°$, angle in same arc

n $s = 49°$, $t = 22°$, angles in same arcs
o $u = 54°$, $v = 45°$, angles in same arcs
p $w = 180°$, angle at centre is double the angle at the circumference
q $x = 45°$, angles in isosceles triangle
r $y = 45°$, angle at centre is double the angle at the circumference

S2.4

1 a $a = 90°$, angle in a semicircle
  b $b = 61°$, angles in a triangle
  c $c = 43°$, angles in a triangle
  d $d = 135°$, opposite angles in cyclic quadrilateral
  e $e = 76°$, opposite angles in cyclic quadrilateral
  f $f = 15°$, angles in a triangle
  g $g = 143°$, $h = 100°$, opposite angles in cyclic quadrilateral
  h $i = 124°$, $j = 54°$, opposite angles in cyclic quadrilateral
  i $k = 56°$, $l = 124°$, angles in a triangle, angles on a straight line
  j $m = 125°$, $n = 250°$, angles at centre are double the angles at the circumference
  k $p = q = 106°$, opposite angles in cyclic quadrilateral, angles on same arc
  l $r = 50°$, angle in a semicircle
  m $s = 35°$, angle in a semicircle
  n $t = 62°$, $u = 118°$, angles on a straight line, opposite angles in a cyclic quadrilateral
  o $v = 74°$, angles on a straight line, opposite angles in a cyclic quadrilateral
  p $w = x = 90°$, opposite angles in a cyclic quadrilateral
  q $y = 45°$, angles in isosceles triangle
  r $z = 100°$

S2.5

1 a $a = 90°$, angle between tangent and radius
  b $b = 33°$, angle between tangent and radius, $c = 57°$, angles in a triangle in a semicircle
  c $d = 42°$, angle between tangent and radius, $e = 48°$, angles in a triangle in a semicircle
  d $f = 26°$, angle between tangent and radius, $g = 26°$, angles in a triangle in a semicircle
  e $h = 51°$, angle between tangent and radius, angles in a triangle in a semicircle
  f $i = 16°$, $j = 4$ cm; 2 tangents drawn from a point to a circle
  g $k = 44°$, angles in a quadrilateral
  h $l = 40°$, $m = 70°$, angles in a quadrilateral and angle at centre is double the angle at the circumference
  i $n = 16°$, $p = 82°$, angles in a quadrilateral and angle at centre is double the angle at the circumference
  j $q = 30°$, $r = 105°$, angles in a quadrilateral and angle at centre is double the angle at the circumference

k $s = 100°$, $t = 50°$, angles in a quadrilateral and angle at centre is double the angle at the circumference

l $u = 116°$, $v = 58°$, angles in a quadrilateral and angle at centre is double the angle at the circumference

S2 Exam review

1 $60°$

2 **a** $27°$, angle between tangent and radius
   b $63°$, angles in a triangle in a semicircle

A3 Before you start …

1 **a** 10, 12   **b** 70, 64   **c** 16, 22
   d −2, −5   **e** 48, 96   **f** $\frac{1}{6}, \frac{1}{7}$

2 **a** $4(n-1)$, $2n + 7$, $15 - n$, $2n^2 = \frac{9}{n} + 15$

3 Square numbers. They form a pattern of squares when drawn.

A3.1

1 **a** 29, 34   **b** 65, 58
   c 16, 21   **d** 999 999, 9 999 999
   e 13, 21   **f** 3.375, 1.6875

2 **a** 7, 13   **b** 8, 16
   c 93, 91, 89   **d** 8, 125

3 **a** 3, 6, 9, 12, 15
   b 2, 4, 8, 16, 32
   c 2, 3, 5, 7, 11
   d 121, 144, 169, 196, 225

4 **a** 6, 10, 15
   b 1, 3, 6, 10, 15, 21, 28, 36, 45, 55
   c They form a pattern of squares when drawn.
   d They form a pattern of cubes when drawn.

5 **a** 110   **b** $\frac{10}{11}$ **c** 100   **d** 100 000

6 **a** $6667^2 = 44\,448\,889$, $66\,667^2 = 4\,444\,488\,889$
   b 4 444 444 444 488 888 888 889
   c 666 666 667

A3.2

1 **a** 10, 18, 26, 34, 42   **b** 1, 6, 11, 16, 21
   c 7, 14, 21, 28, 35   **d** 8, 6, 4, 2, 0
   e −2, 1, 6, 13, 22   **f** 2, 8, 18, 32, 50

2 **a** 8, 7, 6, 5, 4   **b** 6, 12, 20, 30, 42
   c 1, $\frac{1}{2}, \frac{1}{3}, \frac{1}{4}, \frac{1}{5}$ **d** −1, −8, −27, −64, −125
   e −4, −9, 0, 35, 108   **f** 1, 16, 81, 256, 625

3 **a** n^2, n^3 **b** $18 + 2n$, $n(n + 5)$
   c n^3, $n(n + 5)$

4 **a** 10, 14, 18, 22, 26, 30, 34, 38; $T(n) = 4n + 6$ (for $n \leqslant 7$)
   b $T(n) = n^2 + 1$, until $T(n)$ is over 49

5 **a** For example, n^2, $6n - 5$
   b For example, n^2, $6n - 5$
   c No

A3.3

1 **a** $5n - 1$ **b** $2n - 1$ **c** $2n + 8$
   d $0.5n + 0.5$ **e** $2n - 6$ **f** n
   g $13n$ **h** $10n - 6$ **i** $12 - 2n$
   j $105 - 5n$ **k** $50\frac{1}{4} - \frac{1}{4}n$ **l** $79 - 4n$

2 Many possibilities

3 **a** False   **b** False   **c** False

4 **a** $8n - 20$ **b** $3n + 2$ **c** $2n - 46$

5 **a** $\dfrac{2n + 1}{3n + 4}$ **b** $\dfrac{2n + 8}{33 - 3n}$ **c** $\dfrac{n + 6}{n^2}$ **d** $\dfrac{n^3}{13 - 2n}$

6 **a** $\frac{1}{n}$ **b**

   c It gets closer and closer to zero without ever actually reaching it.

A3.4

1 **a** For each square added, 3 sides added, and in first case there is an extra side
   b Number of black tiles is matched by number of white tiles in middle of figure, with 4 extra whites at side
   c Length of rectangle is always one more than width

2 **a** $W = 2B + 6$
   bi $B = 2W + 2$ **bii** $L = 3W + 1$
   ci $P = n(n - 1)$ **cii** $H = \frac{n(n-1)}{2}$

3 **a** $C = 4$, $E = 4(n - 2)$, $M = (n - 2)^2$
   b $C = 4$, $E = 2(m - 2) + 2(n - 2)$, $M = (m - 2)(n - 2)$

A3.5

1 **a** 6, 9, 14, 21, 30   **b** −1, 2, 7, 14, 23
   c 3, 12, 27, 48, 75   **d** 3, 8, 15, 24, 35
   e 1, 7, 17, 31, 49   **f** 4, 13, 28, 49, 76

2 **a** $n^2 + 3$ **b** $n^2 - 3$ **c** $10n^2$
   d $3n^2$ **e** $n^2 + n$ **f** $0.5n^2$

3 **a**

nth term	First five terms	First differences	Second differences
$2n^2$	2, 8, 18, 32, 50, …	6, 10, 14, 18, …	4, 4, 4, …
$3n^2$	3, 12, 27, 48, 75, …	9, 15, 21, 27, …	6, 6, 6, …
$4n^2$	4, 16, 36, 64, 100, …	12, 20, 28, 36, …	8, 8, 8, …
$5n^2$	5, 20, 45, 80, 125, …	15, 25, 35, 45, …	10, 10, 10, …
$10n^2$	10, 40, 90, 160, 250, …	30, 50, 70, 90, …	20, 20, 20, …

   b Second difference is double the coefficient of the n^2 term
   ci $6n^2$ **cii** $2n^2 - 1$ **ciii** $3n^2 + n$

4 **a** $T = h^2 + h$
   b Width of rectangle is one more than the height.

5 a $A = (n + 1)(n + 2)$

   bi The height of the rectangle is always one more than the term, the width is always two more.

   bii Formula equals $n^2 + 3n + 2$, which is quadratic.

A3 Exam review

1 a $2n - 1$ **b** 1, 3   **2 a** $m = 6n$

D2 Before you start …

1

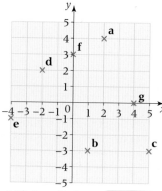

2 ai €2.8   **aii** €6.3   **bi** £2.85   **bii** £5

3 a 40   **b** 60   **c** 80   **d** 15   **e** 50

   f 30   **g** 75   **h** 150   **i** 120   **j** 630

D2.1

1 a

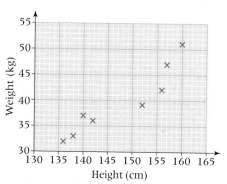

b Positive correlation

c As weight increases, so does height.

2 a

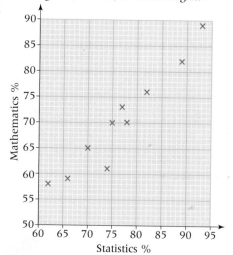

b Positive correlation

c As percentage achieved in statistics increases, so does percentage achieved in maths.

3 a

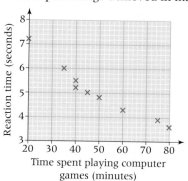

b Negative correlation

c As time spent playing computer games increases, reaction time decreases.

4 a

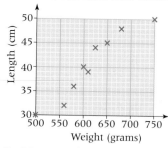

b Positive correlation

c As weight of fish increases, so does length.

D2.2

1 a

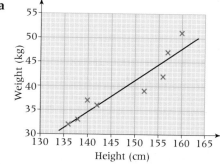

bi 38 kg   **bii** 156 cm

2 a

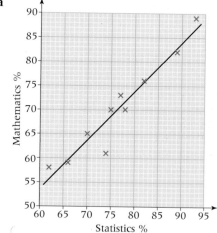

bi 74%   **bii** 86%

c Score of 46% outside range of data collected

3 a

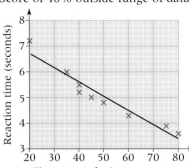

Time spent playing computer games (minutes)

bi 3.9 seconds   **bii** 26 minutes

c Outside range of data collected

4 a

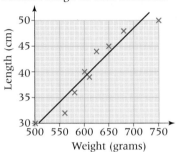

Weight (grams)

bi 48 cm   **bii** 635 g

D2.3

1 a

```
4 | 5  8
5 | 1  1  2  4  9  9
6 | 1  2  3  6  9
7 | 0  1  4  5  7  8
8 | 1  2  9
9 | 3
```

Key: 4|5 means 45%

b Most students scored between 50% and 80%.

ci 48%   **cii** 66%

ciii 54%, 77%   **civ** 23%

2 a

```
2 | 6  8  9
3 | 0  3  7  8  9
4 | 0  1  3  3  4  7  9
5 | 2  5  8  9
6 | 0  1  3  6  7  9
7 | 1  3  5  7
8 | 0  4
```

Key: 3|4 means 34 minutes

b Most people spend 30 to 70 minutes playing computer games each day.

ci 58 minutes

cii 52 minutes

ciii 39 minutes, 67 minutes

civ 28 minutes

3 a

```
0 | 3  5  6  7  8  9
1 | 0  1  2  2  5  6  7  8  9
2 | 0  1  2  4  6  7  7  9  9
3 | 1  2  3  4  5  7
4 | 0  1
```

Key: 2|7 means 27 minutes

b Most people take between 10 and 30 minutes to complete the crossword.

ci 38 minutes

cii 20.5 minutes

ciii 11.5 minutes, 30 minutes

civ 18.5 minutes

4 a

```
3 | 2  6  7  9
4 | 2  5  7  8  9
5 | 1  2  4  6  6  6  7  8  9
6 | 0  1  3  6
7 | 0
```

Key: 3|6 means 36 kg

b Most boys weigh between 40 kg and 60 kg.

ci 38 kg   **cii** 54 kg

ciii 45 kg, 59 kg   **civ** 14 kg

5 a

```
14 | 6  7  9
15 | 0  0  2  2  3  4  5  5  7  8  9
16 | 0  2  3  5  7  8
17 | 1  2  2
```

Key: 14|2 means 142 cm

b Most girls are between 150 cm and 160 cm tall.

ci 26 cm   **cii** 157 cm

ciii 152 cm, 165 cm   **civ** 13 cm

6 a

```
8  | 8  9
9  | 1  2  4  8
10 | 1  3  4  5  6  7  8  8  9
11 | 0  2  4  6  6  7  7  8  9
12 | 1  5  6  7  9
13 | 1  3
```

Key: 11|4 means IQ of 114

b Most students have an IQ between 100 and 120.

ci 45   **cii** 110

ciii 103, 119   **civ** 16

7 a

```
0 | 3  6  8  9  9
1 | 0  1  2  5  6  7  8  9
2 | 1  1  2  3  3  5  7
3 | 0  2  3
```

Key: 3|2 means 32 minutes

b Most people took 10 to 20 minutes

ci 30 minutes

cii 18 minutes

ciii 10 minutes, 23 minutes

civ 13 minutes

D2.4

1 a

```
              A        B
           8  7 │ 3 │
9 7 5 4 4 3 2 1 │ 4 │ 1 4 8
  8 6 6 5 3 2 1 │ 5 │ 3 3 4 6 7 7
           2  1 │ 6 │ 1 3 5 6 9
                │ 7 │ 0 2 2 5 9
```

Key: 2|5|3 means 52% in A, 53% in B

  bi A: 49%, B: 61%

  bii A: 10%, B: 17%

  biii A: 24%, B: 38%

  c Results for test A are generally lower but less varied than for test B.

2 a

```
        Men                       Women
              │ 14 │ 8 9
         9  8 │ 15 │ 0 1 2 3 4 5 5 6 7 8 8
9 9 8 7 7 5 0 │ 16 │ 1 2 2 5 6 7 9
8 8 7 6 5 4 2 1 │ 17 │ 2 4 8
  8 4 3 3 2 0 │ 18 │
```

Key: 2|16|3 means 162 cm for men, 163 cm for women

  bi Men: 174 cm, women: 158 cm

  bii Men: 13 cm, women: 13 cm

  biii Men: 30 cm, women: 29 cm

  c Men are taller than women and heights are evenly spread in both groups.

3 a

```
            X                       Y
              9 │ 8 │ 7
          9 8 6 │ 9 │ 3 7 9
    9 8 6 5 5 4 3 │ 10 │ 2 3 4 6 7 7 9
8 8 7 7 6 5 5 2 0 0 │ 11 │ 3 4 4 4 6 8 8 8 9 9
    9 6 4 3 3 2 1 │ 12 │ 1 1 4 6 6 8 9
            4 1 0 │ 13 │ 0 1 2
```

Key: 7 | 9 | 6 means for 97 in X, 96 in Y

  bi X: 116, Y: 116

  bii X: 18, Y: 18

  biii X: 45, Y: 45

  c Average IQ and spread are same in X and Y.

4 a

```
      Boys                Girls
8 8 8 6 2 2 │ 3 │
   7 5 4 2 │ 4 │ 0 3 4 6 8
9 8 7 6 3 2 │ 5 │ 2 2 5 6 9
  8 6 4 2 1 │ 6 │ 2 3 3 5 6 7
        2 1 │ 7 │ 2 3 4 6 7
            │ 8 │ 0 2
```

Key: 5|4|6 means 4.5 minutes for boys,
4.6 seconds for girls

  bi Boys: 5.3 seconds, girls: 6.3 seconds

  bii Boys: 2.4 seconds, girls: 2.1 seconds

  biii Boys: 4 seconds, girls: 4.2 seconds

  c Boys have a faster reaction time than girls, variation in times is similar.

5 a

```
        P                           Z
        9 8 8 6 │ 0 │ 8 9
9 8 7 7 4 3 2 1 0 │ 1 │ 1 2 4 5 6 7 7 8 9
        9 7 4 2 │ 2 │ 1 3 4 5 8 9
          7 1 │ 3 │ 2 3
```

Key: 4|3|6 means 34 minutes for P, and 36 minutes for Z

  bi P: 17 minutes, Z: 18 minutes

  bii P: 14 minutes, Z: 11 minutes

  biii P: 31 minutes, Z: 25 minutes

  c The average time taken is similar but times for P are more varied than those for Z.

D2.5

1

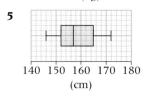

(%)

2

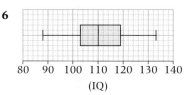

(Minutes per day)

3

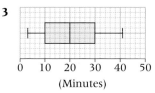

(Minutes)

4

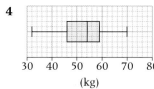

(kg)

5

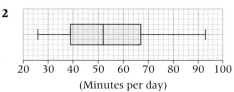

(cm)

6

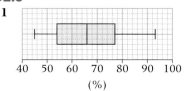

(IQ)

7

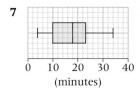

(minutes)

D2 Exam review

1 a

```
4 | 9
5 | 0 2 2 2 3 4 5 5 8 8 9
6 | 0 0 1 1 1 9
7 | 0 2
```

Key: 5|2 means 52 minutes

bi 58 minutes
bii 9 minutes
biii 23 minutes

2 a and **2 c**

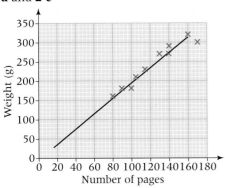

b The weight increases with the number of pages.
di 143 pages **dii** 235 g

A4 Before you start ...

1

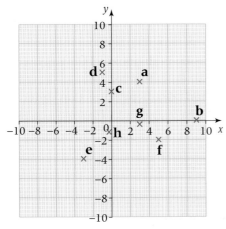

2

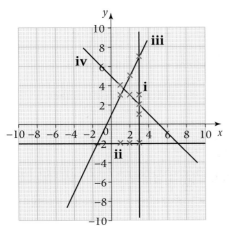

3 a $6\frac{1}{2}$ **b** $5\frac{1}{4}$ **c** $4\frac{5}{7}$ **d** $-9\frac{1}{11}$
 e $\frac{7}{3}$ **f** $\frac{15}{2}$ **g** $\frac{23}{4}$ **h** $-\frac{35}{8}$

4 a $n = 2$ **b** $m = 1$ **c** $p = \frac{1}{2}$

A4.1

1 $y = 2x + 3$, $y = 7 - 3x$, $y = 5x$, $y = 7$, $2x + 7y = 8$, $x = -2$

2 ai

x	0	1	2
y	2	5	8

aii

x	0	2	3
y	−2	0	2

aiii

x	0	5	1
y	2	0	1.6

b

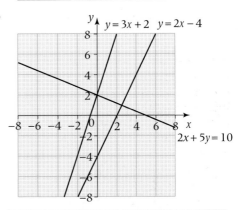

3 a

x	1	2	3	4	5
y	12	19	26	33	40

b

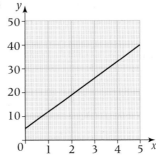

c £26
d $y = 7x + 5$
4 a No, $2 \times 3 + 1 \neq 8$
 b For example (3, 7)

5 a

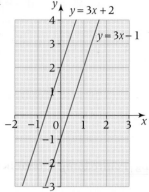

b They are parallel so will never intersect.

c $y = 3x + k$, where k is any number other than -1 or 2.

6 Need to do 40 chores to receive the same amount under both options. If you do less than 40 chores, better to take £5 option; if more, take £3 option.

A4.2

1 a $y = 7$ is horizontal, $x = 9$ and $x = -0.5$ are vertical, $y = 2x - 1$ is diagonal, $y = x^2 + x$ is none of these

2 a $y = 3$ **b** $y = \frac{3}{4}$ **c** $y = -3$

   d $x = -2$ **e** $x = \frac{1}{4}$ **f** $x = 2.5$

3 a Vertical line cutting x-axis at $x = 5$
   b Horizontal line cutting y-axis at $y = 2$
   c Vertical line cutting x-axis at $x = 1.6$
   d Horizontal line cutting y-axis at $y = -3$
   e Horizontal line cutting y-axis at $y = 1$
   f Vertical line cutting x-axis at $x = -1\frac{1}{4}$

4 a $(5, 2)$ **b** $(4, -3)$ **c** $(-2, 9)$ **d** $(-2, -4)$

5 a For example $x = 4$, $x = 7$, $y = 1$, $y = 6$
   b For example $x = 4$, $x = 7$, $y = 3$, $y = 6$
   c For example $x = 4$, $y = 3$, $4y + 3x = 36$

6 a $(1, 2)$
   b For example, below $y = 5$, above $y = 3$, right of $x = 2$, on the line $y = x + 1$

A4.3

1 a $-2, 1$ **b** $3, 2$ **c** $1, -3$ **d** $\frac{1}{2}, -1$

2 a Any line sloping up from left to right.
   b Any line through $(0, 3)$.
   c Any line that goes up 1 unit for every 4 across.
   d Any line that slopes down and goes through $(0, -2)$.
   e The line $y = 3x + 5$.
   f The line $y = \frac{2}{3}x + 1$.

3 a

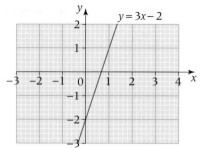

b $3, -2$
c Number before x gives gradient, number after x term gives intercept.

d

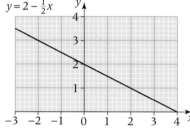

Gradient $-\frac{1}{2}$, intercept 2

4 a $\frac{2}{3}$

   b Divide difference in y-coordinates by difference in x-coordinates.

   c 2

A4.4

1 a $y = 2 - 4x$ **b** $y = 3x + 1$
   c $y = 4x - 2$ **d** $y = x$

2

Equation	Gradient	Direction	Intercept
$y = 4x + 3$	4	Positive	3
$y = 3x + 4$	3	Positive	4
$y = 9x - 2$	9	Positive	-2
$y = 4x - 5$	4	Positive	-5
$2y = 8x + 6$	4	Positive	3

3 ai Gradient = 1   Intercept = 2
   aii $y = x + 2$
   bi Gradient = 2   Intercept = 3
   bii $y = 2x + 3$
   ci Gradient = 3   Intercept = 0
   cii $y = 3x$
   di Gradient = -2   Intercept = 3
   dii $y = -2x + 3$

4 a $y = 3x + k$, where k is any number
   b $y = 3 - 2x$ **c** $y = -3x + 1$

5 a $y = 3x + 4$ **b** $y = 3x + 4$

A4.5

1 $y = 3x + 5$, $y = 5x - 2$, $y = 7 - 2x$, $2y = x + 18$, $4y = -x - 12$, $y = 4$, $y = x$

2 a $y = 4x$, $y = 12 - 2x$ **b** $y = 4x$, $y = 5x - 1$

3 **a** $y = 7x + 5$ **b** $y = \frac{1}{2}x + 3$ **c** $y = 4x - 4$

d $y = 3x - 5$ **e** $y = 5 - 2x$ **f** $y = \frac{1}{4}x - 2$

g $y = 4x + 1$ **h** $y = x + 2$ **i** $y = 8x - 6$

4 **a** 3 **b** $y = 3x + 5$ **c** -2, $y = 16 - 2x$

5 **a** $(0, -2.5)$ **b** $(\frac{5}{9}, 0)$ **c** $(2, 6.5)$

6 $3x + 2y = 12$

A4 Exam review

1 **a**

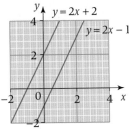

b No, they are parallel

2 $y = 2x + 6$

D3 Before you start …

1 **a** $\frac{3}{5}$ **b** $\frac{3}{7}$ **c** $\frac{5}{8}$ **d** $\frac{7}{11}$

2 **a** $\frac{5}{6}$ **b** $\frac{9}{20}$ **c** $\frac{23}{24}$ **d** $\frac{5}{8}$

3 **a** $\frac{1}{6}$ **b** $\frac{3}{20}$ **c** $\frac{2}{15}$ **d** $\frac{5}{12}$

4 **a** 0.7 **b** 0.75 **c** 0.375

d 0.4 **e** $0.\dot{3}$ **f** 0.0625

D3.1

1 0.53

2 0.65

3 **a** $\frac{5}{18}$ **b** 0 **c** $\frac{5}{9}$ **d** $\frac{1}{6}$ **e** $\frac{5}{6}$

4 **a** $\frac{7}{25}$ **b** $\frac{18}{25}$ **c** $\frac{9}{25}$ **d** $\frac{16}{25}$ **e** $\frac{2}{5}$ **f** $\frac{17}{25}$

5 0.27

6 **a** Gerry will not be chosen.

b 0.4

7 **a** The chance of Rangers winning is double that of Rovers winning.

b 0.2

D3.2

1 **a** $\frac{7}{20}$ **b** $\frac{13}{20}$ **c** $\frac{1}{20}$ **d** $\frac{19}{20}$ **e** $\frac{11}{20}$

f $\frac{9}{20}$ **g** $\frac{8}{20} = \frac{2}{5}$ **h** $\frac{15}{20} = \frac{3}{4}$ **i** $\frac{16}{20} = \frac{4}{5}$

2 **a** $\frac{1}{9}$ **b** $\frac{8}{9}$ **c** $\frac{5}{9}$ **d** $\frac{1}{3}$ **e** $\frac{1}{9}$

f $\frac{8}{9}$ **g** $\frac{2}{9}$ **h** $\frac{7}{9}$ **i** $\frac{4}{9}$

3 **a** 30

bi $\frac{7}{30}$ **bii** $\frac{1}{2}$ **biii** $\frac{2}{5}$

biv $\frac{3}{10}$ **bv** $\frac{3}{10}$ **bvi** $\frac{7}{10}$

4 **a** $\frac{1}{15}$ **b** $\frac{1}{3}$ **c** $\frac{23}{30}$ **d** $\frac{1}{2}$

D3.3

1 **a**

	France	Spain	UK	Total
Girls	3	4	11	18
Boys	3	8	3	14
Total	6	12	14	32

bi $\frac{3}{16}$ **bii** $\frac{13}{16}$ **biii** $\frac{7}{16}$

biv $\frac{2}{3}$ **bv** $\frac{9}{16}$ **bvi** $\frac{5}{6}$

2 **a**

	Orienteering	Paintballing	Quadbiking	Total
Girls	11	8	4	23
Boys	5	8	14	27
Total	16	16	18	50

bi $\frac{8}{25}$ **bii** $\frac{17}{25}$ **biii** $\frac{16}{25}$ **biv** $\frac{17}{25}$

ci 128 **cii** 184

3 **a**

	Science	Humanities	Other subjects	Total
Girls	29	18	32	79
Boys	27	3	11	41
Total	56	21	43	120

bi $\frac{79}{120}$ **bii** $\frac{7}{40}$ **biii** $\frac{77}{120}$ **biv** $\frac{9}{40}$

ci 280 **cii** 105

D3.4

1 140

2 120

3 $\frac{5}{8}$

4 **a** Frequency of 3 is double frequency of 1.

b $\frac{17}{60}$ **c** 50

5 **ai** $\frac{11}{100}$ **aii** $\frac{8}{25}$ **aiii** $\frac{27}{100}$

aiv $\frac{41}{100}$ **av** $\frac{87}{100}$

bi 80 **bii** 120

6 **a** 104 **b** 114 **c** 25

D3.5

1 Expected number of tails is 160, which is not close to 114, so coin is probably not fair.

2 133 is close to expected number of black (140), so spinner is fair.

3 Yes, each outcome occurs with similar frequency.

4 No, 2 occurs more than twice as often as other outcomes.

5 **a** Relative frequency: $\frac{4}{10}, \frac{7}{20}, \frac{11}{30}, \frac{13}{40}, \frac{18}{50}, \frac{22}{60}, \frac{26}{70}, \frac{29}{80}, \frac{32}{90}, \frac{34}{100}$

b $\frac{34}{100}$

c Expected number of heads is 50, which is not close to 34, so coin could be biased.

D3 Exam review

1 ai $\frac{10}{100}$ **aii** $\frac{23}{100}$ **aiii** $\frac{33}{100}$

b Might not be fair as a 6 was rolled more than twice as often as a 2.

2 a

	France	Germany	Spain	Total
Female	2	23	9	34
Male	15	2	9	26
Total	17	25	18	60

b $\frac{5}{12}$

N4 Before you start …

1 a 40 **b** 630 **c** 90 **d** 200

2 £8.50

3 a 40 mph **b** 42.5 mph **c** 54 mph
  d 32 mph

4 ai £49.50 **aii** 66 mm
  aiii 52.8 km **aiv** 4.4 hours
  bi 4.5 miles **bii** 43.5 minutes
  biii 10.5 kg **biv** £46.5

N4.1

1 a 75 g **b** 4.5 g **c** 100 g **d** 125 g

2 a 90 ml **b** 228 ml **c** 1.2 litres **d** 168 cm³

3 a 60 g **b** 36 mm **c** 380 g **d** 31.2 km

4 a $\frac{1}{10}$ **b** $\frac{3}{40}$ **c** $\frac{1}{5}$ **d** $\frac{3}{16}$

5 a 10% **b** 7.5% **c** 20% **d** 18.75%

6 ai $\frac{3}{20}$ **aii** 15%

  bi $\frac{1}{25}$ **bii** 4%

  ci $\frac{2}{5}$ **cii** 40%

  di $\frac{8}{125}$ **dii** 6.4%

7 a 73.536 mm **b** £315 **c** £9.88 **d** 840 kg

8 a $\frac{3}{25}$ **b** 12%

9 a 200 ml **b** 3 litres **c** 1.8 litres

10 a £2100 **b** $\frac{8}{15}$, 53.3%

N4.2

1 a £8.25 **b** £12.38 **c** £18.84 **d** £75.76

2 a £17.25 **b** £11.90 **c** £7.76 **d** £16.73

3 12 kg

4 75 kg

5 a 60p **b** £1.80

6 a 40p **b** £2.80

7 a 3p **b** 0.52 g

8 a £1.88 **b** £1.48 **c** £1.73
  d £2.04 **e** £1.86

9 £3.64

10 a Pack A = 1.21p per pin. Pack B = 1.15p per pin. Pack B is the better value.

11 Regular, at 0.7p per sheet, is better value than Super, at 0.77p per sheet.

N4.3

1 a $31 **b** $54.25 **c** $16.28 **d** $59.75

2 a €7.45 **b** €44.70 **c** €87.91 **d** €393.36

3 a £1.07 **b** £19.22 **c** £127.67 **d** £305.31

4 a £2 **b** £10 **c** £1.46 **d** £3.66
  e £31.71 **f** £2.90

5 a CAN$40.91 **b** $61.11

6 a €501 **b** €760.87

7 1.760 pints

8 a 12.701 kg **b** 1016.047 kg
  c 170.324 kg **d** 17.293 kg

9 a 40.2 km to 3 sf **b** 6.21 miles to 3 sf
  c 24 900 miles to 3 sf **d** 7.24 km to 3 sf

10 a 157.5 litres **b** 11.1 gallons to 3 sf
  c 56.25 litres to 3 sf **d** 8.62 gallons to 3 sf

N4.4

1 9 mph

2 £2.10 per metre

3 32 mph

4 89.25 mph

5

Speed (km/h)	Distance (km)	Time
105	525	5 hours
48	106	2 hours 12.5 minutes
$37\frac{1}{3}$	84	2 hours 15 minutes
86	215	2 hours 30 minutes
37.1	65	1 hour 45 minutes

6 a 5 g/cm³ **b** 87.88 g

7 a 7.59 g/cm³ to 3 sf **b** 105

8 a 9.46 kg to 3 sf **b** 6.15 litres to 3 sf

9 1358 m

N4.5

1 a 25 g **b** 200 cm **c** 7.5 h **d** 625 kg

2 a 400 g **b** 20 sec **c** 30° **d** 48 h

3

Original number	Proportional change	Result
42	Decrease by $\frac{1}{4}$	31.5
110	Increase by $\frac{1}{5}$	132
250	Increase by $\frac{1}{10}$	275
450	Decrease by $\frac{2}{5}$	270
965	Increase by $\frac{1}{10}$	1061.5

4 90 g butter, 4.5 tsp sugar, 300 ml milk

5 a 280 g **b** 350 g **c** $233\frac{1}{3}$ g **d** 875 g

6 a £126 **b** £252 **c** £525 **d** £75.60

7 a £495 **b** £510 **c** £954 **d** £658.60

8 Andrew's pay increases by a larger amount (£10.80) than Bella's (£10.40)

9 £203.34 to 2 dp

N4 Exam review

1 a 750 g **b** 187.5 g

2 £9720

S3 Before you start …

1 A(3, 2), B(5, 5), C(4, −1), D(−2, −3), E(1, 6)

2 a $x = 3$ **b** $x = 5$ **c** $y = 2$
  d $y = -1$ **e** $y = x$

3

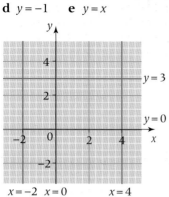

$x = -2$ $x = 0$ $x = 4$

4 a and **b**

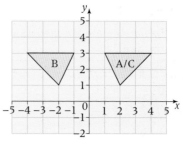

c A and C are the same.

S3.1

1

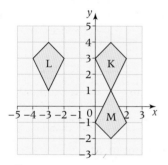

2

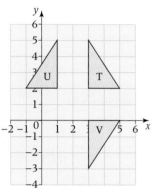

3

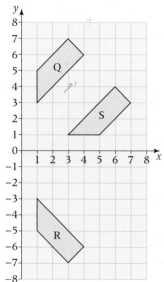

S3.2

1

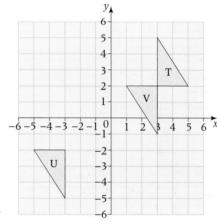

2

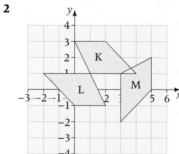

3

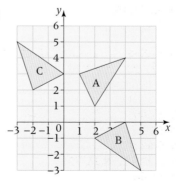

4 a and b

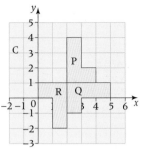

c It maps to R. Two 90° rotations are the same as one 180° rotation.

S3.3

1

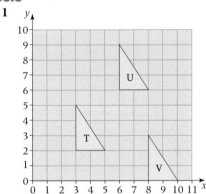

2

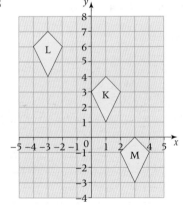

3 a

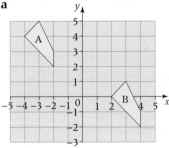

b B translates back to A. The translation has been reversed.

S3.4

1 a Reflection in *y*-axis
  b Rotation by 180° about (0, 0)
  c Reflection in *x*-axis
  d Rotation by 180° about (0, 0)
  e Reflection in *y*-axis
  f Reflection in *y*-axis

2 a Translation by $\begin{pmatrix} 16 \\ 2 \end{pmatrix}$ **b** Translation by $\begin{pmatrix} 5 \\ 3 \end{pmatrix}$

  c Translation by $\begin{pmatrix} 9 \\ 8 \end{pmatrix}$ **d** Translation by $\begin{pmatrix} -4 \\ -5 \end{pmatrix}$

  e Translation by $\begin{pmatrix} 7 \\ -6 \end{pmatrix}$ **f** Translation by $\begin{pmatrix} -7 \\ 6 \end{pmatrix}$

3 a Rotation by 90° clockwise about (0, 0)
  b Rotation by 180° about (0, 0)
  c Rotation by 90° anticlockwise about (0, 0)
  d Rotation by 90° clockwise about (0, 0)
  e Rotation by 90° clockwise about (0, 0)
  f Rotation by 180° about (0, 0)

S3.5

1

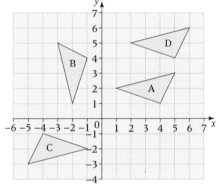

c Rotation 180° about (0, 0)

e Translation by $\begin{pmatrix} 1 \\ 3 \end{pmatrix}$

2

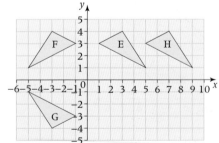

c Rotation 180° about (0, 0)

e Translation by $\begin{pmatrix} 4 \\ 1 \end{pmatrix}$

3

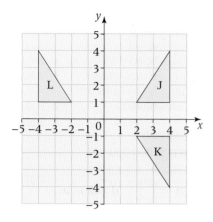

c Reflection in *y*-axis

4 a and **b**

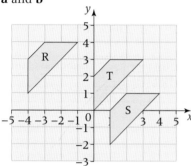

c Translation by $\begin{pmatrix} -4 \\ 1 \end{pmatrix}$

S3 Exam review

1

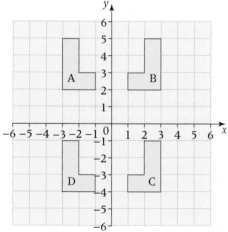

b Translation by $\begin{pmatrix} -2 \\ -6 \end{pmatrix}$

d Reflection in the line $x = 0$

2 a

b Rotation by 180° about (0, 1)

N5 Before you start …

1 a 3^2 **b** 4^5 **c** 6^3 **d** 5^4
2 a $3 \times 3 \times 3$ **b** 6×6
   c $4 \times 4 \times 4 \times 4 \times 4$ **d** $8 \times 8 \times 8$
   e $7 \times 7 \times 7 \times 7$ **f** $5 \times 5 \times 5 \times 5 \times 5$
3 a 16   **b** 27   **c** 16
   d 125   **e** 128   **f** 7

N5.1

1 a 3^2 **b** 2^3 **c** 3^3 **d** 5^4
   e 7^3 **f** 10^3 **g** 6^4 **h** 5^3
2 a $3 \times 3 \times 3 \times 3$
   b 5×5
   c $7 \times 7 \times 7 \times 7$
   d $10 \times 10 \times 10 \times 10 \times 10$
   e $4 \times 4 \times 4 \times 4 \times 4 \times 4 \times 4 \times 4 \times 4$
   f $6 \times 6 \times 6$
   g $2 \times 2 \times 2 \times 2 \times 2$
   h $9 \times 9 \times 9$
3 a 16   **b** 64   **c** 32   **d** 100
   e 1000   **f** 27   **g** 8   **h** 9

4

Index form	Product	Value
10^6	$10 \times 10 \times 10 \times 10 \times 10 \times 10$	1 000 000
10^5	$10 \times 10 \times 10 \times 10 \times 10$	100 000
10^4	$10 \times 10 \times 10 \times 10$	10 000
10^3	$10 \times 10 \times 10$	1 000
10^2	10×10	1 00
10^1	10	1 0

5

Index form	Product	Value
2^8	$2 \times 2 \times 2 \times 2 \times 2 \times 2 \times 2 \times 2$	256
2^7	$2 \times 2 \times 2 \times 2 \times 2 \times 2 \times 2$	128
2^6	$2 \times 2 \times 2 \times 2 \times 2 \times 2$	64
2^5	$2 \times 2 \times 2 \times 2 \times 2$	32
2^4	$2 \times 2 \times 2 \times 2$	16
2^3	$2 \times 2 \times 2$	8
2^2	2×2	4
2^1	2	2

6 a 9^2 **b** 5^3 **c** 2^7 **d** 10^5
   e 3^4 **f** 7^3

7 a 3^6 **b** 2^9 **c** 4^8 **d** 5^5
   e 8^8 **f** 3^8 **g** $2^7 \times 3^4$ **h** $5^4 \times 7^7$

8 a 2^4 **b** 11^2 **c** 3^3 **d** 5^3
   e 13^2 **f** 5^4 **g** 3^5 **h** 2^8

9 a $2^2 \times 13$ **b** $2^2 \times 3^2$ **c** 2×5^2 **d** $2^3 \times 3$
   e 2×3^2 **f** $2^4 \times 3$ **g** $2^2 \times 3 \times 5$
   j $2^4 \times 3^2$

10 a 16, 448   **b** 19, 266

N5.2

1 a 7^3 **b** 3^3 **c** 5^3 **d** 6^4
   e 5^2 **f** 8^5 **g** 9^6 **h** 8^8

2 a 6^5 **b** 4^9 **c** 2^{13} **d** 11^7
   e 1^{30} **f** 7^{12} **g** 3^{12} **h** 9^{10}

3 a 7^2 **b** 8^4 **c** 3 **d** 9^3
   e 4^6 **f** 1 **g** 12^2 **h** 1

4 a 8^5 **b** 5^5 **c** 2^6 **d** 9^2
   e 8^8 **f** 7^7 **g** 4^{10} **h** 11

5 a 3 **b** 5^{10} **c** 4^2 **d** 7^2
   e 8^{11} **f** 9^{10}

6 a 4^2 **b** 6^2 **c** 9^2 **d** 8
   e 5^3 **f** 6^5 **g** 8^2 **h** 10^0

7 a $3^3 \times 4^4$ **b** $7^2 \times 8^3$ **c** $5^6 \times 6^4$ **d** $3^6 \times 4^9$
   e $2^3 \times 5^2$ **f** $7^4 \times 9^7$ **g** $5^3 \times 8^5$ **h** $3^8 \times 8^7$
   i $2^2 \times 9^5$

8 a $5^2 \times 8^3$ **b** $6^2 \times 7^2$ **c** $5^2 \times 6^2$ **d** $5^3 \times 7^6$
   e $3^3 \times 8^2$ **f** $4^5 \times 5^2$ **g** $6^4 \times 7^2$ **h** $4^4 \times 7^5$

N5.3

1 a 5 **b** 6 **c** 1 **d** 1
   e 1 **f** 41 **g** 1 **h** 0

2 a 10 **b** 4 **c** 7 **d** 2
   e 8 **f** 3 **g** 11 **h** 8
   i 12 **j** 2 **k** 3 **l** 10

3 a 6 **b** 36 **c** 9 **d** 81

4 a 3^{-1} **b** 5^{-1} **c** 7^{-1} **d** 11^{-1}
   e 2^{-1} **f** 5^{-1} **g** 10^{-1} **h** 3^{-1}

5 a $\frac{1}{9}$ **b** 1 **c** 3 **d** 9
   e 81 **f** 729 **g** $59\,049$ **h** $\frac{1}{729}$

6 a 7^{-2} **b** 9^{-2} **c** 2^{-2} **d** 2^{-3}
   e 2^{-5} **f** 3^{-4} **g** 5^{-3} **h** 6^{-4}

7 a $\frac{12}{8}$ **b** $\frac{13}{7}$ **c** $\frac{12}{5}$ **d** $\frac{14}{9}$
   e $\frac{12}{3}$ **f** $\frac{13}{9}$ **g** $\frac{15}{4}$ **h** $\frac{16}{6}$

8 a $\frac{1}{16}$ **b** $\frac{1}{4}$ **c** 1 **d** 2
   e 4 **f** 16 **g** 64 **h** $\frac{1}{2}$

9 a $3^{-\frac{1}{2}}$ **b** $5^{-\frac{1}{2}}$ **c** $7^{-\frac{1}{2}}$ **d** $11^{-\frac{1}{2}}$

10 a 2^4 **b** 2^6 **c** 3^6 **d** 4
   e 5^8 **f** 4^{-6} **g** 7^{12} **h** 5^{-4}

11 a 2^{-2} **b** 3^{-1} **c** 4^{-6} **d** $3^{-1} \times 5^{-1}$
   e $5^4 \times 7^{-4}$

N5.4

1 a 10^2 **b** 10^1
   c 10^5 **d** 10^0

2 a 2×10^2 **b** 8×10^2 **c** 9×10^3
   d 6.5×10^2 **e** 6.5×10^3 **f** 9.52×10^2
   g 2.358×10 **h** 2.5585×10^2

3 a 500 **b** 3000 **c** $100\,000$
   d 250 **e** 4900 **f** $3\,800\,000$
   g $750\,000\,000\,000$
   h $8\,100\,000\,000\,000\,000\,000$

4 a 6×10^2 **b** 4.5×10^4
   c 6.5×10^0 **d** 5×10^6

5 a 4×10^5 **b** 9×10^7
   c 2.5×10^8 **d** 2.4×10^{13}

6 a 2×10^2 **b** 2×10^4
   c 5×10 **d** 7.5×10^2

7 a 9.75×10^9 **b** 1.37×10^4
   c 4.01×10^{11} **d** 2.06×10^8

8

Planet	Mean distance from Sun (m)	Light travel time
Mercury	5.79×10^{10}	3 minutes 13 seconds
Earth	1.50×10^{11}	8 minutes 20 seconds
Mars	2.28×10^{11}	12 minutes 40 seconds
Jupiter	7.78×10^{11}	43 minutes 13 seconds
Pluto	5.90×10^{12}	5 hours 27 minutes 47 seconds

N5.5

1 a 3×10^{-1} **b** 4.7×10^{-3}
   c 7.8×10^{-5} **d** 4.485×10^{-1}

2 a 2.8×10^{-1} **b** 4×10^{-2}
   c 1.35×10^{-3} **d** 1.2×10^{-7}

3 a 1×10^{-2} km   **b** 2×10^{-3} g
   c 5×10^{-6} m   **d** 1.1×10^{-2} litre

4 a 5×10^{-1} **b** 9.2×10^{-8}
   c 2×10^{-2} **d** 4.2×10^{-8}

5 a 1.15×10^{-7} **b** 1.83×10^{-1}
   c 4.85×10^{-6} **d** 5.01×10^{-1}

6 a 5.2×10^{-1} **b** 4.6×10^{-2}
   c 2.09×10^{-2} **d** 1.3×10^{-2}

7 See Q6

8 9.11×10^{-8} m^3

9 5.3×10^{-3} kg

10 About 385 atoms

N5 Exam review

1 a $\frac{1}{3}$ **b** 3 **c** 1

2 a 1×10^{-9} **b** 2×10^8

A5 Before you start …

1 $A = lw$, area of a rectangle; $A = \pi r^2$, area of a circle; $a^2 + b^2 = c^2$, length of sides in a right-angled triangle; $V = lwh$, volume of a cuboid; $A = \frac{1}{2}bh$, area of a triangle

2 a $x = 5\frac{2}{3}$ **b** $x = 2\frac{1}{2}$ **c** $x = \pm 5$ **d** $x = 16$
e $x = 2$ **f** $x = 2$

3 a $2 + 4 = 6$ **b** $(-2)^2 = 4$
c $2 + 3 = 5$ **d** $3 \times 2 = 6$

A5.1

1

Identities	Equations	Formulae
$3x(x+1) = 3x^2 + 3x$	$3x + 1 = 10$	$C = 2\pi r$
$y + y = y^2$	$2x + 5 = 3 - 7x$	$A = \frac{1}{2}(a+b)h$
$20 - x = -(x - 20)$	$2x^2 = 50$	$a^2 + b^2 = c^2$

2 a £17 **b** 9
c You can't have half a person.

3 a 53.6 °F **b** 0 **c** 4
d 108, 5 or −5 **e** −14 **f** 1

4 a 33 cm²
b Any a, b such that both are positive and $a + b = 10$

A5.2

1 a $P = 4s$ **b** $A = \frac{1}{8}\pi r^2$
c $P = 26 + 2y$ **d** $T = 10 + 35p$

2 a, b, c are lengths of the base of the triangle, height of the triangle and length of the prism. Correct as multiplies area of cross-section by length.

3 a $V = \pi r^2 h$
b $2\pi r^2$ represents the combined area of the top and bottom circles. The $2\pi rh$ term represents the curved section; if opened out flat it would be a rectangle measuring $2\pi r$ in length (the circumference of the circle) and h in height.

A5.3

1 a $x = \frac{C - b}{a}$ **b** $x = M + b + c$
c $x = Kt + qt$ **d** $x = \frac{W - t}{y}$
e $x = Hp - z$ **f** $x = q + \frac{D}{p}$ or $\frac{D + pq}{p}$
g $x = AB + ct$ **h** $x = \frac{Y - c}{m}$

2 Sebastian expanded the brackets to begin with but James didn't.

3 a $y = \sqrt{c}$ **b** $y = 4(k + 2)$
c $y = \frac{M + t}{xz}$ **d** $y = 4x^2$
e $y = \sqrt[3]{p - 2}$ **f** $y = \sqrt{\frac{T}{k}}$
g $y = \frac{3R}{az}$ **h** $y = p^3$

4 He should have divided by m before taking cube root.

5 $\frac{8(D + k)}{ab} = c$
$8(D + k) = abc$
$D + k = \frac{1}{8}abc$
$d = \frac{1}{8}abc - k$

6 a 86 °F **b** $C = \frac{5(F + 40)}{9} - 40$ **c** $-35\frac{5}{9}$ °C

A5.4

1 a $p = \frac{m + q}{x}$ **b** $p = \sqrt{w + r}$
c $p = (m - h)^3$ **d** $p = t(h + g)$
e $p = 2(q - r)$ **f** $p = \sqrt{\frac{k}{b}}$
g $p = \frac{z}{aw}$ **h** $p = (2x + y)^2$

2 a $x = k - w$ **b** $x = \frac{t - p}{a}$
c $x = \frac{b - y}{t}$ **d** $x = a - \frac{m}{n}$ or $\frac{an - m}{n}$
e $x = \frac{k}{w}$ **f** $x = \frac{t}{m}$
g $x = \frac{h}{g - p}$ **h** $x = \sqrt{\frac{p}{k}}$

3 a 36 mph **b** $t = \frac{d}{S}$
c 1 hour 26 minutes to nearest minute

4 $k = \frac{t}{p - q}$

5 D $x = \frac{2(p - y)}{ab}$
A $abx = 2(p - y)$
E $\frac{1}{2}abx = (p - y)$
B $y + \frac{1}{2}abx = p$
C $y = p - \frac{1}{2}abx$

6 a $g = \frac{4\pi^2 p}{T^2}$ **b** 9.8

7 a $h = \frac{v}{\pi r^2}$ **b** $r = \sqrt{\frac{V}{\pi h}}$

8 $b = \frac{2A}{h} - a$

A5.5

1 a $21 - 7 = 14$, which is even
b 5 squared is 25, which is odd
c $3 \times 2 = 6$, which is even, and 2 is prime
d $3 \times 4 = 12$, which is even

2 a For example, $2 + 3 + 4 + 5 + 6 = 20$ which is 4×5
b $n + (n + 1) + (n + 2) + (n + 3) + (n + 4)$
$= 5n + 10 = 5(n + 2)$, which is clearly a multiple of 5 for any n

3 a Even numbers $2m$ and $2n$, then $2m + 2n = 2(m + n)$ which is a multiple of 2 and hence even for any m, n
b $(2n)^2 = 4n^2$, which is always in the four times table for any n

c $2n + (2m + 1) = 2(m + n) + 1$, which is odd for any m, n

d $k(k + 1) - k = k^2 + k - k = k^2$, which is a square number for any k

4 a Any number between 0 and 1 inclusive

b Any number between 0 and 1 inclusive

5 a $k(k + 2) = k^2 + 2k$,
$(k + 1)^2 = k^2 + 2k + 1 = k(k + 2) + 1$ for any k

b $(n + 4)(n + 7) - (n + 3)(n + 8)$
$= n^2 + 11n + 28 - n^2 - 11n - 24 = 4$

c $m^2 - n^2 = m^2 - mn + mn - n^2 = (m - n)(m + n)$ for any m, n

6 If angle ACB is x then angle OAC is also x (isosceles triangle) so angle AOC is $180 - 2x$ (angles in a triangle).

If angle ABC is y then angle OAB is also y (isosceles triangle) so angle AOB is $180 - 2y$ (angles in a triangle).

Hence, $180 - 2x + 180 - 2y = 180$ (angles on a straight line)

$180 = 2x + 2y$

$90 = x + y$

i.e. angle CAB is $90°$

A5 Exam review

1 a $P = 3x - 1$

bi $x = \frac{P + 1}{3}$ **bii** 3 cm

2 When $n = 2$, $n^2 + 3 = 7$ which is not even.

S4 Before you start ...

1

2 a 49 **b** 52 **c** 34 **d** 48 **e** 45 **f** 80

S4.1

1 a Yes **b** No **c** Yes

2 a

2 b First is a line of symmetry, second is not.

3

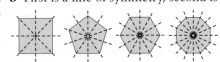

4 Cuboid: planes through middle of sides and through diagonally opposite edges.
Triangular prism: planes vertically through middle and from edges through opposite side.

5 One horizontal plane of symmetry halfway up the cylinder and infinitely many vertical planes of symmetry through the diameters of the circle on top of the cylinder.

S4.2

1 22.5 cm²

2 a Square with sides 7.1 cm **b** 50 cm²

3 a All 42 cm²

b Halve the product of the diagonals.

4 60 cm²

5 Both shapes have been split into congruent triangles; all sides equal in both cases.

6 a True; all squares have 2 pairs of sides of equal length and 4 equal angles.

b False; not all kites have diagonals which bisect each other.

c False; not all rhombuses have 4 equal angles.

S4.3

1 a 5 cm **b** 17 cm **c** 13 cm **d** 10.3 cm
e 10.8 cm **f** 9.90 cm **g** 12.5 cm **h** 7.3 cm

2 3, 4, 5
8, 15, 17
5, 12, 13

3 a 8 cm **b** 19.8 cm **c** 10 cm **d** 9 cm
e 7.5 cm **f** 5.4 cm **g** 5.5 cm **h** 7.1 cm

4 a 6, 8, 10; 10, 24, 26; 9, 12, 15

b 6, 8, 10 is 3, 4, 5 doubled; 10, 24, 26 is 5, 12, 13 doubled; 9, 12, 15 is 3, 4, 5 trebled

S4.4

1 a 8.6 cm **b** 7.9 cm

2 9.6 cm

3 11.3 cm

4 5.7 cm

5 16.7 cm

6 13.6 cm

7 5.4 m

S4.5

1 a (4, 5.5) **b** (0.5, 1) **c** (2, 3) **d** (1.5, 2)

2 a 5 units **b** 8.6 units **c** 10.0 units
d 10.6 units

3 J (2, 4, 5), K (3, 1, 5), L (−1, 2, 5), M(4, −3, 0)
N (0, 2, 0), P (−3, 2, −1), Q (0, 0, 2), R (2, 0, −3)

4 ai (1, 3, 3.5) **aii** (3.5, 3, 2.5)
aiii (1.5, 3.5, 1.5) **aiv** (3, 2.5, 4.5)

b ai Length of AB = 3.61 (2 dp)
aii Length of CD = 4.24 (2 dp)

c BC = $\sqrt{3}$ = 1.72 (2 dp)
If rounded, $\sqrt{2}$ = 1.41 so BC = 1.73 (2 dp)

S4 Exam review

1 a 4.1 units **b** (2.5, 5)

2 ai D **aii** B **aiii** C **b** D

N6 Before you start ...

1 a 14 **b** 2 **c** 15 **d** 16

2 a 2 **b** 16 **c** 5 **d** 10

3 a 45.6 **b** 9.9 **c** 104.0 **d** 16.1
4 a $20 \times 30 = 600z$ **b** $350 + 60 = 410$
 c $1200 - 800 = 400$ **d** $7000 + 6000 = 13\,000$

N6.1

1 a 59 **b** 37 **c** 2 **d** 26
2 a 1 **b** 10 **c** 43 **d** 110
3 a + **b** − **c** × **d** ×
4 a $(11 - 1) \times 5 = 50$ **b** $(12 + 3) \div 3 = 5$
 c $12 - (4 - 1) = 9$ **d** $8 \div (4 + 4) + 1 = 2$
5 a 2 **b** 5 **c** 110 **d** 14
6 a 196 **b** 4 **c** 18 **d** 289
7 a 121 **b** 39 **c** 18 **d** 5.29
8 a 10 **b** 18
9 a ≈ 14.72 **b** ≈ 102.38 **c** ≈ 286.63 **d** ≈ 9.90

N6.2

1 a $2\sqrt{3}$ **b** $2\sqrt{5}$ **c** 5 **d** $3\sqrt{7}$
2 a 2π **b** 3π **c** $2 + 2\pi$ **d** 10π
 e 18π **f** 16π
3 a $4 + \sqrt{3}$ **b** $3 + \sqrt{2}$ **c** $7 + 2\sqrt{7}$ **d** $14 + \sqrt{17}$
4 a 32π **b** $7 + 2\pi$ **c** $28 + 4\sqrt{2}$ **d** 62π
5 a 2 **b** 5 **c** $3 + 3\sqrt{3}$ **d** $8 + 2\sqrt{3}$
6 a 5.66 **b** 3.24 **c** 4.24 **d** 106.10
7 a 20.9 **b** 44.7 **c** 12.3 **d** 32.2
8 a $2\sqrt{5}$ **b** $5\sqrt{5}$ **c** $3\sqrt{2}$ **d** $7\sqrt{2}$
9 a 3.146 **b** 3.162 **c** 11.18 **d** 4.897
10 a 3.146 **b** 3.162 **c** 11.18 **d** 4.899
11 a $14 + 7\sqrt{2}$ **b** $17 + 7\sqrt{5}$ **c** $15 - 3\sqrt{3}$ **d** 20

N6.3

1 a 1 **b** 1.2 **c** 1.5 **d** 8
 e 1.4 **f** 7.5 **g** 16 **h** 15
2 a 40 **b** 10 **c** 20 **d** 80
 e 0.2 **f** 10 **g** 10 **h** 5000
3 a 0.8 **b** 0.8 **c** 21 **d** 0.21
 e 4.8 **f** 40 **g** 0.9 **h** 4
4 as Q3
5 a $4 \div 2$ **b** $6 \div 5$ **c** $12 \div 5$ **d** $2 \div 1000$
6 a 4×2 **b** 6×4 **c** 16×100 **d** 15×20
7 a 32 **b** 30 **c** 35 **d** 1
 e 0.32 **f** 200 **g** 1 **h** 720
 i 0.375 **j** 28 **k** 1 **l** 7
8 a 80 **b** 1000 **c** 520 **d** 500
9 a 78.72 to 4 sf **b** 1063 to 4 sf
 c 509.2 **d** 548.4 to 4 sf
10 a 9.3 **b** 700 **c** 12.2 **d** 10.6
 e 500 **f** 130 **g** 570 **h** 0.65
11 a 9.3 **b** 700 **c** 12.2 **d** 10.6
 e 463.3 to 4 sf **f** 130 **g** 570
 h 0.65

N6.4

1 a 38.76 **b** 2.61 **c** 8.201 **d** 22.52
2 a 2.12 **b** 4.19 **c** 2.04 **d** 5.7
3 a 155.09 **b** 0.45 **c** 24.32 **d** 14.72

4 a 3.97 **b** 0.095 **c** 12.44 **d** 58.34
5 a 2.81 **b** 0.757 **c** 88.01 **d** 1126.72
6 See Q1 to Q5
7 a 63.6 **b** 5.37 **c** 13.52 **d** 0.003 36
8 a 1.7 **b** 0.43 **c** 48 **d** 269
9 a 16.72 **b** 10.26 **c** 793.8 **d** 7.918
10 a 8.82 **b** 203 **c** 886 **d** 21.9
11 See Q7 to Q10

N6.5

1 a 5.6 **b** 73 **c** 0.85 **d** −35
 e 110 **f** 38
2 a 141 kg **b** 8.4 m^2 **c** £13.63 **d** 45 sec
3 a 4.745 m, 4.755 m **b** 12.55 s, 12.65 s
 c 149.5 cm, 150.5 cm **d** 24.45 kg, 24.55 kg
 e 8.065 g, 8.075 g **f** 4.325 s, 4.335 s
4 a 2.545 m, 2.555 m **b** 1.65 s, 1.75 s
 c 1.548 m/s to 4 sf **d** 1.454 m/s to 4 sf
5 a 56.3 m^2, 72.3 m^2 **b** 40.3 cm^2, 41.6 cm^2
 c 1.09 m^2, 1.11 m^2 **d** 8.70 mm^2, 9.30 mm^2
 e 14.0 m^2, 14.1 m^2 **f** 99.7 cm^2, 99.9 cm^2
6 a 36.765, 38.645 **b** 0.3, 0.5
 c 0.3618, 0.3864 **d** 0.1075, 0.1485 to 4 sf

N6 Exam review

1 ai 1250 **aii** 125 **b** 500
2 a 17.9867 **b** $(1.6 + 3.8 \times 2.4) \times 4.2$

A6 Before you start …

1 a $x = 9$ **b** $x = 3\frac{1}{2}$ **c** $x = -14\frac{1}{2}$
 d $x = \frac{1}{4}$ **e** $x = 3$ **f** $x = 14$
2 a $x^2 + 7x + 10$ **b** $y^2 + 4y - 21$
 c $w^2 + 8w + 16$ **d** $y^2 - 15y + 54$
 e $6w^2 + 14w + 4$ **f** $16y^2 - 40y + 25$
3 a $2(2x + 1)$ **b** $y(x + y)$
 c $5a(1 + 2b)$ **d** $4(x - 2y)$
 e $3x^3(3 + x^2)$ **f** $2x(y + 2z - 3x)$
4 a $(x + 2)(x + 3)$ **b** $(x - 6)(x + 4)$
 c $(x - 3)(x - 4)$ **d** $(x - 3)(x - 3)$
 e $(x + 10)(x - 10)$ **f** $2(x + 2)(x + 1)$

A6.1

1 a $x = 6$ **b** $y = 14$ **c** $x = 6$ **d** $p = 8$
 e $t = 2$ **f** $b = 2$
2 a $x = 1.1$ **b** $y = -1\frac{5}{22}$ **c** $z = 1\frac{3}{4}$ **d** $a = 5\frac{1}{8}$
 e $a = 4.3$ **f** $x = 5$ **g** $x = -13$ **h** $y = 1$
3 a $x = 2$ **b** $y = 3$ **c** $z = \frac{3}{8}$ **d** $a = -\frac{5}{7}$
 e $b = -1$ **f** $c = -2.5$
4 a $40 - 3x = 22, x = 6$
 b $2(x + 6) = 6(x - 5), x = 10.5$
 c $(x + 7)^2 = (x - 5)^2, x = -1$
5 a $(x - 4)(x + 11) = (x + 3)^2, x = 53$
 b $26 - 4x = 15 - 2x, x = 5.5$

A6.2

1 a $2x^2 = 50$, $x^2 = 100$, $x^2 = -49$, $x^2 = 169$
 b $2x^2 = 50$, $x = 5$ or -5
  $x^2 = 100$, $x = 10$ or -10
  $x^2 = 169$, $x = 13$ or -13

2 a $(x+2)(x+6)$ **b** $(x+3)(x+8)$
 c $(x+9)(x+4)$ **d** $(x+11)(x+5)$
 e $(x-3)(x-4)$ **f** $(x-4)(x-6)$
 g $(x+7)(x-4)$ **h** $(x+3)(x-5)$

3 a $x = -3$ or -4 **b** $x = -2$ or -6
 c $x = -5$ **d** $x = -7$ or 2
 e $x = 5$ or -1 **f** $x = 2$ or 3
 g $x = 3$ or 7 **h** $x = 8$ or -5

4 a $x = 0$ or 8 **b** $x = 0$ or -4
 c $x = 0$ or 6 **d** $y = 0$ or -5
 e $x = 0$ or 9 **f** $x = 0$ or 12
 g $x = 0$ or -4 **h** $x = 0$ or 6

5 No two integers add up to 7 and multiply to give 11.

6 a She should make the right-hand side equal to zero first.
 b $(x+4)(x-2) - 7 = 0$, $x^2 + 2x - 15 = 0$, $(x+5)(x-3) = 0$, $x = -5$ or 3

7 $x = 1$ or 5

A6.3

1 6.2

2 a $x(x+4) - 77 = 0$, length and width are 7 cm and 11 cm
 b $h(h+1)(h+2) - 990 = 0$; 9 cm by 10 cm by 11 cm

3 a $x = -5$ or 3
 b Factorisation is more efficient

4 a 8.8 **b** 4.6 **c** 9.51 **d** 4.28

5 a $x(x+1)(x+2) = 85\ 140$; 43, 44, 45
 b $x(x+2) = 57$; 6.6 cm
 c $3x^3 = 500$, 5.5 mm

6 0 or 6.03

A6.4

1 a, b, e

2 a $x = 4$, $y = 3$ **b** $x = 2$, $y = -3$
 c $x = -1$, $y = 4$ **d** $x = 3$, $y = 0.5$
 e $m = 5$, $n = 2$ **f** $x = -2$, $y = 1$

3 a $x = 7$, $y = -1$ **b** $x = 3$, $y = 0.5$
 c $a = 3$, $b = -1$ **d** $x = 2$, $y = 5$
 e $x = 3.25$, $y = -\frac{2}{7}$ **f** $p = 2$, $q = 7$

4 a $x = 5$, $y = -2$ **b** $x = 2$, $y = 3$
 c $a = -2$, $b = 3$ **d** $v = 4$, $w = 2$
 e $p = 2$, $q = 2$ **f** $x = 5$, $y = 2$

5 a $a = 9\frac{2}{3}$, $b = 1\frac{1}{3}$ **b** $v = 7$, $w = 4$

6 a 17p **b** 6.4 cm

A6.5

1 a $x = 12$, $y = -16$ **b** $x = 7$, $y = -1$
 c $a = 3$, $b = -2$ **d** $v = 3$, $w = 2$
 e $p = 0$, $q = 3$ **f** $x = 5$, $y = 2$

2 $2x + y = 12$, $y - x = 15$; $x = -1$, $y = 14$
  $2x + y = 12$, $3x - 4y = 7$; $x = 5$, $y = 2$
  $y - x = 15$, $3x - 4y = 7$; $x = -67$, $y = -52$

3 a $x = 6$, $y = 2$ **b** $a = 8$, $b = -1$
 c $p = 8$, $q = -1$

4 a 17, 24 **b** 17, 23
 c 4 large, 1 small **d** 105°, 37.5°, 37.5°
 e £5 for a paperback, £10 for a hardback

A6 Exam review

1 $x = 4$, $y = 2$ **2** $x = 4.2$

D4 Before you start …

1 a 10 **b** 25 **c** 20 **d** 12.5
2 a 16 **b** 22.5 **c** 14 **d** 18.5
3 a 6 **b** 4.5 **c** 7 **d** 10.5
4 a 12 **b** 33 **c** 21 **d** 10.5

D4.1

1 ai 5 **aii** 6 **aiii** 5.8 **aiv** 4 **av** 2
 bi 5 **bii** 5 **biii** 4.96 **biv** 4 **bv** 2
 ci 4 **cii** 6 **ciii** 6.59 **civ** 6 **cv** 4
 di 6 and 7 **dii** 6 **diii** 5.77 **div** 6 **dv** 3

2 42.5

3 2.46

D4.2

1 ai $10 < t \leqslant 15$ **aii** $15 < t \leqslant 20$ **aiii** 16.1
 bi $10 < t \leqslant 20$ **bii** $20 < t \leqslant 30$ **biii** 23.5
 ci $5 < t \leqslant 10$ **cii** $10 < t \leqslant 15$ **ciii** 14.7
 di $15 < t \leqslant 25$ **dii** $25 < t \leqslant 35$ **diii** 30

2 a $20 < b \leqslant 30$ **b** £24.17 to 3 sf

3 a $165 \leqslant h < 170$ **b** 164.3 cm **c** $165 \leqslant h < 170$

D4.3

1 a

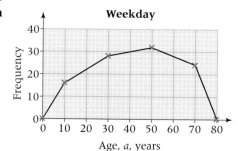

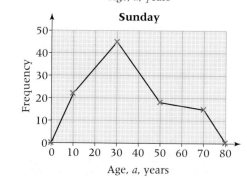

b Weekday: $40 < a \leqslant 60$, Sunday: $20 < a \leqslant 40$

c Generally, people visiting on a weekday are older.

2 a

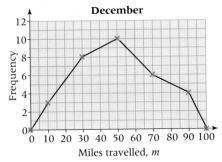

b December: $40 < m \leqslant 60$, January: $40 < m \leqslant 60$

c Less variation in miles travelled in January, less short journeys. The most common journey length does not change.

3 a

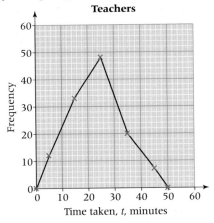

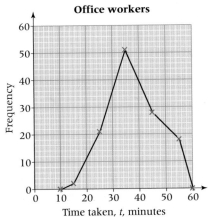

b Teachers: $20 < t \leqslant 30$, office workers: $30 < t \leqslant 40$

c On average, office workers take longer travelling home.

D4.4

1

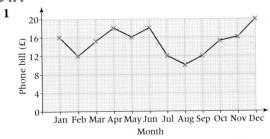

2

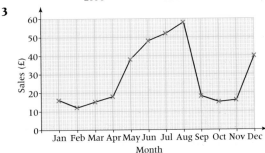

3

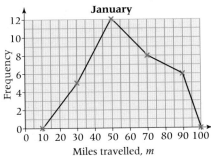

4

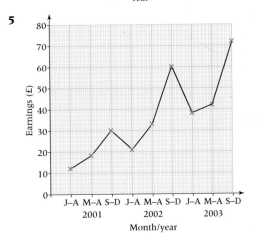

5

6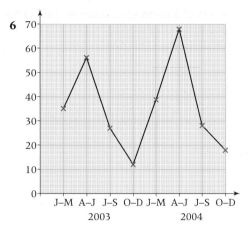

D4.5

1 £14.33, £15, £16.33, £17.33, £15.33, £13.33, £11.33, £12.33, £14.33, £17

2 **a** £32, £34.25, £35.25, £36, £36.50

   b
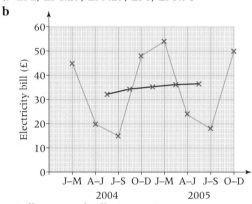

   c Bills are gradually increasing

3 £15.25, £20.75, £29.75, £39, £49, £44, £35.75, £26.75, £22.25

4 **a** 16%, 21%, 26%, 26.5%, 22.5%, 19%, 20%

   b 21.8%, 23%, 23%, 22.6%

5 **a** £20, £23, £28, £38, £44, £47, £51

   b
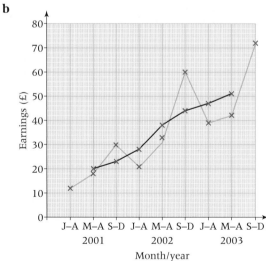

Month/year

   c Money earned is increasing over time.

6 **a** £32.50, £33.50, £36.50, £37, £38.50

   b
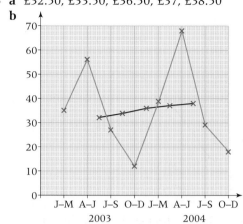

   c Expenses are gradually increasing.

7 **a**
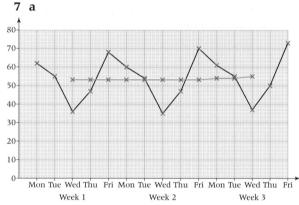

   b It will help her spot a trend over time

   c 53.6, 53.2, 53, 52.8, 52.8, 53.2, 53.4, 53.6, 54, 54.6, 55.2

   d See part **a**

   e The number of cups of coffee sold is changing very little over time, perhaps a slight increase.

D4 Exam review

1 1.65

2 **a** $150 < C \leqslant 250$

   b Incorrect, the median (21st customer) is in $150 < C \leqslant 250$

   c £6500

S5 Before you start ...

1 34°, 72°, 105°

2 **a** Circle radius 3 cm   **b** Arc radius 5 cm

3

S5.1

1 **a** 050°   **b** 320°

2 **a** Bearing of 070°   **b** Bearing of 155°

   c Bearing of 340°   **d** Bearing of 260°

3 a 316° **b** 265° **c** 068°
4 a 284° **b** 263° **c** 117°
5 b 170°
6 aii 328° **bii** 357°

S5.2

1 a Triangle with sides 8 cm, 4 cm, 7 cm
  b Triangle with sides 3 cm and 4 cm, and 30° angle
  c Triangle with sides 10 cm, 7.5 cm, 6 cm
  d Triangle with sides 8 cm and 2 cm, and 90° angle
  e Triangle with sides 6 cm, 9 cm, 5 cm
  f Triangle with two 45° angles and a 4 cm side
2 a 4 + 3 < 9, two short sides will never meet.
  b 4 + 5 = 9 so triangle is a straight line.
3 a Yes   **b** Yes   **c** Yes   **d** No
  e No   **f** No   **g** Yes   **h** Yes
4 a Triangle with sides 7 cm, 7 cm, 5 cm
  b Triangle with sides 5 cm, 5 cm, 7 cm
5 a Equilateral triangle with sides 5 cm
  b Rhombus with sides 5 cm
  c Rhombus with sides 3.5 cm
6 a Triangle with sides 5 cm, 12 cm, 13 cm
  b Right-angled triangle
7 a Triangle with sides 3 cm, 4 cm, 5 cm
  b Rectangle with sides 3 cm, 4 cm
  c Rectangle
8 a, b

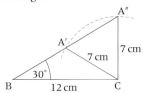

  c No, SSA triangles are not unique since you can construct two different triangles with the given side and angle measurements.

S5.3

1 Angle bisectors
2 a Equilateral triangle with sides 5 cm
  b Angles bisectors of triangle in part **a**
  c Bisectors meet at a point and cut the midpoints of the opposite sides.
3 a Equilateral triangle with sides 4 cm
  b Angles bisector of triangle in part **a**
4 a Perpendicular bisector of 6 cm line
  b Perpendicular bisector of 9 cm line
  c Perpendicular bisector of 5.6 cm line
  d Perpendicular bisector of 10 cm line
  e Perpendicular bisector of 11.2 cm line
5 a Equilateral triangle with sides 5 cm
  b Perpendicular bisectors of sides of triangle in part **a**
  c Perpendicular bisectors and angle bisectors of equilateral triangles are the same.

6 d Both sets of lines meet at common points but not the same points.
7 c 90°, 135°, 180°, 225°, 270°, 315°

S5.4

1 Perpendiculars from points to lines
2 Perpendiculars from points on lines

S5.5

These sketches not drawn to scale.

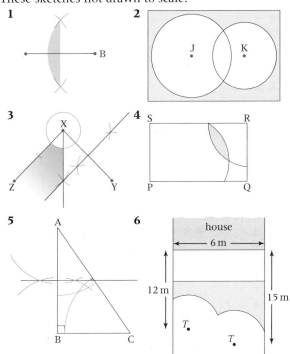

S5 Exam review

1 a 197°
  bi

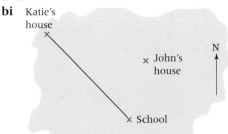

  bii 105°
2

N7 Before you start …

1 a 30% **b** 65% **c** 72.5%
  d 105% **e** 6%
2 a 0.15 **b** 0.065 **c** 0.125
  d 0.975 **e** 1.08
3 a £35 **b** 20 m **c** 54°
  d 120 cm **e** £54 **f** 90 mm

N7.1

1 a $3\frac{1}{2}$ **b** $1\frac{3}{5}$ **c** $3\frac{1}{3}$ **d** $\frac{2}{3}$ **e** $2\frac{2}{5}$
2 a 10 **b** 21 **c** 12 **d** 13 **e** 22
3 a 24 m **b** 36 m
4 a 15 litres **b** 180 cl **c** 48 cl **d** 250 ml
5 a 30 m **b** 8 km **c** 16 mm **d** 90 m
6 a $1\frac{2}{3}$ **b** $5\frac{1}{2}$ **c** $4\frac{1}{2}$ **d** $\frac{2}{3}$ **e** $9\frac{1}{3}$
7 a $2\frac{2}{3}$ miles **b** $33\frac{1}{3}$ miles
  c $12\frac{1}{2}$ miles **d** $3\frac{3}{4}$ miles
8 a 6.67 m **b** 5.33 mm
  c 8.73 cm **d** 22.3 km
9 40.83 m³ to 2 dp
10 a 5 hours 15 minutes **b** 13 hours 45 minutes
  c 16 hours 48 minutes **d** 24 hours 30 minutes
11 a £4.67 **b** £16.67 **c** £12.75 **d** £29.33.

N7.2

1 a 10.5 **b** 126 **c** 240 **d** 300
  e 132 **f** 99
2 a £225 **b** 990 m **c** 2520 kg **d** €144
3 a 135 **b** 91 **c** 744 **d** 18
  e 387 **f** 192
4 a 1104 mm **b** 1952 kg
  c €1443 **d** £493
5 a 0.5 **b** 0.6 **c** 0.25 **d** 0.51
  e 0.64 **f** 0.22 **g** 0.15 **h** 0.7
  i 0.07 **j** 0.085
6 a 5.7 **b** 200 **c** 15.93 **d** 99.84
  e 16.81 **f** 20
7 a £3.84 **b** £53.55 **c** £13.95 **d** £170.10
  e £28.16 **f** £15.90
8 £425
9 £73.70
10 649
11 £49 090.60

N7.3

1 a 108 **b** 72 **c** 49.5 **d** 618
  e 600 **f** 700
2 a 25.2 **b** 36 **c** 51 **d** 37.5
  e 228 **f** 64.35
3 a 342 **b** 264 **c** 1008 **d** 325
  e 1372.5 **f** 1109.2
4 a 672 **b** 532 **c** 272 **d** 301.5
  e 136.5 **f** 124.8
5 a 1.2 **b** 1.3 **c** 1.45 **d** 1.85
  e 1.065

6 a 0.6 **b** 0.4 **c** 0.65 **d** 0.28
  e 0.815
7 a 56.71 **b** 40.32 **c** 719.2 **d** 246
8 a £33 600 **b** £19 317.15
  c £27 348 **d** £56 021
9 a £351 **b** £559 **c** £725 **d** £714.10
10 They could have more than doubled in price, which gives a price increase of more than 100%.
11 A price fall by more than 100% leads to a negative cost, which is impossible.

N7.4

1 a £5 **b** £24
  c £315 **d** £227.50
2 a £262.50 **b** £416.16
  c £1348.32 **d** £1061.21
3 a 1.06 **b** 1.1236
4 a £530 **b** £561.80
5 a 1.157 625 **b** 1.370 086 663
6 a £5788.13 **b** £1096.07
7 Option **a**
8 a 0.85 **b** Multiply by 0.85 again

N7.5

1 a 80% **b** 28% **c** 12% **d** 50% **e** 50%
2 a 60% **b** 48% **c** 60%
3 a 51.35% **b** 28.57% **c** 3.83%
4 a Science: 88.3%, Maths: 86.7% (both to 3 sf)
  b Jason dropped a greater proportion of the total marks in maths than he did in science.
5 £5
6 a 50 **b** 40 **c** 80 **d** 110
7 a £9 **b** £80
8 a Francesca **b** £78.03
9 She needs to find 20% of the original price, which is not the same as 20% of the sale price.

N7 Exam review

1 a £32 **b** £12 **ci** £36 **cii** $\frac{9}{20}$ **ciii** 45%
2 a 6 hours **b** £255.84

D5 Before you start …

1 a 62 **b** 70 **c** 60 **d** 45
  e 102 **f** 108
2 a 350 **b** £70
3 2 days

D5.1

1 a

Height, h, cm	Cumulative frequency
$h < 145$	0
$h < 150$	7
$h < 155$	32
$h < 160$	78
$h < 165$	95
$h < 170$	100

b
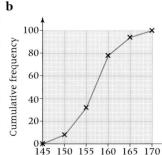

c $155 \leqslant h < 160$

2 a

Age, A, years	Cumulative frequency
$A < 20$	0
$A < 30$	18
$A < 40$	55
$A < 50$	106
$A < 60$	134
$A < 70$	150

b

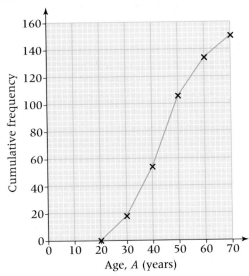

c $40 \leqslant A < 50$

3 a

Time, t, minutes	Cumulative frequency
$t < 10$	4
$t < 20$	15
$t < 30$	44
$t < 40$	81
$t < 50$	108
$t < 60$	120

b

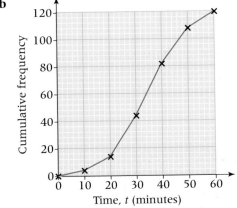

c $30 \leqslant t < 40$

4 a

Weight, w, grams	Cumulative frequency
$w < 1500$	0
$w < 2000$	9
$w < 2500$	31
$w < 3000$	68
$w < 3500$	88
$w < 4000$	100

b

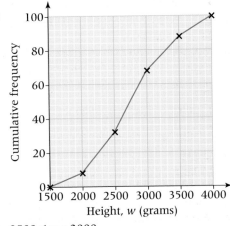

c $2500 \leqslant w < 3000$

5 a

Height, h, cm	Cumulative frequency
$h < 40$	0
$h < 60$	2
$h < 80$	19
$h < 100$	47
$h < 120$	86
$h < 140$	110
$h < 160$	120

b

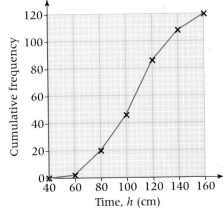

c $100 \leqslant h < 120$

6 a

Amount, £p	Cumulative frequency
$p < 10$	16
$p < 20$	30
$p < 30$	53
$p < 50$	70
$p < 70$	85
$p < 100$	100

b

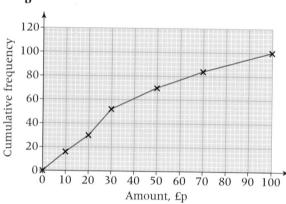

c $20 \leqslant p < 30$

D5.2

1 a	157 cm	**b**	6 cm	**ci** 18	**cii**	12
2 a	44 cm	**b**	17 cm	**ci** 36	**cii**	30
3 a	34 minutes	**b**	18 minutes	**ci** 30	**cii**	26
4 a	2750 g	**b**	750 g	**ci** 18	**cii**	10
5 a	107 cm	**b**	36 cm	**ci** 98	**cii**	86
6 a	£28	**b**	£39	**ci** 30	**cii**	10

D5.3

1 The boys' results are higher than the girls', on average. The middle half of the girls' results is less varied than that of the boys. The range is the same for the girls and the boys.

2 Farmer Jenkins' sunflowers are shorter than Farmer Giles', on average.
The middle half of Farmer Jenkins' sunflowers vary in height more than those of Farmer Giles. The range of heights of Farmer Jenkins' sunflowers is greater than that of Farmer Giles'.

3 Boys have higher mobile phone bills, on average. The middle half of the mobile phone bills varies the same for girls and boys. The range of mobile phone bills is the same for girls and boys.

D5.4

1 Min 145 cm, LQ 153.5 cm, median 157 cm, UQ 159.5 cm, max 170 cm

2 Min 20 cm, LQ 35 cm, median 44 cm, UQ 52 cm, max 70 cm

3 Min 0 minutes, LQ 25 minutes, median 34 minutes, UQ 43 minutes, max 60 minutes

4 Min 1500 g, LQ 2350, median 2750 g, UQ 3100 g, max 4000 g

5 Min 40 cm, LQ 88 cm, median 107 cm, UQ 124 cm, max 160 cm

6 Min £0, LQ £17, median £28, UQ £56, max £100

7 a Min 11 years, LQ 13.1 years, median 14.3 years, UQ 15.4 years, max 18 years

b Min 0 minutes, LQ 23 minutes, median 31 minutes, UQ 42 minutes, max 70 minutes

D5.5

1 On average, waiting times are higher at the dentist. The range of waiting times is greater at the doctor. The middle half of waiting times varies more at the doctor than at the dentist. The waiting times at the doctor are symmetrical, but those at the dentist are negatively skewed.

2 The average reaction time is the same for boys and girls. The range of reaction times is greater for girls. The middle half of reaction times varies less for girls than for boys. Reaction times for girls are symmetrical, but for boys they are negatively skewed.

3 On average, results are the same in the English and French tests. The ranges of results are the same. The middle half of results varies more in the English test. The English test results are negatively skewed, but the French test results are symmetrical.

4 On average, 17-year-old girls make longer phone calls than 13-year-olds. The range of the length of calls made is greater for 17-year-olds. The middle half of the calls made varies more for 13-year-olds than for 17-year-olds. The lengths of calls made by 13-year-olds is negatively skewed, but those for 17-year-olds are symmetrical.

D5 Exam review

1 a

Height	Cumulative frequency (girls)	Cumulative frequency (boys)
$h < 140$	0	0
$h < 145$	4	2
$h < 150$	9	6
$h < 155$	12	11
$h < 160$	14	14
$h < 165$	15	17

b

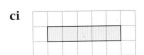

c Girls are on average shorter (lower median). The range of heights are the same. The interquartile ranges are the same.

2 a Median 33 seconds

b Min 10 seconds, LQ 16 seconds, median 33 seconds, UQ 44.5 seconds, max 60 seconds

c The average times for boys and girls are similar. The range of times is greater for boys.

S6 Before you start ...

1 a Cuboid　　**b** Square-based pyramid
　c Triangular prism

2 For example,

3 a 21 cm²　**b** 59.3 cm²　**c** 40 cm²

S6.1

1 ai

aii

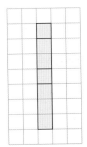

bi

bii

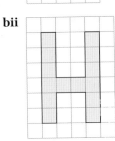

ci

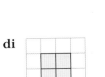

cii

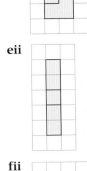

di

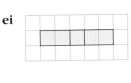

dii

ei

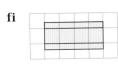

eii

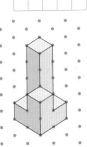

fi

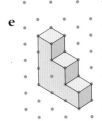

fii

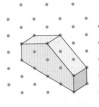

2 a

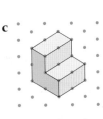

b

c

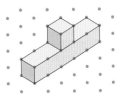

d

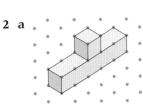

e

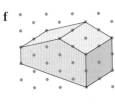

f

S6.2

1 a 105 cm³　　**b** 60 cm³　　**c** 72 cm³
　d 36 cm³　　**e** 64 cm³　　**f** 28 mm³
2 a 75.4 cm³　　**b** 628.3 cm³　**c** 201.1 cm³
　d 160.8 cm³
3 a 270 cm³　　**b** 600 mm³　　**c** 1080 cm³

1 a 27 cm³   **b** 343 cm³
    c 64 cm³   **d** 0.125 m³
2 a 155.9 cm³   **b** 86.6 cm³
    c 304.8 cm³   **d** 125.7 mm³
3 a 384 cm²   **b** 600 cm²
    c 216 cm²   **d** 6 cm²
4 a 1 cm × 1 cm × 80 cm, 2 cm × 2 cm × 20 cm,
    4 cm × 4 cm × 5 cm
    bi 112 cm²   **bii** 322 cm²
5 a 1 cm × 1 cm × 24 cm, 2 cm × 2 cm × 6 cm;
    56 cm²; 98 cm²
    b 1 cm × 1 cm × 64 cm, 2 cm × 2 cm × 16 cm,
    4 cm × 4 cm × 4 cm; 96 cm²; 258 cm²

S6.4

1 a 400 m   **b** 630 m   **c** 4200 cm
    d 12 m   **e** 80 km   **f** 45 m
    g 50 mm   **h** 300 mm   **i** 60 cm
    j 70 mm   **k** 4 cm   **l** 18 050 m
2 a 2.6 m²   **b** 70 000 m²   **c** 4.5 cm²
    d 1200 cm²   **e** 80 cm²   **f** 4.5 cm²
    g 8 400 000 mm²   **h** 300 m²
    i 20 000 cm²   **j** 1 000 000 m²
3 a 3000 mm³   **b** 200 cm³   **c** 4800 cm³
    d 3000 cm³   **e** 10 m³   **f** 50 m³
4 a Area   **b** Area   **c** Area
    d Volume   **e** Length   **f** Length
    g Area   **h** None of these   **i** Area
    j Volume   **k** None of these   **l** Length
5 a The formula has only two dimensions; it
    should have three.
    b Area
6 a $\pi r + 2h + 2r$ **b** $2hr + \frac{1}{2}\pi r^2$

7 a $a^2 + 2ab + a\sqrt{a^2 + b^2}$ **b** $\frac{1}{2}a^2 b$

S6.5

1 a 2.5 litres/s   **b** 1.6 litres/s
2 1200 litres
3 a 30 ml   **b** 24 ml   **c** 12.6 ml
4 ai 96 km   **aii** 12 km   **aiii** 16 km
    bi 3 hours   **bii** 1 hour 30 minutes
    biii 10 minutes
5

Distance	Time taken	Average speed
120 km	1.5 hours	80 km/h
250 miles	4 hours	62.5 mph
4 km	15 minutes	16 km/h
120 m	24 seconds	5 m/s
300 m	15 seconds	20 m/s
0.4 km	160 seonds	2.5 m/s or 12 km/h
3 km	125 seconds	24 m/s
20 km	20 minutes	60 km/h

6 a 57 040 kg   **b** 34 000 kg   **c** 61 824 kg
    d 125 cm³   **e** 2500 cm³   **f** 0.002 m³
7 a 8000 kg/m³   **b** 4500 kg/m³   **c** 2700 kg/m³

S6 Exam review

1

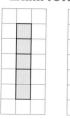
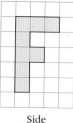

Front   Side   Plan

    b 8 cm³
2 170 g

A7 Before you start …

1 a 9   **b** 27   **c** 18   **d** 30
    e 24   **f** 60   **g** 63   **h** 87
2 a 4   **b** −8   **c** 8   **d** −12
    e 14   **f** −20   **g** 8   **h** 2
3 a Vertical, cutting x-axis at (4, 0)
    b Horizontal, cutting y-axis at (0, 5)
    c Vertical, cutting x-axis at (−3, 0)
    d Gradient 2, cuts y-axis at (0, −1)
4 a

x	1	2	3
y	2	8	11

b

x	−1	0	6
y	14	12	0

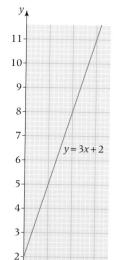

$y = 3x + 2$

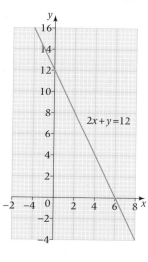

$2x + y = 12$

A7.1

1 a

Straight line	Parabola
$y = 3x - 2$	$y = x^2 - 2$
$3x + 2y = 8$	$y = 10 + x^2$
$y = x$	$y = x^2 + 2x + 1$

2 b

x	−4	−3	−2	−1	0	1	2	3	4
x^2	16	9	4	1	0	1	4	9	16
$y = x^2 - 2$	14	7	2	−1	−2	−1	2	7	14

c

$y = x^2 - 2$

3 ai

x	−4	−3	−2	−1	0	1	2	3	4
$y = x^2 + 3$	19	12	7	4	3	4	7	14	19

aiii

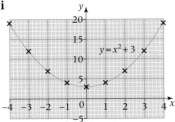

$y = x^2 + 3$

aiv (0, 3)

bi

x	−4	−3	−2	−1	0	1	2	3	4
$y = 2x^2$	32	18	8	2	0	2	8	18	32

biii

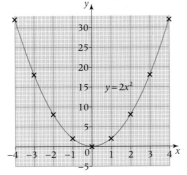

$y = 2x^2$

biv (0, 0)

ci

x	−4	−3	−2	−1	0	1	2	3	4
$y = 3x^2 - 1$	47	26	11	2	−1	2	11	26	47

ciii

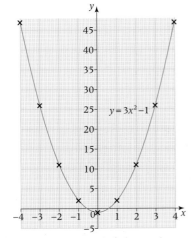

$y = 3x^2 - 1$

civ (0, −1)

di

x	−4	−3	−2	−1	0	1	2	3	4
$y = x^2 + x$	12	6	2	0	0	2	6	12	20

diii

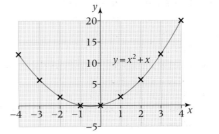

$y = x^2 + x$

div (−0.5, −0.25)

4 False, since $4^2 - 5 \neq 10$.

5 a graph will be other way up

b

x	−3	−2	−1	0	1	2	3
$y = 10 - x^2$	1	6	9	10	9	6	1

c

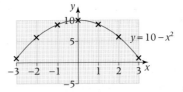

$y = 10 - x^2$

6 a

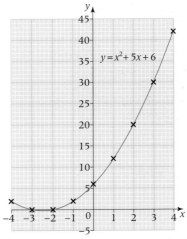

$y = x^2 + 5x + 6$

b (−2, 0), (−3, 0)

c $x = -2$ or -3

d The coordinates give the solutions of the equation, as x-axis is the line $y = 0$

A7.2

1 From left to right: $y = x^3$, $y = 3 - 2x$, $y = x^2 - x - 6$, $x = \frac{1}{2}$. Remaining graphs:

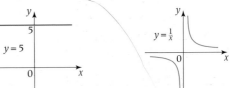

$y = 5$

$y = \frac{1}{x}$

2 b

x	−3	−2	−1	0	1	2	3
y	−26	−7	0	1	2	9	28

c

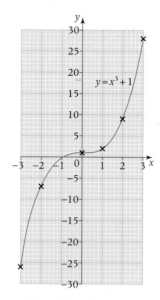

$y = x^3 + 1$

c

x	-2	-1	0	1	2	3
x^3	-8	-1	0	1	8	27
$x + 1$	-1	0	1	2	3	4
y	-9	-1	1	3	11	31

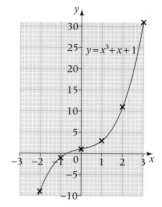

$y = x^3 + x + 1$

di 4.375   **dii** 1.125

3 a

x	-2	-1	0	1	2	3
x^3	-8	-1	0	1	8	27
$x^3 - 4$	-12	-5	-4	-3	4	23

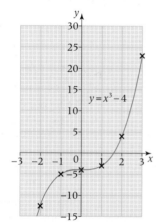

$y = x^3 - 4$

d

x	-2	-1	1	2	3
$\frac{3}{x}$	$-\frac{3}{2}$	-3	3	$1\frac{1}{2}$	1
y	$-\frac{1}{2}$	-2	4	$2\frac{1}{2}$	2

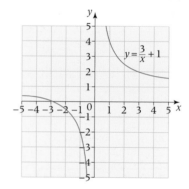

$y = \frac{3}{x} + 1$

b

x	-2	-1	1	2	3
$\frac{1}{x}$	$-\frac{1}{2}$	-1	1	$\frac{1}{2}$	$\frac{1}{3}$
y	-1	-2	2	1	$\frac{2}{3}$

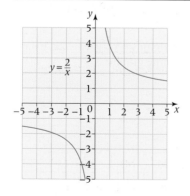

$y = \frac{2}{x}$

4 a   **b** (1.9, 1.4)

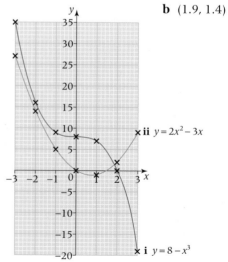

ii $y = 2x^2 - 3x$

i $y = 8 - x^3$

A7.3

1 ai $x=2, y=3$ **aii** $x=1, y=4$
aiii $x=1, y=1$ **aiv** $x=\frac{1}{4}, y=1\frac{3}{4}$
b The lines are parallel so they never intersect.

2 a $x=3, y=7$ **b** $x=1, y=1$ **c** $x=3, y=-1$

3 a $x=1, y=\frac{1}{2}$ **b** $x=1, y=\frac{1}{2}$
c They are the same

4 a 2, 0   **b** James is 3, Isla is 1

5 a The lines are parallel so they never intersect.
b Not if both are lines, but you could have a curve and a line intersecting twice or more.

6 a $y=4x+3, y=6x+1$ **b** $x=1, y=7$ **c** $(1, 7)$

A7.4

1 a $x=-1.5$ or 2.5 **b** $x=-0.6$ or 1.6
c $x=-1$ or 3

2 a $x=-1.2$ or 3.2 **b** $x=0$ or 2
c $x=1$ **d** $x=-0.3$ or 3.3
e $x=-1.6$ or 2.6 **f** $x=-1$ or 3

3 a $x=-2.6$ or 2.6 **b** $x=-1.3$
c $x=0$ or 0.5 **d** $x=1.7$

A7.5

1 a $x=-3$ or 1 **b** $x=-2.6$ or 1.6
c $x=-3.4$ or 1.4 **d** $x=-1$ or 0

2 a $y=3$ **b** $y=0$ **c** $y=2x+1$
d $y=4$ **e** $y=2$ **f** $y=10x-4$

3 a $x=1.1$ **b** $x=1.2$ **c** $x=-1.5$
d $x=-1.6, 0$ or 0.6 **e** $x=-2.5$ **f** $x=0.8$

A7 Exam review

1 a

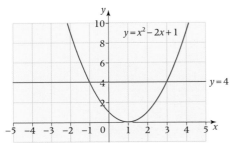

b $x=3, x=-1$; the x-coordinate of where the graphs cross gives the solutions

2 a

x	-2	-1	0	1	2	3
y	-1	1	3	5	7	9

b

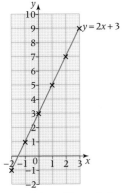

ci $y=0.4$ **cii** $x=1.2$

N8 Before you start …

1 Gill gets £400, Paul gets £100
2 33.3% decrease
3 a Batch B, it has a greater proportion of black paint
b 9 litres black, 21 litres white

N8.1

1 a $17:13$ **b** $11:19$ **c** $14:15$
2 11A: $7:23$, 11B: $13:17$, 11C: $16:13$
3 a $1:1$ **b** $2:5$ **c** $2:3$ **d** $3:4$
e $1:2$ **f** $1:1$ **g** $1:3$ **h** $4:3$
4 a $3:2$ **b** $4:1$ **c** $1:2$ **d** $1:4$
e $2:3$ **f** $3:2$ **g** $2:3$ **h** $3:2$
5 a 24, 16   **b** 27, 18   **c** 54, 36
d 60, 40   **e** 150, 100   **f** 1200, 800
g 4500, 3000   **h** 5550, 3700
6 a 16 hours, 8 hours   **b** 120°, 60°
c 30 minutes, 15 minutes   **d** €240, €120
e 164 cm, 82 cm   **f** 80 kg, 40 kg
g 54 km, 27 km   **h** 36 miles, 18 miles
7 a £180, £180   **b** £120, £240
c £240, £120   **d** £135, £225
8 a $5:3$ **b** Karla: £13.75, Wayne: £8.25
9 a £200   **b** Annie: £120, Ben: £180

N8.2

1 a $1:2$ **b** $1:2:2$ **c** $3:1:2$
d $5:10:4$ **e** $2:5:3$ **f** $3:7:2$
2 a £80, £160   **b** £96, £144
c £150, £90   **d** £40, £80, £120
e £150, £30, £60   **f** £140, £40, £60
3 a 50 m, 200 m   **b** 100 m, 150 m
c 71.43 m, 178.57 m   **d** 214.29 m, 35.71 m
e 115.38 m, 134.62 m   **f** 160.71 m, 89.29 m
4 Peter: £2700, Bob: £1500, Yasmin: £3300
5 Ann: £377.78, Charles: £283.33, Edward: £188.89

6

Quantity	Ratio	Share 1	Share 2	Share 3
200 km	2:3:5	40 km	60 km	100 km
38 kg	1:2:3	6.33 kg	12.67 kg	19 kg
450 cm	2:3:8	69.2 cm	104 cm	277 cm
£720	4:5:10	£151.58	£189.47	£378.95
95 litres	1:6:7	6.79 litres	40.71 litres	47.5 litres

7 Mango: 444 ml, pineapple: 333 ml, passion fruit: 222 ml
8 a £119.05, £116.67, £214.29
b £125, £166.67, £208.33
9 Robert: £12 660, Kathleen: £16 880

N8.3

1 a $2:3$ **b** $\frac{2}{5}$
2 a $1:2$ **b** $\frac{1}{3}$
3 a $1:3:2$ **bi** $\frac{1}{6}$ **bii** $\frac{1}{2}$ **biii** $\frac{1}{3}$
4 a $\frac{1}{2}$ **b** $\frac{1}{3}$ **c** $\frac{2}{5}$ **d** $\frac{3}{8}$ **e** $\frac{4}{7}$ **f** $\frac{5}{7}$
5 a $\frac{1}{2}$ **b** $\frac{2}{3}$ **c** $\frac{3}{5}$ **d** $\frac{5}{8}$ **e** $\frac{3}{7}$ **f** $\frac{2}{7}$

6 2 : 3

7 a £600   **b** 3 : 7

8 £770

9 $\frac{1}{2}$

10 £5333.33, $\frac{8}{21}$

N8.4

1 10%

2 a 4 : 7   **b** 36.4%

3 Nickeline: 80%, US nickel coinage: 75%, Medal bronze: 93%

4 60%

5 44.4%

6 a 80%   **b** 3 : 1

7 a £1140, £2660   **b** 30%

8 a $324.47, $175.53   **b** 35.1%

9 a 333 g, 104 g, 63 g (to 0 dp)

   b 66.7%, 20.8%, 12.5%

N8.5

1 a £125   **b** £560   **c** 46 m   **d** 54 cm

2 a £45   **b** 20 cm   **c** 360 cm   **d** 102 mm

3 a 1.2   **b** 0.85   **c** 1.06   **d** 0.95

   e 0.94   **f** 0.83

4 a £59.52   **b** 51.3 kg   **c** 10.088 seconds

   d 198 m   **e** €260.40   **f** $300.90

5 a £50   **b** £30.94

6 £560

7 a £321.63   **b** £1749.60

N8 Exam review

1 a Shop B   **b** Shop A

2 ai £1250   **aii** $\frac{1}{25}$

   b 12 : 5

S7 Before you start ...

1

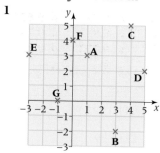

2 a 1 : 3   **b** 1 : 3   **c** 2 : 1   **d** 5 : 2

3 8 : 3

4 a $x = 14$ **b** $x = 15$ **c** $x = 10$ **d** $x = 18$

S7.1

1

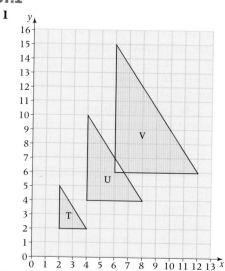

2

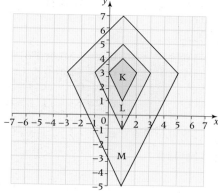

3

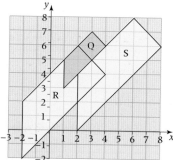

d Twice as large

4

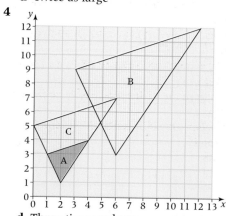

d Three times as large

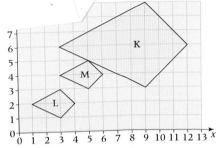

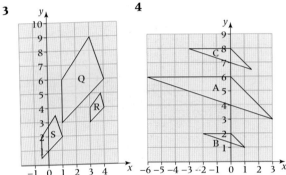

S7.4

1 $a = 6$ cm, $b = 4.5$ cm
2 $c = 6$ cm, $d = 2.5$ cm
3 $e = 5$ cm, $f = 2.4$ cm
4 15 cm
5 **a** 3 : 2   **b** $5\frac{1}{3}$ cm   **c** 22 cm, $14\frac{2}{3}$ cm
   d 3 : 2, same as ratio of lengths of sides

S7.5

1 **a** 4.5 cm   **b** 9 cm
2 **a** 2.67 mm   **b** 4.5 mm
3 **a** 6 cm   **b** 5 cm   **c** 15 cm
4 **a** MN = 6.3 cm, JN = 6 cm   **b** 15.5 cm
5 **a** 19.35 cm   **b** 32.1 cm

S7 Exam review

1 **a**

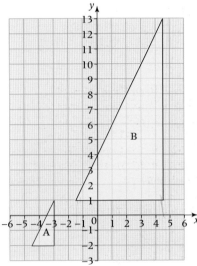

   b Enlargement scale factor $\frac{1}{4}$, centre $(-5.5, -3)$
2 **a** AD = 7.5 cm   **b** BC = 3 cm

D6 Before you start …

1 **a** 0.55   **b** 0.04   **c** 0.72   **d** 0.625
2 **a** 0.6   **b** 0.34   **c** 0.9   **d** 0.75
3 **a** 0.18   **b** 0.17   **c** 0.192   **d** 0.12
4 **a** $\frac{1}{6}$ **b** $\frac{4}{5}$ **c** $\frac{2}{9}$ **d** $\frac{3}{4}$
5 **a** $\frac{13}{15}$ **b** $\frac{11}{12}$
6 **a** $\frac{5}{9}$ **b** $\frac{8}{45}$

D6.1

1 **a** $\frac{4}{10}$ **b** $\frac{6}{10}$ **c** $\frac{7}{10}$ **d** $\frac{4}{10}$ **e** $\frac{8}{10}$ **f** $\frac{9}{10}$
2 241
3 102
4 **a** $\frac{13}{28}$ **b** $\frac{15}{28}$ **c** $\frac{11}{14}$ **d** $\frac{3}{14}$
5 **ai** $\frac{9}{16}$ **aii** $\frac{3}{8}$ **aiii** $\frac{13}{16}$ **aiv** $\frac{9}{16}$ **av** $\frac{5}{16}$ **avi** $\frac{1}{8}$
   b the black triangle
6 **ai** $\frac{1}{4}$ **aii** $\frac{3}{4}$ **aiii** 1 **aiv** 0   **b** 64

S7.3

1 **a** Enlargement with centre (0, 0) and scale
       factor 3
   b Enlargement with centre (0, 0) and scale
       factor $\frac{1}{3}$
2 **ai** Enlargement with centre (−3, 0) and scale
       factor 2
   aii Enlargement with centre (−3, 0) and scale
       factor $\frac{1}{2}$
   b 16 units
3 **a** Enlargement with centre (0, 0) and scale
       factor 2
   b Enlargement with centre (0, 0) and scale
       factor $\frac{1}{2}$
4 **ai** Enlargement with centre (0, 0) and scale
       factor $\frac{1}{2}$
   aii Enlargement with centre (0, 0) and scale
       factor 2
   aiii Enlargement with centre (0, 0) and scale
       factor 4
   b 4 times

D6.2

1 a

Coin/Dice	1	2	3	4	5	6
Head	H1	H2	H3	H4	H5	H6
Tail	T1	T2	T3	T4	T5	T6

bi $\frac{1}{12}$ **bii** $\frac{1}{12}$ **biii** $\frac{1}{12}$

2 a

Red/blue	1	2	3	4	5	6
1	1, 1	1, 2	1, 3	1, 4	1, 5	1, 6
2	2, 1	2, 2	2, 3	2, 4	2, 5	2, 6
3	3, 1	3, 2	3, 3	3, 4	3, 5	3, 6
4	4, 1	4, 2	4, 3	4, 4	4, 5	4, 6
5	5, 1	5, 2	5, 3	5, 4	5, 5	5, 6
6	6, 1	6, 2	6, 3	6, 4	6, 5	6, 6

bi $\frac{1}{36}$ **bii** $\frac{1}{36}$ **biii** $\frac{1}{12}$ **biv** $\frac{1}{18}$

3 $\frac{1}{6}$

4 a $\frac{1}{5}$ **b** $\frac{1}{20}$ **c** $\frac{3}{20}$ **d** $\frac{1}{10}$ **e** $\frac{1}{20}$

5 a $\frac{1}{100}$ **b** $\frac{1}{50}$ **c** 0 **d** $\frac{1}{25}$ **e** $\frac{3}{50}$

6 a

10p/2p	Head	Tail
Head	HH	HT
Tail	TH	TT

b $\frac{1}{4}$

7 $\frac{1}{4}$

D6.3

1 First choice   Second choice

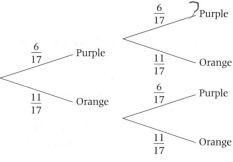

2 First choice   Second choice

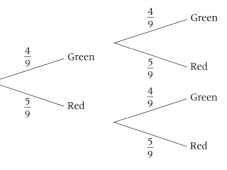

3 First choice

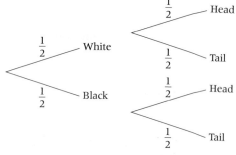

4 10p coin   2p coin

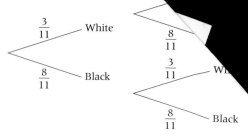

5 Bag   Coin

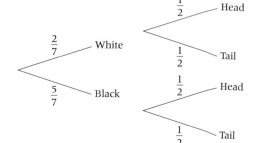

6 First choice   Second choice

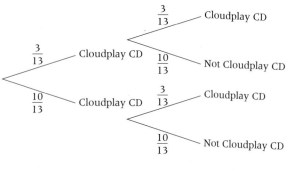

D6.4

1 a $\frac{9}{49}$ **b** $\frac{16}{49}$ **c** $\frac{12}{49}$

2 a $\frac{16}{81}$ **b** $\frac{25}{81}$

3 a $\frac{9}{121}$ **b** $\frac{64}{121}$

4 a $\frac{1}{4}$ **b** $\frac{1}{4}$ **c** $\frac{1}{4}$

5 a

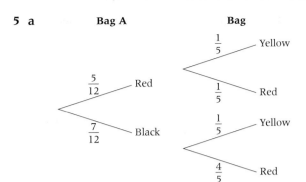

Bag A   Bag

bi $\frac{7}{60}$ **bii** $\frac{1}{3}$

D6.5

1 a 0.2

b

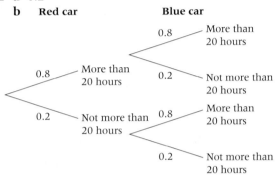

Red car   Blue car

ci 0.64   **cii** 0.32   **ciii** 0.96

2 a

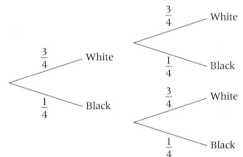

First choice   Second choice

bi $\frac{1}{16}$ **bii** $\frac{3}{8}$ **biii** $\frac{7}{16}$

3 a $\frac{19}{20}$

b

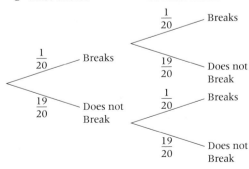

First choice   Second choice

ci $\frac{361}{400}$ **cii** $\frac{38}{400}$ **ciii** $\frac{39}{400}$

4 a

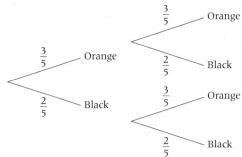

First spin   Second spin

bi $\frac{4}{25}$ **bii** $\frac{16}{25}$ **biii** $\frac{12}{25}$

5 a 0.9

b Grey plane   Orange plane

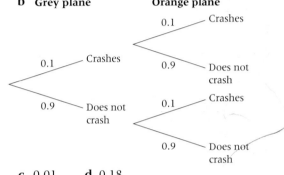

c 0.01   **d** 0.18

D6 Exam review

1 ai $\frac{1}{6}$ **aii** $\frac{5}{6}$ **aiii** $\frac{5}{12}$

    b 0, there are no purple balls in the bag

2 a From Julie's experiment P(6) = $\frac{1}{3}$.

For a fair dice P(6) = $\frac{1}{6}$.

The probability of a 6 inJulie's experiment
is twice that expected from a fair dice.
This suggests that the dice is not fair.

b

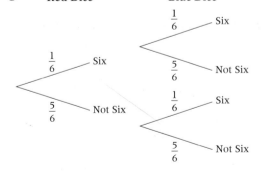

Red Dice   Blue Dice

A8 Before you start …

1 a 20 mph **b** 80 mph **c** 90 mph **d** 24 mph

2

Equation	Gradient	y-axis intercept	Direction
$y = 3x + 4$	3	(0, 4)	up
$y = 10 - 4x$	−4	(0, 10)	down
$2y = 8x + 10$	4	(0, 5)	up
$2y - 4x = 15$	2	(0, 7.5)	up
$y = 7$	0	(0, 7)	horizontal

3 a $\frac{3}{2}$ **b** −1 **c** $-\frac{1}{2}$ **d** $\frac{1}{3}$ **e** $\frac{3}{5}$

4 a 5 **b** $\frac{5}{4}$ **c** 11

A8.1

1 a 60 km **b** 1 hours 30 minutes
  c 60 km/h **d** 11:30 am and 12 noon
  e 30 km/h

2 Claire and Christina: Claire's vertical line means she travelled a distance in no time, and Christina's line sloping backwards means she has travelled backwards in time.

3 a

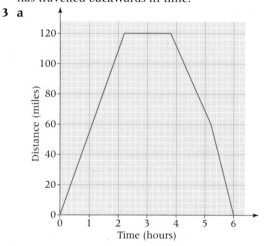

b

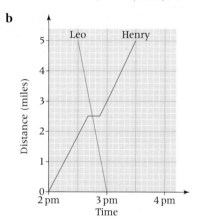

4

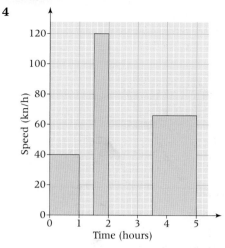

A8.2

1 A3, B1, C4, D2

2 Growth spurt at around 5, steadies down until next growth spurt at about 13–17 years (puberty) then steadies down again.

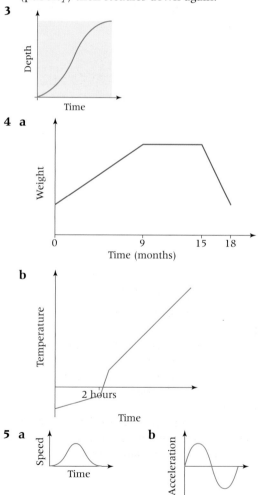

3

4 a

b

5 a   **b**

A8.3

1 ai 32 km **aii** 37.5 miles
  b Charlie **c** $y = 1.6x$

2 a

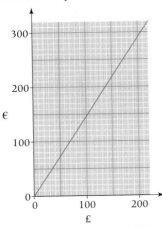

  bi Approximately €138 **bii** Approximately £59

3 a

Number of people	1	2	3	4	5	6	7	8	9	10	11	12	13	14	15
Cost (£)	18	21	24	27	30	33	36	39	42	45	48	51	54	57	60

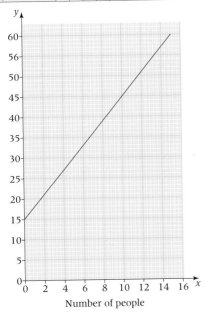

Number of people

  b 7
  c This would mean that a fraction of a person stayed in the tent.
  d $c = 15 + 3p$, where c = cost per night and p = number of people
  e £96

4 Power Up! is cheaper for less than 10 units; they cost the same for 10 units; Sparks Are Us! is cheaper for more than 10 units.

A8.4

1 A: $y = 4x + 3$, B: $y = 2x + 1$, C: $y = \frac{1}{2}x + 3$,
  D: $y = \frac{1}{3}x + 1$

2 a $y = 6x + 50$; m is 6, meaning that for every year you age, you grow by 6 cm; c is 50, meaning that when you are born you are 50 cm tall, this is reasonable.

  b $y = \frac{1}{2}x + 20$; m is $\frac{1}{2}$, meaning that for every two years a person needs one more driving lesson; it is not sensible to interpret c as no one takes a driving lesson at birth.

3 a Pocket money is not given to very young children or adults, so only makes sense between $x = 4$ and $x = 21$
  b Does make sense to interpret the equation for any value of x, as temperature can take any value.

A8.5

1 $(-\frac{1}{4}, -4\frac{1}{8})$

2 a

x	−3	−2	−1	0	1	2	3
x^2	9	4	1	0	1	4	9
y	7	3	1	1	3	7	13

  b

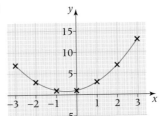

  c Minimum is 0.75 at $x = -0.5$

3 a

Time (x)	0	1	2	3	4	5
$20x$	0	20	40	60	80	100
$4x^2$	0	4	16	36	64	100
Height (y)	0	16	24	24	16	0

  b

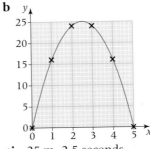

  ci 25 m, 2.5 seconds
  cii 0.7 seconds, 4.3 seconds
  ciii 3.2 seconds

A8 Exam review

1 a 12 m   **b** 50 m

2 a Speed = 40 km/h

b

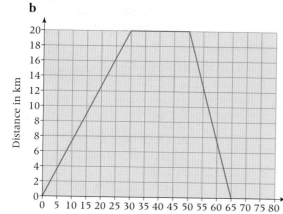

S8 Before you start ...

1 a $x = 6y$ **b** $x = 5y$ **c** $x = 10y$

   d $x = \frac{2}{y}$ **e** $x = \frac{5}{y}$ **f** $x = \frac{8}{y}$

2 a $a = 8.1$ units   **b** $b = 9.8$ units

   c $c = 8.5$ units   **d** $d = 13.7$ units

S8.1

1 a 3.63 cm   **b** 11.3 cm   **c** 10.8 cm

   d 3.02 cm   **e** 6.38 cm   **f** 3.79 cm

   g 30.5 cm   **h** 14.2 cm   **i** 74.5 cm

   j 6.06 cm   **k** 7.00 cm   **l** 12.3 cm

2 Right-angled isosceles triangles; They are the same as the other shorter side in each triangle.

S8.2

1 a 6.89 cm   **b** 8.60 cm   **c** 7.88 cm

   d 4.62 cm   **e** 4.77 cm   **f** 39.7 cm

   g 10.6 cm   **h** 9.56 cm   **i** 34.4 cm

   j 13.5 cm   **k** 11.8 cm   **l** 6.15 cm

   m 14.2 cm   **n** 7.88 cm   **o** 5.12 cm

S8.3

1 a 26.4°   **b** 25.7°   **c** 56.2°   **d** 67.4°

   e 17.6°   **f** 23.6°   **g** 69.2°   **h** 55.1°

   i 19.8°   **j** 35.7°   **k** 27.7°   **l** 40.1°

   m 73.0°   **n** 41.0°   **o** 45°   **p** 30°

   q 60°

S8.4

1 a $a = 8$ cm, $b = 14.4$ cm

   b $c = 9.57$ cm, $d = 16.0$ cm

   c $e = 13.2$ cm, $f = 10.8$ cm, $g = 13.1$ cm

2 a 28.9°   **b** 35.5°   **c** 61.2°

3 a 15.2 cm   **b** 19.5°

4 a 8.5 cm   **b** 58.4°

5 a 15.2 cm   **b** 6.9 cm

S8.5

1 49.4 cm^2

2 63.8 cm^2

3 31.5 cm^2

4 91.2°

5 10.3 cm

6 14.7 cm

7 38.9°

8 7.5 km

9 a 130° or 050°   **b** 20.8 km or 5.4 km

10 6.6 m

11 Yes, both give roughly the same height for phone mast.

S8 Exam review

1 bi 63.7 paces   **bii** 279.3°

2 a DG = 11.7 m (to 3 sf)   **b** 36.9° (to 1 dp)

Index